PHYSICAL
PROPERTIES OF
CRYSTALS

PHYSICAL PROPERTIES OF CRYSTALS

THEIR REPRESENTATION BY
TENSORS AND MATRICES

By

J. F. NYE, F.R.S.

CLARENDON PRESS · OXFORD

Oxford University Press, Walton Street, Oxford OX2 6DP

London New York Toronto
Delhi Bombay Calcutta Madras Karachi
Kuala Lumpur Singapore Hong Kong Tokyo
Nairobi Dar es Salaam Cape Town
Melbourne Auckland
and associated companies in
Beirut Berlin Ibadan Mexico City Nicosia

Oxford is a trade mark of Oxford University Press

Published in the United States
by Oxford University Press, New York

© Oxford University Press 1957, 1985

First published in hardback 1957

Reprinted with corrections 1960, 1964, 1967, 1969, 1972, 1976, 1979

First published in paperback with corrections and new material 1985

British Library Cataloguing in Publication Data
Nye, J. F.
Physical properties of crystals.
1. Crystals
I. Title
548'.8   QD931
ISBN 0-19-851165-5

Library of Congress Cataloging in Publication Data
Nye, J. F.
Physical properties of crystals.
Bibliography: p.
Includes indexes.
1. Crystals.   2. Calculus of tensors.   3. Matrices.
I. Title
QD931.N9   1985   548'.8   84-19026
ISBN 0-19-851165-5 (pbk.)

Printed in Great Britain by
J. W. Arrowsmith Ltd, Bristol

PREFACE TO FIRST EDITION

THE purpose of this textbook is to formulate the physical properties of crystals systematically in tensor notation—to explain what the tensors are, and how they are used. The book is not concerned with the explanation of particular crystal properties in terms of structure; it aims rather at a unified presentation of the tensor properties in terms of two themes: the common mathematical basis of the properties, and the thermodynamical relations between them.

The plan has been to start with the mathematical groundwork, treating at first tensors up to the second rank only—this forms Part 1—and then to show how the various properties in turn find their places within the tensor scheme. Tensors of higher rank than the second, and matrix methods, are introduced later as natural developments of the theory. This mathematical arrangement of the properties goes hand in hand with a threefold division on a thermodynamical and optical basis represented by Parts 2, 3, and 4.

The book originated in a course in crystal physics given by the author at Cambridge to second-year undergraduates reading crystallography (as part of the main subject Mineralogy and Crystallography) in Part I of the Natural Sciences Tripos. The level of treatment is, for the most part, well within the grasp of honours students in physics in their second or later years; but the author has also had in mind the post-graduate student, and the research worker in solid-state physics or metallurgy who needs an introductory text on the use of tensors and matrices in his subject. The mathematics has been kept as simple as possible, particularly in the early chapters where the tensor notation and the formal manipulations are explained in detail. The reader who already knows something of crystal symmetry will be at an advantage, but the essential symmetry theory needed for following the argument is summarized in an appendix. The chapter on thermoelectricity may be found a little more difficult than the others; it was not possible to give a satisfactory treatment of thermoelectricity at the level of the rest of the book—and yet the property fits so naturally into the systematic development that it seemed right to include it.

The references to published work are not intended to be complete: they are simply pointers to further reading and acknowledgements of sources.

The rationalized metre-kilogram-second system of units is used throughout. It was chosen primarily because it is a rationalized system, and thereby avoids the awkward factors of 4π which otherwise spoil the simplicity of many of the electrical and magnetic formulae. Any rationalized system would have served, but the well-known special advantages of the m.k.s. system, and its growing acceptance by physicists, made it an obvious choice.

I have the pleasant task of expressing my gratitude to Dr. R. C. Evans and Dr. N. F. M. Henry, of the Department of Mineralogy and Petrology at Cambridge, for encouraging me to write this book, and for criticizing in detail the first drafts of all the chapters. Their advice has been of the greatest assistance at all times. I have also to thank the other friends and colleagues who have read and commented upon the manuscript: Professor F. C. Frank, F.R.S. (for reading Chapters I to X), Professor K. Lonsdale, F.R.S. (for reading the first drafts of Chapters I, II, III, VIII and IX), and Sir Edward Bullard, F.R.S. Dr. D. Polder has helped me with Chapter XII and Dr. F. G. Fumi with Chapter XIV. The book was begun while I was on the staff of Bell Telephone Laboratories, Murray Hill, New Jersey, and I appreciate very much the generous co-operation I received there. Three colleagues at Murray Hill gave me particular help: Mr. W. L. Bond, who showed me his unpublished 'least-squares octagonal-disc' method and supplied the numerical illustrations of matrix methods in § 7 of Chapter IX; Dr. A. N. Holden, who likewise gave me the use of unpublished work, reproduced in the *résumé* on pp. 181–2, and who read Chapter X; and Dr. Conyers Herring, who helped me with the troublesome subject of magnetic and electrical energy.

The dot-and-ring notation used for showing the form of the crystal-property matrices in the thirty-two crystal classes is taken, with slight variations, from the unpublished *Manual of Piezoelectric Data* by K. S. Van Dyke and G. D. Gordon, by kind permission of the authors.

Some of the exercises distributed through the book have been taken or adapted, by permission, from the collections used by the teaching staff of the Department of Mineralogy and Petrology at Cambridge, to whom I tender my thanks.

In referring to the department of which I was formerly a member, I should like to acknowledge my particular debt to Dr. W. A. Wooster, for the present book is an outcome of the interest aroused by the course which he initiated there.

J. F. N.

Bristol
July 1955

PREFACE TO THE 1985 EDITION

SINCE the first edition of this book artificial crystals of many more substances have been successfully grown, and study of their physical properties has produced a formidably extensive literature. For example, Landolt-Börnstein lists some 4700 references on the properties of the elastic dielectric alone, and that was six years ago. There has been great progress on the atomistic and experimental sides, but amidst all this activity the principles of the macroscopic continuum treatment of the subject have remained the same, and that is why it now seems worthwhile to produce this new edition.

The only change in the main text worth special note concerns the role of body-torques in the formulation of elasticity and related properties, the confusion that previously surrounded this topic having now cleared away. At the same time I have taken the opportunity of adding at the end of the book an up-to-date bibliography with notes, which I hope may serve as an indication of the main new developments that have taken place in the subject. I thank Dr. J. W. Steeds and Professor F. G. Fumi for their ready help in guiding me to relevant literature.

J. F. N.

Physics Department
University of Bristol
April 1984

CONTENTS

NOTE

FIG. 13.7 is reproduced, by permission, from Hartshorne and Stuart's *Crystals and the polarising microscope* (Edward Arnold (Publishers) Ltd.).

NOTATION

ORDINARY letters in bold-face type, $\mathbf{P}$, $\mathbf{h}$, etc., denote vectors.

Bold sans serif letters, P, h, etc., and bold Greek letters, $\boldsymbol{\alpha}$, $\boldsymbol{\beta}$, etc., denote matrices.

The range of values of all letter suffixes is 1, 2, 3 unless some other range is specified.

INTRODUCTION

THE physical properties of crystals are defined by relations between measurable quantities. Density, for example, is defined from a relation between mass and volume. Now both mass and volume may be measured without reference to direction, and, accordingly, density is a property that does not depend on direction. On the other hand, a crystal property such as electrical conductivity is defined as a relation between two measurable quantities (the electric field and the current density) both of which have to be specified in direction as well as in magnitude. We therefore have to allow for the possibility that a physical property of this sort will depend upon the direction in which it is measured—and, as an experimental fact, the electrical conductivity of many crystals does indeed vary with direction. In such cases the crystals are said to be *anisotropic* for the property in question.

The problem then arises of how to specify the value of a crystal property that can depend upon direction—clearly, a single number will not suffice. There is also the problem of how the specification, when we have it, is related to the symmetry of the crystal. The answers to these two questions, and some of their implications, form the subject of this book.

Electrical conductivity is one of many crystal properties that can depend upon the direction of measurement. A few further examples are: the flow of heat produced by a temperature gradient (thermal conductivity); the polarization produced in a dielectric by an electric field (dielectric susceptibility); the polarization of a crystal that may be produced by mechanical stress (piezoelectricity); the deformation caused by a mechanical stress (elasticity); and the birefringence that can be set up by an electric field (electro-optical effect) or by a stress (photoelastic effect).

For a few properties, such as density, all crystals are isotropic. Cubic crystals happen to be isotropic for certain other properties as well, such as conductivity and refractive index, and this sometimes leads to the misconception that they are isotropic for all properties. Nevertheless, the symmetry elements of a cubic crystal are not the same as those of a completely isotropic body, and, in fact, cubic crystals are anisotropic, often markedly so, for elasticity, photoelasticity, and certain other properties. We must therefore regard cubic crystals as *potentially* anisotropic, and then we can go on to prove that, for *certain* properties, they are isotropic. All crystals are anisotropic for some of their properties.

In this book, then, we study how to specify the physical properties of crystals; a large number of the properties are represented by mathematical quantities called *tensors*, and only these properties will concern us. A list of them is given in Appendix C; to help put the subject in proper perspective a further list is given there of some other properties that are not directly represented by tensors. It is, of course, part of the task of physics to explain the values of these tensors for any particular crystal in terms of its atomic and crystalline structure. That is, in a sense, the next stage. Here we are less ambitious; we concern ourselves more with the form and general significance of the tensors than with their actual numerical values. For our purpose it will be sufficient to regard a crystal as simply an anisotropic continuum, without structure, having certain properties of symmetry. Moreover, except in dealing with one property, namely thermoelectricity, we shall assume *homogeneity*: that the properties of a crystal are the same at all points.

Plan of the book

Tensors are classified by their *rank*. Chapter I introduces the concept of a tensor and shows how tensors of zero, first and second rank may be used for studying crystal properties. Chapter II continues the mathematical development. In Chapters III to VI the tensor method is applied to various physical properties in turn; then, in Chapters VII and VIII, tensors of the third and fourth rank are introduced and are used for representing piezoelectricity and elasticity. Chapter IX describes the alternative method of representing crystal properties by matrices, a technique which is particularly useful for carrying out numerical calculations. All the properties dealt with up to this point are describable by reference to equilibrium states and thermodynamically reversible changes; a unified treatment of the properties and their thermodynamical inter-relations is given in Chapter X. In Chapters XI and XII we pass on to conduction and thermoelectricity; these are treated after the equilibrium properties, because, being transport phenomena, they are irreversible and their thermodynamics needs special consideration. The two final chapters, XIII and XIV, are devoted to crystal optics, and especially to the electro-optical effect, photoelasticity and optical activity.

Selection for first reading

For a first reading of the book the following selection, which is more or less self-contained, is suggested.

Chapter I.   The groundwork of crystal physics
Chapter III.   Paramagnetic and diamagnetic susceptibility
Chapter IV.   Electric polarization
Chapter V.   The stress tensor
Chapter VI.   The strain tensor and thermal expansion
Chapter VII.   Piezoelectricity. Third-rank tensors
Chapter VIII.   Elasticity. Fourth-rank tensors
Chapter XI.   Thermal and electrical conductivity (up to the end of § 4)

PART 1

GENERAL PRINCIPLES

I

THE GROUNDWORK OF CRYSTAL PHYSICS

1. Scalars, vectors and tensors of the second rank

(i) *Scalars.* In physics we are accustomed to dealing with certain quantities, such as the density or the temperature of a body, which are not connected in any way with direction. With the usual definitions of density and temperature it is meaningless to speak of measuring these quantities in any particular direction. Such non-directional physical quantities are called *scalars*, and we note that the value of a scalar is completely specified by giving a single number. For a reason that will appear later, a scalar is also called a *tensor of zero rank*.

(ii) *Vectors.* In contrast to scalars there are physical quantities of a different type, called *vectors*, which can only be defined with reference to directions. Mechanical force is a well-known example. To specify completely a force acting at a point we need to give both its magnitude and its direction. It may be conveniently represented by an arrow of definite length and direction. Other examples of vectors are: the strength of an electric field at a point, the moment of a magnetic dipole, and the temperature gradient at a point. In this book we denote vectors by bold-face type: thus $\mathbf{E}$ denotes the strength of an electric field at a point. The magnitude, or length, of a vector $\mathbf{p}$ is denoted by p.

As an alternative to specifying a vector by giving its magnitude and direction we may, instead, choose three mutually perpendicular axes Ox_1, Ox_2, Ox_3 and give the *components* of the vector along them. The components are simply the projections of the vector on the axes. If the components of $\mathbf{E}$ are E_1, E_2, E_3, we write

$$\mathbf{E} = [E_1, E_2, E_3].$$

Thus, when the axes of reference have been chosen, a vector is completely specified by giving the values of its three components along the axes. For a reason that will appear later, a vector is also called a *tensor of the first rank*.

The methods of manipulation of vectors form the subject of *vector analysis*. We shall not have to use this much in what follows, but we shall occasionally draw on it. It will be assumed that the reader is already familiar with the notions of scalar and vector product, the gradient of a scalar, the divergence of a vector and the curl of a vector—see, for example, Abraham and Becker (1937). A summary of vector notation and formulae is given in Appendix A.

(iii) *Tensors of the second rank.* We now have to introduce an extension of the idea of a vector. Consider the following example. If an electric field given by the vector **E** acts in a conductor, an electric current flows. The current density (current per unit cross-section perpendicular to the current) is denoted by the vector **j**. If the conductor

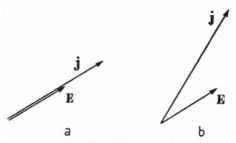

Fig. 1.1. The relation between the electric current density **j** and the electric field **E** in (*a*) an isotropic conductor and (*b*) an anisotropic conductor.

is isotropic and obeys Ohm's Law, **j** is parallel to **E** (Fig. 1.1 *a*), and the magnitude of **j** is proportional to the magnitude of **E**. We write

$$\mathbf{j} = \sigma \mathbf{E}, \tag{1}$$

where σ is the conductivity. If, with axes Ox_1, Ox_2, Ox_3, $\mathbf{j} = [j_1, j_2, j_3]$ and $\mathbf{E} = [E_1, E_2, E_3]$, it follows that

$$j_1 = \sigma E_1, \quad j_2 = \sigma E_2, \quad j_3 = \sigma E_3. \tag{2}$$

Each component of **j** is proportional to the corresponding component of **E**.

Now, if the conductor is a crystal the relation between the components of **j** and **E** is not as simple as this, for crystals are, in general, anisotropic in their conducting properties. (It happens that cubic crystals form a special group of crystals whose conductivity is isotropic; but, for the moment, as mentioned in the Introduction, we shall treat cubic crystals as potentially anisotropic, like all other crystals, and then later (§ 5.1) we shall prove that in fact they are isotropic for conductivity.) For crystals, relations (2) are replaced by

$$\left. \begin{array}{l} j_1 = \sigma_{11} E_1 + \sigma_{12} E_2 + \sigma_{13} E_3 \\ j_2 = \sigma_{21} E_1 + \sigma_{22} E_2 + \sigma_{23} E_3 \\ j_3 = \sigma_{31} E_1 + \sigma_{32} E_2 + \sigma_{33} E_3 \end{array} \right\}, \tag{3}$$

where σ_{11}, σ_{12},... are constants. Each component of $\mathbf{j}$ is now linearly related to all three components of $\mathbf{E}$. It follows that $\mathbf{j}$ is no longer in the same direction as $\mathbf{E}$ (Fig. 1.1 b).

Each of the coefficients σ_{11}, σ_{12},... in (3) can be given a physical

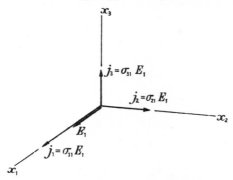

Fig. 1.2. The components of current density when a field is applied along Ox_1.

meaning. For instance, if the field is applied along x_1† (Fig. 1.2), $\mathbf{E} = [E_1, 0, 0]$, and equations (3) become

$$\left.\begin{aligned} j_1 &= \sigma_{11} E_1 \\ j_2 &= \sigma_{21} E_1 \\ j_3 &= \sigma_{31} E_1 \end{aligned}\right\}.$$

Thus, there are now components of $\mathbf{j}$ not only along x_1, but along the other axes as well. The direct component is given by σ_{11} and the two transverse components by σ_{21} and σ_{31}. In a similar way σ_{23}, for example, measures the component of $\mathbf{j}$ parallel to x_2 when a field is applied parallel to x_3.

In order to specify the conductivity of a crystal, then, we have to specify the nine coefficients σ_{11}, σ_{12},.... They can be conveniently written down in a square array, thus:

$$\begin{bmatrix} \sigma_{11} & \sigma_{12} & \sigma_{13} \\ \sigma_{21} & \sigma_{22} & \sigma_{23} \\ \sigma_{31} & \sigma_{32} & \sigma_{33} \end{bmatrix}. \tag{4}$$

This array enclosed by square brackets symbolizes a *tensor of the second rank.*‡ σ_{11}, σ_{12},... are the *components* of the tensor. It will be seen

† We sometimes use the symbols x_1, x_2, x_3 instead of Ox_1, Ox_2, Ox_3, to denote the directions of the axes when no confusion can arise with the coordinates of a point, which are also denoted by x_1, x_2, x_3.

‡ The word *tensor* was first used by Voigt to describe mechanical stress. The word *rank* is preferred to *order* in this context, so that we may reserve *order* for its usual meaning in phrases like 'second-order term', 'second-order effect'.

that the first suffix gives the row and the second the column in which the component appears. σ_{11}, σ_{22}, σ_{33} are the components on the *leading diagonal*.

We may now compare the three sorts of quantity that have been introduced so far:

(i) A tensor of zero rank (a scalar) is specified by a single number unrelated to any axes of reference.

(ii) A tensor of the first rank (a vector) is specified by three numbers, or components, each of which is associated with one of the axes of reference.

(iii) A tensor of the second rank is specified by nine numbers, or components, each of which is associated with a pair of axes (taken in a particular order).

Our notation emphasizes these distinctions. A scalar such as density is written without subscripts (for example, density ρ); the components of a vector have one subscript (for example, E_2); and the components of a second-rank tensor have two subscripts (for example, σ_{12}). *The number of subscripts equals the rank of the tensor.*

In the present chapter we shall not deal with tensors of higher rank than the second. Later on, however, in discussing physical properties such as piezoelectricity and elasticity (Ch. IX and X) we shall introduce tensors of the third and fourth ranks. These are natural extensions of the concepts of zero, first- and second-rank tensors.

Besides conductivity, which we have chosen as our first example, crystal physics introduces many other second-rank tensors. In general, if a property T relates two vectors $\mathbf{p} = [p_1, p_2, p_3]$ and $\mathbf{q} = [q_1, q_2, q_3]$ in such a way that

$$\left.\begin{aligned} p_1 &= T_{11}q_1 + T_{12}q_2 + T_{13}q_3 \\ p_2 &= T_{21}q_1 + T_{22}q_2 + T_{23}q_3 \\ p_3 &= T_{31}q_1 + T_{32}q_2 + T_{33}q_3 \end{aligned}\right\}, \qquad (5)$$

where T_{11}, T_{12},... are constants, T_{11}, T_{12},... are said to form a second-rank tensor

$$\begin{bmatrix} T_{11} & T_{12} & T_{13} \\ T_{21} & T_{22} & T_{23} \\ T_{31} & T_{32} & T_{33} \end{bmatrix}. \qquad (6)$$

Table 1 contains some more examples of second-rank tensor properties.

A full list of all the crystal properties in this book represented by tensors is given in Appendix C.

TABLE 1

Some examples of second-rank tensors relating two vectors

Tensor property	Vector given or applied	Vector resulting or induced
Electrical conductivity	electric field	electric current density
Thermal conductivity	(negative) temperature gradient	heat flow density
Permittivity	electric field	dielectric displacement
Dielectric susceptibility	,, ,,	,, polarization
Permeability	magnetic field	magnetic induction
Magnetic susceptibility	,, ,,	intensity of magnetization

1.1. The dummy suffix notation. It is convenient now to shorten our notation. Equations (5) may be written

$$
\left.
\begin{aligned}
p_1 &= \sum_{j=1}^{3} T_{1j}\, q_j \\
p_2 &= \sum_{j=1}^{3} T_{2j}\, q_j \\
p_3 &= \sum_{j=1}^{3} T_{3j}\, q_j
\end{aligned}
\right\}, \tag{7}
$$

or, more compactly, as

$$
p_i = \sum_{j=1}^{3} T_{ij}\, q_j \quad (i = 1, 2, 3). \tag{8}
$$

We now leave out the summation sign:

$$
p_i = T_{ij}\, q_j \quad (i,j = 1, 2, 3), \tag{9}
$$

and introduce the Einstein summation convention: *when a letter suffix occurs twice in the same term, summation with respect to that suffix is to be automatically understood.* With this convention, (7), (8) and (9) are equivalent ways of writing the original equations (5). j in equations (9) is called a *dummy suffix*, and clearly it is immaterial what letter is used for this purpose, provided we do not use a letter that occurs elsewhere in the same term. Thus, for example,

$$
p_i = T_{ij}\, q_j = T_{ik}\, q_k.
$$

What matters is the *position* of the repeated suffixes. i in these equations is a *free suffix*.†

In an equation written in this notation the free suffixes must be the same

---

† The arrangement of the suffixes in equation (9) follows the accepted rule: the suffix of the q term is the same as the *second* suffix of T_{ij}, so that when q is written after T the dummy suffixes adjoin one another. We defined the T's in equations (5) so as to ensure this. Readers familiar with Wooster's *Crystal Physics* should note that he reverses this traditional order of suffixes: in his notation equation (9) would appear as $p_i = T_{ji}\, q_j$.

in all the terms on both sides of the equation; while the dummy suffixes must occur as pairs in each term. This is a logical necessity from the meaning of the notation. For example, we might have an equation of this sort:

$$A_{ij} + B_{ik} C_{kl} D_{lj} = E_{ik} F_{kj};$$

i, j are free suffixes here and k, l are dummy suffixes. Note that the order of the members of a product does not matter in this notation; for example, the second term on the left in the above equation could be alternatively written $C_{kl} B_{ik} D_{lj}$. It is often convenient to keep the dummy suffixes together in pairs, as we have done, but this is purely a matter of choice. [We shall later have to do with the matrix notation (Ch. IX) where the order of members in a product is significant.]

In this book the range of values of all letter suffixes is 1, 2, 3 unless some other range is specified.

Returning now to equation (9), we see that it sums up the relation between **p** and **q** in a crystal. The corresponding equation for an isotropic body would be

$$p_i = T q_i, \tag{10}$$

where T is a single constant. We shall write the tensor (6) briefly as $[T_{ij}]$, but where no ambiguity can arise we shall sometimes, following normal practice, use T_{ij}, without brackets, to denote the complete tensor. In a similar way the vector $\mathbf{p} = [p_1, p_2, p_3]$ is written as $[p_i]$ when we want to emphasize its components, or often simply as p_i. It is, of course, immaterial what letter is used for the suffix in the symbol $[p_i]$ or p_i.

2. Transformations

In equations (5) the constant coefficients T_{ij} determine how the three components p_i vary as different values are given to the components q_i. It is important to notice that the directions of the axes were chosen arbitrarily, except for the fact that they were mutually orthogonal. If we had begun with a different set of axes we should have found a different set of coefficients in equations (5). However, *both* sets of coefficients equally well represent the *same* physical quantity—in our illustrative example, the conductivity. There must therefore be some relation between them. If the coefficients are given for one particular choice of axes the property is completely defined. The coefficients for any other set of axes whose relation to the first set is known are therefore determinate. In short, when we change the axes of reference it is only our method of representing the property that changes; the property itself remains the same.

Our next task is to find out how the values of the nine components T_{ij} transform when the axes are transformed. To do this we have first to specify the transformation of axes (§ 2.1) and then to discuss how the components of the vectors $[p_i]$ and $[q_i]$ transform under the change of axes (§ 2.2).

2.1. Transformations of axes. By a transformation of axes we shall mean a change from one set of mutually perpendicular axes to another set with the same origin. The unit of measurement along each axis is always the same. We call the first set x_1, x_2, x_3 and the second

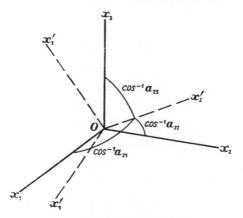

FIG. 1.3. Transformation of axes.

set x_1', x_2', x_3' (Fig. 1.3). The angular relations between the axes may be specified by drawing up a table of direction cosines:

'Old'

		x_1	x_2	x_3
	x_1'	a_{11}	a_{12}	a_{13}
'New'	x_2'	a_{21}	a_{22}	a_{23}
	x_3'	a_{31}	a_{32}	a_{33}.

(11)

Thus, for example, the direction cosines of x_2' with respect to x_1, x_2, x_3 are a_{21}, a_{22}, a_{23}, and the direction cosines of x_3 with respect to x_1', x_2', x_3' are a_{13}, a_{23}, a_{33}. The first subscript in the a's refers to the 'new' axes and the second to the 'old'; a_{ij} is the cosine of the angle between x_i' and x_j.† The array of a_{ij}, denoted collectively by (a_{ij}), is an example of a *matrix*. The nine a_{ij} are not independent of one another; we derive the relations between them later (Ch. II, § 1.1). Note that, in general, $a_{ij} \neq a_{ji}$.

† Some authors reverse the order of suffixes in a_{ij}; there is no general agreement.

2.2. Transformation of vector components. Now suppose we have a certain vector **p** whose components with respect to x_1, x_2, x_3 are p_1, p_2, p_3 (Fig. 1.4). Let the components with respect to x_1', x_2', x_3' be p_1', p_2', p_3'. p_1' is obtained by resolving p_1, p_2, p_3 (thought of as vectors along x_1, x_2, x_3 respectively) along the new direction x_1'. Thus

$$p_1' = p_1 \cos \widehat{x_1 \, x_1'} + p_2 \cos \widehat{x_2 \, x_1'} + p_3 \cos \widehat{x_3 \, x_1'}.$$

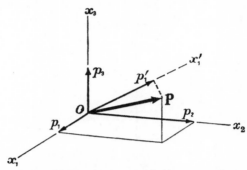

FIG. 1.4.   Transformation of the components of a vector.

Hence, from (11), $p_1' = a_{11} p_1 + a_{12} p_2 + a_{13} p_3,$ (12.1)

and similarly, $p_2' = a_{21} p_1 + a_{22} p_2 + a_{23} p_3,$ (12.2)

$$p_3' = a_{31} p_1 + a_{32} p_2 + a_{33} p_3,$$ (12.3)

which may be written in the dummy suffix notation (§ 1.1) as

'new' in terms of 'old', $p_i' = a_{ij} p_j.$ (13)

By repeating the argument for the reverse transformation and by referring again to (11) we have

$$\left. \begin{aligned} p_1 &= a_{11} p_1' + a_{21} p_2' + a_{31} p_3' \\ p_2 &= a_{12} p_1' + a_{22} p_2' + a_{32} p_3' \\ p_3 &= a_{13} p_1' + a_{23} p_2' + a_{33} p_3' \end{aligned} \right\},$$ (14)

or, briefly,   'old' in terms of 'new', $p_i = a_{ji} p_j'.$ (15)

Similarly, for another vector **q**,

$$q_i = a_{ji} q_j'.$$ (16)

Note that in the transformation, 'new' in terms of 'old' (13), the dummy suffixes occur in neighbouring places. In the reverse transformation (15) they are separated.

2.3. Transformation of the coordinates of a point. The co-ordinates (x_1, x_2, x_3) of a point P with respect to Ox_1, Ox_2, Ox_3 are the components of the vector **OP**. It follows from equations (12.1 to 12.3) that the coordinates (x'_1, x'_2, x'_3) with respect to Ox'_1, Ox'_2, Ox'_3 are given by

$$\left.\begin{aligned} x'_1 &= a_{11} x_1 + a_{12} x_2 + a_{13} x_3 \\ x'_2 &= a_{21} x_1 + a_{22} x_2 + a_{23} x_3 \\ x'_3 &= a_{31} x_1 + a_{32} x_2 + a_{33} x_3 \end{aligned}\right\}, \tag{17}$$

or
$$x'_i = a_{ij} x_j. \tag{18}$$

We also have, from (14),

$$\left.\begin{aligned} x_1 &= a_{11} x'_1 + a_{21} x'_2 + a_{31} x'_3 \\ x_2 &= a_{12} x'_1 + a_{22} x'_2 + a_{32} x'_3 \\ x_3 &= a_{13} x'_1 + a_{23} x'_2 + a_{33} x'_3 \end{aligned}\right\}, \tag{19}$$

or
$$x_i = a_{ji} x'_j. \tag{20}$$

2.4. Transformation of the components of a second-rank tensor. We return now to the relation between the vectors **p** and **q** expressed by the three equations (5). We repeat that the particular values of the coefficients T_{ij} in these equations depend on the set of axes x_1, x_2, x_3 that we happened to choose. Now let us choose a new set of axes x'_1, x'_2, x'_3 related to the old axes by the direction cosines (11). The vectors **p** and **q** now have components p'_i and q'_i. We want to find out what the relations are between these new components; to do this we use the following series of equations

$$p' \xrightarrow{(13)} p \xrightarrow{(9)} q \xrightarrow{(16)} q',$$

where $\rightarrow$ here stands for 'in terms of'. Remembering that any letter may be used for the dummy suffix, (13) may be written

$$p'_i = a_{ik} p_k. \tag{13'}$$

Using k as the free suffix and l as the dummy suffix, (9) takes the form

$$p_k = T_{kl} q_l; \tag{9'}$$

and, similarly, by changing the free suffix from i to l, (16) becomes

$$q_l = a_{jl} q'_j. \tag{16'}$$

Then, combining the equations,

$$p'_i = a_{ik} p_k = a_{ik} T_{kl} q_l = a_{ik} T_{kl} a_{jl} q'_j;$$

or
$$p'_i = T'_{ij} q'_j, \tag{21}$$

where T'_{ij} is written for the coefficient of q'_j in the ith equation. Thus, finally,

$$\boxed{T'_{ij} = a_{ik} a_{jl} T_{kl}.} \tag{22}$$

Equations (21) give the relation between the p'_i and the q'_i; they are to be compared with (9). Just as the components p_i and q_i transform to p'_i and q'_i when the axes are changed, so we may say that the nine coefficients T_{ij} transform to the nine coefficients T'_{ij}. We have proved that the relation between them, *the transformation law for second-rank tensors*, is (22). This is a most important result; it is essential that its meaning and significance be properly understood, and we shall discuss it fully in § 3.

In (22) it will be noticed that k and l are dummy suffixes, while i and j are free. The reader may find it helpful, at first, to expand such expressions in two stages, first for one suffix and then for the other— the order is immaterial. Thus, expanding for l,

$$T'_{ij} = a_{ik}a_{j1}T_{k1} + a_{ik}a_{j2}T_{k2} + a_{ik}a_{j3}T_{k3}.$$

k is now a dummy suffix in each term and so a further expansion is indicated:
$$T'_{ij} = a_{i1}a_{j1}T_{11} + a_{i1}a_{j2}T_{12} + a_{i1}a_{j3}T_{13} +$$
$$+ a_{i2}a_{j1}T_{21} + a_{i2}a_{j2}T_{22} + a_{i2}a_{j3}T_{23} +$$
$$+ a_{i3}a_{j1}T_{31} + a_{i3}a_{j2}T_{32} + a_{i3}a_{j3}T_{33}. \tag{23}$$

To each pair of values of i and j there corresponds one such equation. It will be seen that throughout all the terms the pattern of the suffixes, typified in (22), is preserved (for example, the second and fifth suffixes are always the same). The economy of the dummy suffix notation is apparent here, for (22) represents a set of nine equations each with nine terms on the right-hand side.

The equation giving the old components in terms of the new may be obtained by considering the reverse transformation, which, as we have seen, is given by the same scheme of a_{ij}'s but with the suffixes reversed in order. Thus

$$T_{ij} = a_{ki}a_{lj}T'_{kl}. \tag{24}$$

It is left as an exercise for the reader to prove equations (24) from the series of equations
$$p \to p' \to q' \to q,$$

where $\to$ means 'in terms of'.

The following is an aid to memory: *in the transformation* (22), *where the new components are expressed in terms of the old, the dummy suffixes occur as close together as possible; while in the reverse transformation* (24) *the opposite is true.*

(22) and (24) may be compared with the vector transformation laws (13) and (15). We repeat the laws in Table 2 and include the transformation law for a scalar ϕ. For completeness we add also the transformation laws for third- and fourth-rank tensors, but we shall not

discuss these further until Chapters VII and VIII. The table could be continued indefinitely with tensors of higher rank.

These transformation laws are of fundamental importance and, as we shall see (§ 3), they form the basis of an exact definition of a tensor. The resemblance between them is the reason for regarding scalars and vectors as tensors.

TABLE 2

Transformation laws for tensors

Name	Rank of tensor	Transformation law	
		New in terms of old	Old in terms of new
Scalar	0	$\phi' = \phi$	$\phi = \phi'$
Vector	1	$p'_i = a_{ij}p_j$ (13)	$p_i = a_{ji}p'_j$ (15)
—	2	$T'_{ij} = a_{ik}a_{jl}T_{kl}$ (22)	$T_{ij} = a_{ki}a_{lj}T'_{kl}$ (24)
—	3	$T'_{ijk} = a_{il}a_{jm}a_{kn}T_{lmn}$	$T_{ijk} = a_{li}a_{mj}a_{nk}T'_{lmn}$
—	4	$T'_{ijkl} = a_{im}a_{jn}a_{ko}a_{lp}T_{mnop}$	$T_{ijkl} = a_{mi}a_{nj}a_{ok}a_{pl}T'_{mnop}$

2.5. Transformation law for products of coordinates. We found (pp. 10, 11) that the transformation law for a first-rank tensor,

$$p'_i = a_{ij}p_j, \tag{13}$$

was the same as that for the coordinates of a point,

$$x'_i = a_{ij}x_j. \tag{18}$$

In the same way, the transformation law for a second-rank tensor turns out to be the same as the transformation law for *products* of coordinates. This may be proved as follows.

Consider the products $x'_i x'_j$ formed from the coordinates x'_i of a certain point with respect to the axes Ox'_i. The coordinates of the point with respect to the axes Ox_i are x_i. Now the products $x'_i x'_j$ are expressible in terms of the x_i by (18). Thus we have

$$x'_i x'_j = a_{ik}x_k . a_{jl}x_l = a_{ik}a_{jl}x_k x_l \tag{25}$$

as the connexion between the products of the primed coordinates and the products of the unprimed coordinates. The reader should verify (25) by using the full equations (17) and writing out the expression for one of the products, say $x'_1 x'_2$. As a transformation law, (25) is formally the same as (22); we may say that *the tensor component T_{ij} transforms like the product $x_i x_j$*. This statement needs one qualification. When the right-hand side of (25) is written out in full, the terms with $k \neq l$ could be grouped together in pairs such as:

$$a_{i1}a_{j2}x_1 x_2 + a_{i2}a_{j1}x_2 x_1 = (a_{i1}a_{j2} + a_{i2}a_{j1})x_1 x_2.$$

A corresponding grouping does not take place in (22) unless it happens that $T_{kl} = T_{lk}$ (so that we are dealing with a symmetrical tensor, § 3.2). In general, products only transform like tensor components if this grouping is not performed—if, in fact, $x_1 x_2$ is regarded as different from $x_2 x_1$, and so on.

The analogy between tensor components and products extends to tensors of higher rank than the second (Chs. VII and VIII); we shall make much use of it in analysing the effect of crystal symmetry on tensor components.

3. Definition of a tensor

The importance of the transformation laws for tensors in Table 2 is that they may be used as definitions. We now *define* a vector as a quantity which, with respect to a set of axes x_1, has three components p_i that transform according to equations (13). Let us examine this definition more closely (Eddington 1923). We have three numbers p_1, p_2, p_3 which we associate with a certain set of axes. We ask what these numbers become when we change the axes. If this were all we were told the question would be meaningless; unless we do something to the numbers when the axes are changed they stay as they were. If, however, we begin by saying that these three numbers are always to be calculated as the components, along the axes, of a certain physical quantity—an electric field, for example—then we have a rule that tells us how to find the numbers for any given position of the axes. It is then a legitimate question to ask whether or not the numbers transform according to equations (13). The essential point is that the components of a vector are the components of a *physical quantity* which we conceive to retain its identity however the axes may be changed.

In a similar way we now *define* a tensor of the second rank as a physical quantity which, with respect to a set of axes x_i, has nine components T_{ij} that transform according to equations (22). In speaking of the transformation of components it is implied that there is some rule, independent of (22), by which we can find the components for any given set of axes. For electrical conductivity, which we took as our example on p. 4, we should have to measure the components of **j** when **E** was directed along each of the axes in turn; see equations (3). A tensor of the second rank, like a vector, represents a *physical quantity* (in this instance it is the conductivity) that we think of as existing in its own right, and quite independent of the particular choice of axes that we happen to make. We repeat that, when we change the axes, the physical quantity does not change, but only our method of representing it. Similar definitions based on transformation properties apply to tensors of any rank.

If we use equations (22) now as a definition, it may be asked what becomes of the apparent proof that we gave of them. The proof now amounts to saying that, if, for any set of axes, the nine coefficients T_{ij} connect the components of two vectors p_i and q_i in linear relationships, then, on changing to another set of axes, the T_{ij} transform according to equation (22), and *hence* form a second-rank tensor.

EXERCISE 1.1. If p_i and q_i are vectors, then $p_1/q_1, p_2/q_2, p_3/q_3$ are three numbers whose values are defined for any set of axes. Are they the components of a vector?

3.1. Difference between the transformation matrix (a_{ij}) and the tensor $[T_{ij}]$.
Although both (a_{ij}) and $[T_{ij}]$ are arrays of nine numbers, this is virtually their only point of resemblance. In other respects they are quite different. This is clearly brought out in their definitions: (a_{ij}) is an array of coefficients relating two sets of axes; $[T_{ij}]$ is a physical quantity that, for one given set of axes, is represented by nine numbers. (a_{ij}) straddles, as it were, two sets of axes; $[T_{ij}]$ is tied to one set at a time. One cannot speak, for instance, of transforming (a_{ij}) to another set of axes, for it would not mean anything.

3.2. Symmetrical and antisymmetrical tensors.
A tensor $[T_{ij}]$ is said to be *symmetrical* if $T_{ij} = T_{ji}$. Thus

$$\begin{bmatrix} 5 & 2 & -3 \\ 2 & 8 & 4 \\ -3 & 4 & 12 \end{bmatrix}$$

is a symmetrical tensor.

A tensor $[T_{ij}]$ is said to be *antisymmetrical* or *skew-symmetrical* if $T_{ij} = -T_{ji}$. This implies that $T_{11} = T_{22} = T_{33} = 0$. Thus

$$\begin{bmatrix} 0 & -\gamma & \beta \\ \gamma & 0 & -\alpha \\ -\beta & \alpha & 0 \end{bmatrix}$$

is the typical antisymmetrical tensor (the reason for the distribution of minus signs will appear in Chapter II, § 2).

It is important to notice that the property of a tensor of being symmetrical or antisymmetrical is independent of the axes of reference. Thus, if $T_{ij} = \pm T_{ji}$, then $T'_{ij} = \pm T'_{ji}$. The proof is straightforward and is left to the reader.

4. The representation quadric

We have seen, from the example of conductivity and the other examples in Table 1, that many crystal properties whose values depend on the direction in which they are measured are represented by second-rank tensors. Other crystal properties are represented by tensors of higher or lower rank (and some not by tensors at all). It is natural, then, to look for some geometrical interpretation of a tensor, and, in this section, we give a geometrical representation of a second-rank tensor.

Consider the equation
$$S_{ij} x_i x_j = 1,$$ (26)

where the S_{ij} are coefficients; we do not at this stage say that the S_{ij} are the components of a tensor. Performing the summations with respect to i and j in (26), we have

$$S_{11} x_1^2 \ + S_{12} x_1 x_2 + S_{13} x_1 x_3 +$$
$$+ S_{21} x_2 x_1 + S_{22} x_2^2 \ + S_{23} x_2 x_3 +$$
$$+ S_{31} x_3 x_1 + S_{32} x_3 x_2 + S_{33} x_3^2 \ = 1.$$

If we now let $S_{ij} = S_{ji}$ and collect the terms, we obtain

$$S_{11} x_1^2 + S_{22} x_2^2 + S_{33} x_3^2 + 2S_{23} x_2 x_3 + 2S_{31} x_3 x_1 + 2S_{12} x_1 x_2 = 1.$$

This is the general equation of a second-degree surface (a quadric) referred to its centre as origin. It may, in general, be an ellipsoid or a hyperboloid.†

Equation (26) may be transformed to new axes Ox_i' by using equations (20) in the forms

$$x_i = a_{ki} x_k' \quad \text{and} \quad x_j = a_{lj} x_l'.$$

We obtain
$$S_{ij} a_{ki} a_{lj} x_k' x_l' = 1,$$
which may be written

$$S_{kl}' x_k' x_l' = 1, \quad \text{where} \quad S_{kl}' = a_{ki} a_{lj} S_{ij}.$$

If this is compared with the second-rank tensor transformation law

$$T_{ij}' = a_{ik} a_{jl} T_{kl},$$ (22)

it is seen to be identical so far as the relative position of the suffixes is concerned, which is all that is significant. Now we have put $S_{ij} = S_{ji}$. It follows that the *coefficients* S_{ij} *of the quadric* (26) *transform like the components of a symmetrical tensor of the second rank.* The theory of the transformation of a symmetrical second-rank tensor is thus identical with the theory of the transformation of a quadric; to find how the

† For the properties of quadrics (or conicoids) see Bell (1937) or Eisenhart (1939).

components of such a tensor transform we need only examine the transformation of the corresponding quadric. For this reason the surface (26) is called the *representation quadric* for the tensor S_{ij}.†

All the second-rank tensors given in Table 1, p. 7, are symmetrical; and, indeed, all except one of the second-rank tensors describing crystal properties that are mentioned in this book are symmetrical (the exception is the thermoelectric tensor). Thus, for example, in the electrical conductivity tensor we have

$$\sigma_{ij} = \sigma_{ji}. \tag{27}$$

The proof of symmetry always involves thermodynamical considerations, and we postpone it until later.‡ Meanwhile, it is worth noting that the relation is not an obvious one. In the case of conductivity, for instance, it states that the conductivity coefficient found by applying **E** in the x_1 direction and measuring the component of **j** in the x_2 direction (σ_{21}) will be the same as that found by applying **E** in the x_2 direction and measuring the component of **j** in the x_1 direction (σ_{12}).

Our conclusion is: *a representation quadric can be used to describe any symmetrical second-rank tensor, and, in particular, it can be used to describe any crystal property which is given by such a tensor.*

4.1. Principal axes. An important property of a quadric is the possession of *principal axes*. These are three directions at right angles such that, when the general quadric (26) is referred to them as axes, its equation takes the simpler form

$$S_1 x_1^2 + S_2 x_2^2 + S_3 x_3^2 = 1. \tag{28}$$

We shall not prove that a transformation which will accomplish this simplification exists,‖ but in Chapter II, § 3, we indicate how the directions of the axes and their lengths can be found (a practical method is illustrated in Chapter IX, § 7.2). Just as any quadric takes its simplest form when referred to its principal axes, so likewise does a symmetrical second-rank tensor. Thus, when

$$[S_{ij}] = \begin{bmatrix} S_{11} & S_{12} & S_{31} \\ S_{12} & S_{22} & S_{23} \\ S_{31} & S_{23} & S_{33} \end{bmatrix}$$

---

† The physical dimensions of x_1, x_2, x_3 in equation (26) are determined by those of the S_{ij}; they are in fact $1/\sqrt{}$(dimensions of the S_{ij}). This need cause no difficulty when it is recalled that one is accustomed in physics to plotting graphs in which distances along the axes represent not only pure numbers but mass, velocity, temperature, etc.

‡ For thermodynamically reversible properties see pp. 57–60, 74 and 182; for irreversible properties, see pp. 205–12.

‖ For a rigorous proof of the existence of principal axes see Eisenhart (1939), pp. 256–62.

is transformed to its principal axes it becomes

$$\begin{bmatrix} S_1 & 0 & 0 \\ 0 & S_2 & 0 \\ 0 & 0 & S_3 \end{bmatrix},$$

this array being formed from the coefficients in equation (28). S_1, S_2, S_3

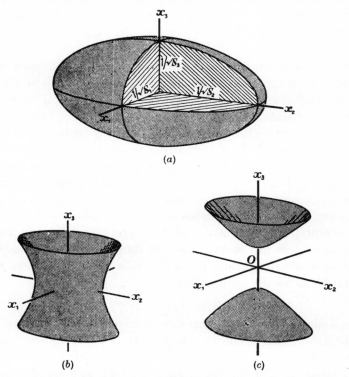

(a)

(b)   (c)

FIG. 1.5. The representation quadric for the tensor $[S_{ij}]$, as (a) an ellipsoid,
(b) a hyperboloid of one sheet, and (c) a hyperboloid of two sheets.

are called the *principal components* of the tensor $[S_{ij}]$ or of the property
that it represents.

By comparing (28) with the standard equation

$$\frac{x^2}{a^2} + \frac{y^2}{b^2} + \frac{z^2}{c^2} = 1,$$

it is clear that the semi-axes of the representation quadric are of lengths
$1/\sqrt{S_1}$, $1/\sqrt{S_2}$, $1/\sqrt{S_3}$ (Fig. 1.5 a). If S_1, S_2, S_2 are all positive, the surface
(28) is an ellipsoid (Fig. 1.5 a). If two coefficients are positive and one
negative, it is a hyperboloid of one sheet (Fig. 1.5 b), for it can be seen

that two principal sections (central sections perpendicular to the axes) are hyperbolae and one is an ellipse. If one coefficient is positive and two are negative, (28) is a hyperboloid of two sheets (Fig. 1.5c), for two principal sections are hyperbolae and one is an imaginary ellipse. If all three coefficients are negative, the surface is an imaginary ellipsoid.

[In a symmetrical tensor referred to arbitrary axes the number of independent components is six. If the tensor is referred to its principal axes, the number of independent components is reduced to three; the number of 'degrees of freedom' is nevertheless still six, for three independent quantities are needed to specify the directions of the axes, and three to fix the magnitudes of the principal components.]

4.2. Simplification of equations when referred to the principal axes. We now return to equation (9) in which the tensor $[T_{ij}]$ relates the vector **p** to the vector **q**. If $[T_{ij}]$ is symmetrical, we may replace it by $[S_{ij}]$ to give

$$p_i = S_{ij} q_j. \tag{29}$$

If $[S_{ij}]$ is now referred to its principal axes, equations (29) written out in full are simply

$$p_1 = S_1 q_1, \quad p_2 = S_2 q_2, \quad p_3 = S_3 q_3; \tag{30}$$

compare equations (5).

To illustrate the results of this simplification let us consider again the example of electrical conductivity that we began with on p. 4. We have

$$j_1 = \sigma_1 E_1, \quad j_2 = \sigma_2 E_2, \quad j_3 = \sigma_3 E_3, \tag{31}$$

where σ_1, σ_2, σ_3 are the principal components of the conductivity tensor, or, shortly, the *principal conductivities*. If **E** is parallel to Ox_1, so that $E_2 = E_3 = 0$, then $j_2 = j_3 = 0$. Hence, in this case, **j** is also parallel to Ox_1. Thus, *when* **E** *is directed along any of the three principal axes the situation is particularly simple*: **j** *is parallel to* **E** *but the conductivity is different for the three axes.*

We now consider a case in which **E** is not parallel to a principal axis. Suppose that $\mathbf{E} = [E_1, E_2, 0]$. Then $j_1 = \sigma_1 E_1, j_2 = \sigma_2 E_2, j_3 = 0$. Fig. 1.6

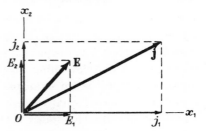

Fig. 1.6. Two-dimensional illustration of the fact that when $\sigma_1 \neq \sigma_2$, **j** is not parallel to **E**.

shows the relation between **j** and **E**. Starting with **E** we construct E_1 and E_2. Multiplying these by σ_1 and σ_2 gives j_1 and j_2, which then compound to give **j**. The fact that **j** is not parallel to **E** is a direct consequence of the fact that $\sigma_1 \neq \sigma_2$. In a three-dimensional case, with $E_3 \neq 0$, we should again naturally find that $\mathbf{j}$ is not parallel to $\mathbf{E}$. In Fig. 1.6, $\sigma_1 > \sigma_2$; Ox_1 may thus be thought of as a 'direction of easy conductivity' and $\mathbf{j}$ has a tendency to swing towards it.

5. The effect of crystal symmetry on crystal properties

We must now leave our discussion of second-rank tensor properties for a while and broaden our outlook to include all crystal properties. We have to examine the question of how the symmetry of a crystal is related to the symmetry of its physical properties. The key to this question is a fundamental postulate of crystal physics, known as *Neumann's Principle*. It may be stated as follows:

The symmetry elements of any physical property of a crystal must include the symmetry elements of the point group of the crystal.

The point group of a crystal is the group of macroscopic symmetry elements that its structure possesses. It is the basis for the division of crystals into the 32 crystal classes. In what follows the reader is assumed to be familiar with the notions of 'symmetry elements', 'crystal class', and 'crystal system'. A survey of these and other essential concepts in the theory of crystal symmetry is given in Appendix B, which also includes, on pp. 284–8, a table of the symmetry elements of the 32 crystal classes.

It is important to notice that Neumann's Principle does not state that the symmetry elements of a physical property are the same as those of the point group. It merely says that the symmetry elements of a physical property must *include* those of the point group. The physical properties may, and often do, possess more symmetry than the point group. As an example of the principle we have the fact that cubic crystals are optically isotropic. The physical property in this case is completely isotropic, and so it certainly possesses the symmetry elements of all the cubic point groups, as the principle demands. Again, consider as an example the optical properties of a crystal of class *3m* of the trigonal system, say tourmaline. It is known that the variation of refractive index with direction is represented by the indicatrix, which in this case is an ellipsoid of revolution about the triad axis (the optic axis). This ellipsoid possesses the vertical triad axis and the three vertical planes of symmetry of the point group *3m*, as the principle demands. However, as the principle allows, the ellipsoid also possesses a centre of symmetry and several other symmetry elements which are not in the point group.

It is worth examining more closely what is meant by 'the symmetry of a physical property'. A physical property of a crystal consists of a relation between certain measurable quantities associated with the crystal. For example, the elasticity of a crystal is a certain relation between a homogeneous stress and a homogeneous strain in the crystal. Now suppose we wish to know whether a property has a certain symmetry element or not. First we measure the property relative to some fixed axes. Then we operate with the potential symmetry element on the crystal,† and again investigate the relation between the measured quantities, taking our measurements in the same directions as before, relative to the fixed axes. If the relation between the measured quantities is unchanged, we say that the property in question, in this particular crystal, possesses this symmetry element.

It is clear that it does not matter whether we operate with the symmetry element on the crystal or on the quantities we measure. Thus, for example, if we were measuring elasticity and wished to test for a centre of symmetry, we could invert the stresses and strains through a centre, instead of inverting the crystal. If we inverted the stresses and strains no change would be apparent in them, because a state of homogeneous stress or strain is already centrosymmetrical. The measurements would therefore be the same as before, and so would the elasticity. This would be so whatever the symmetry of the crystal. We say, then, that elasticity (which is given by a fourth-rank tensor) is a centrosymmetrical property.

All second-rank tensor properties are centrosymmetrical. For, if we take the defining equation, $p_i = T_{ij} q_j$, and reverse the directions of **p** and **q**, the signs of all the components p_i and q_i will change. The equation will then still be satisfied by the same values of T_{ij} as before, and so the value of the property represented by T_{ij} will be unchanged.

In summary, then, a physical property may have certain inherent symmetry—that is, symmetry that is displayed whatever the symmetry, or lack of it, possessed by the crystal. In accordance with Neumann's Principle, however, a physical property in a given crystal must also have any extra symmetry elements that are possessed by the crystal. Thus, the elasticity of, say, a hexagonal crystal is not only centrosymmetrical, but also possesses all the symmetry elements of the crystal.

† The philosophically-minded may object that we cannot actually do this for operations involving a reflection or an inversion. This is true and, to this extent, our definition is not strictly an 'operational' one. However, if we are to link physics to the mathematical theory of symmetry it is difficult to avoid the use of such unperformable operations.

EXERCISE 1.2. What is the fallacy in the following?

The property of diffracting X-rays is a physical property of a crystal. By means of X-ray patterns the *space group* of a crystal can be deduced. But Neumann's Principle states that the symmetry of the property depends on the *point group*. Therefore Neumann's Principle does not apply to X-ray diffraction.

5.1. The effect of crystal symmetry on second-rank tensor properties. The principles of the last section apply to all the physical properties of crystals. We return now to those properties represented by symmetrical second-rank tensors. These tensors, as we have seen, have six independent components when they are referred to arbitrary axes. When the crystal has symmetry, however, the number of independent components is reduced. This is most easily investigated by considering the representation quadric. This surface, whose equation contains as many independent coefficients as there are independent components in a symmetrical second-rank tensor, represents the tensor property completely. Its symmetry is the inherent symmetry of the crystal property. The quadric is to be thought of as fixed in orientation relative to the crystal. We have, therefore, to discuss how the representation quadric is related to the symmetry elements of the crystal. By Neumann's Principle our procedure must be to make sure that each symmetry element of the crystal is also shown by the representation quadric. We do this now for each crystal class in turn. (The procedure is identical with that used in crystal optics in relating the indicatrix to the crystal symmetry.) The results are collected in Table 3. The tabulation is simplified because the restrictions imposed by the crystal symmetry happen to be the same within any one crystal *system*. This is not the case with an unsymmetrical second-rank tensor (p. 227), or with tensors of the first rank (p. 79) or of higher rank than the second (pp. 123–4, 140–1, 247–8, 250–1).

(i) *Cubic system.* The only way in which the representation quadric can have four triad axes is by being a sphere. It then possesses all the symmetry elements that can be shown by cubic crystals (Table 21, p. 288). The property is entirely determined when the radius of the sphere is given. The particular axes to which the tensor is referred are of no consequence, for it takes the same form, that shown in Table 3, for all choices of axes.

(ii) *Uniaxial group of systems (tetragonal, hexagonal and trigonal).* A general quadric possesses three rotation diad axes at right angles, three planes of symmetry perpendicular to the diad axes, and a centre of symmetry (the symmetry of the orthorhombic class *mmm*). The only way in which it can possess a 4-, 6- or 3-fold axis is for this axis to be a principal axis, and for the quadric to be a surface of revolution about it. The quadric will then automatically possess all the other symmetry elements that are found in the various classes of these three systems: rotation diad axes normal to the principal axis, and planes

TABLE 3

The effect of crystal symmetry on properties represented by symmetrical second-rank tensors

Optical classi- fication	System	Characteristic symmetry (see p. 280)†	Nature of repre- sentation quadric and its orientation	Number of inde- pendent coefficients	Tensor referred to axes in the conventional orientation‡
Isotropic (anaxial)	Cubic	4 3-fold axes	*Sphere*	1	$\begin{bmatrix} S & 0 & 0 \\ 0 & S & 0 \\ 0 & 0 & S \end{bmatrix}$
Uniaxial	Tetragonal Hexagonal Trigonal	1 4-fold axis 1 6-fold axis 1 3-fold axis	*Quadric of revo- lution* about the principal sym- metry axis $(x_3)(z)$	2	$\begin{bmatrix} S_1 & 0 & 0 \\ 0 & S_1 & 0 \\ 0 & 0 & S_3 \end{bmatrix}$
Biaxial	Orthorhom- bic	3 mutually perpendicular 2-fold axes; no axes of higher order	*General quadric* with axes $(x_1, x_2, x_3) \parallel$ to the diad axes (x, y, z)	3	$\begin{bmatrix} S_1 & 0 & 0 \\ 0 & S_2 & 0 \\ 0 & 0 & S_3 \end{bmatrix}$
	Monoclinic	1 2-fold axis	*General quadric* with one axis $(x_2) \parallel$ to the diad axis (y)	4	$\begin{bmatrix} S_{11} & 0 & S_{31} \\ 0 & S_2 & 0 \\ S_{31} & 0 & S_{33} \end{bmatrix}$
	Triclinic	A centre of symmetry or no symmetry	*General quadric.* No fixed rela- tion to crystal- lographic axes	6	$\begin{bmatrix} S_{11} & S_{12} & S_{31} \\ S_{12} & S_{22} & S_{23} \\ S_{31} & S_{23} & S_{33} \end{bmatrix}$

  † Axes of symmetry may be rotation axes or inversion axes.
  ‡ The setting of the reference axes x_1, x_2, x_3 in column 6 in relation to the crystallo- graphic axes x, y, z and to the symmetry elements is that shown in column 4. For further notes on axial conventions, see Appendix B.

of symmetry parallel and perpendicular to the principal axis. Two numbers, the lengths of the major and minor axes, are thus sufficient to determine the quadric completely. For writing down the tensor com- ponents we choose the x_3 axis parallel to the principal axis of symmetry of the crystal. These systems are the ones that are optically uniaxial.

(iii) *Orthorhombic system.* A general quadric already possesses three mutually perpendicular diad axes, and we simply have to arrange the axes of the quadric parallel to the crystallographic diad axes. Moreover, since the symmetry of the quadric is that of the holosymmetric class *mmm*, it possesses all the symmetry elements that are found in the classes of the orthorhombic system. The lengths of the three axes of the quadric determine the property completely. If the tensor is referred to the *x*, *y*, *z* crystallographic axes it takes the form shown in Table 3.

(iv) *Monoclinic system*. Here we must arrange one of the diad axes of the general quadric parallel to the crystallographic diad axis, and we notice that, when this is done, the quadric has all the symmetry elements that are found in the classes of the monoclinic system. Apart from the one fixed axis, though, the quadric is free to take up any orientation. We need three numbers to specify the lengths of the quadric axes, that is, the shape of the quadric, and one more to decide the orientation (say the angle through which it has to be turned from some arbitrary position). We set the reference axes so that x_2 is parallel to the crystallographic diad axis.† Then the equation of the quadric will contain no terms involving x_2 except the x_2^2 term. The tensor components (coefficients of the quadric) will therefore be as shown in the table. There are four independent components. Of course, by choosing principal axes we could reduce the number of different components from four to three, but we should then need an additional number to tell us where the principal axes lay with respect to the crystallographic *x* or *z* axis. This point is clearly brought out when we consider that for different properties in the same monoclinic crystal the two principal axes perpendicular to the diad have different directions. There is analogous behaviour in crystal optics: when a monoclinic crystal is heated, the principal directions of the indicatrix rotate about the *y* axis; they also rotate as the wavelength is changed, and so cause dispersion of the optic axes.

(v) *Triclinic system*. Class *1*, which is completely unsymmetrical, evidently imposes no restrictions on the representation quadric. The same is true of class $\bar{1}$, for its centre of symmetry is already possessed by the quadric. The number of independent constants needed to specify the property is therefore equal to the number of independent components of the tensor. They can be thought of as being three to give the lengths of the axes of the quadric and three to specify its orientation.

6. The magnitude of a property in a given direction

6.1. Definition. In discussing second-rank tensor properties of crystals one often uses phrases like 'the conductivity in the direction

---

† An alternative setting with x_3 parallel to the diad is sometimes used.

[100]' or 'the susceptibility in the direction [112]' or whatever direction it may be. The concept of the magnitude of a property in a given direction is one that needs careful definition, because of the lack of parallelism between the vectors involved. Take electrical conductivity as an example (Fig. 1.7). We apply a field **E** in a crystal. There will then, in general, be components of **j** both parallel and transverse to **E**. *The conductivity σ in the direction of* **E** *is then defined to be the component of* **j** *parallel to* **E** *divided by* E, *that is,* $j_{\parallel}/E$. If unit field is applied ($E = 1$), it is simply $j_{\parallel}$. This definition is appropriate because it is $j_{\parallel}$ that is measured in simple experiments. Con-

sider the experiment, for instance, in which a slab of crystal is placed between parallel plate electrodes made of highly conducting material. If we divide the current flowing through the faces by the potential difference between the faces, we evidently obtain a conductance, and hence a conductivity. It may be proved that

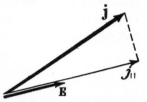

FIG. 1.7. The conductivity in the direction of **E** is $j_{\parallel}/E$.

what is actually measured is the conductivity perpendicular to the slab, in the sense we have just defined. The theory of the corresponding thermal conductivity experiment is worked out explicitly in Chapter XI, § 2 (i); many similar examples occur throughout the book.

In general, if $p_i = S_{ij} q_j$, *the magnitude S of the property* $[S_{ij}]$ *in a certain direction is obtained by applying* **q** *in that direction and measuring* $p_{\parallel}/q$, *where* $p_{\parallel}$ *is the component of* **p** *parallel to* **q**.

6.2. Analytical expressions. Taking conductivity as a typical example we now derive analytical expressions for its magnitude in a given direction.

(i) *Referred to principal axes.* Take the principal axes of conductivity as reference axes and let the direction in question have direction cosines l_1, l_2, l_3. l_1, l_2, l_3 form the components of a vector of unit length (unit vector) l_i. Applying **E** in this direction we have

$$\mathbf{E} = [l_1 E,\ l_2 E,\ l_3 E].$$

Hence, from (31), $\mathbf{j} = [\sigma_1 l_1 E,\ \sigma_2 l_2 E,\ \sigma_3 l_3 E]$.

The component of **j** parallel to **E** is the sum of the components of **j** each resolved along **E**, namely,

$$j_{\parallel} = l_1^2 \sigma_1 E + l_2^2 \sigma_2 E + l_3^2 \sigma_3 E.$$

The magnitude σ of the conductivity in the direction l_i is thus

$$\sigma = l_1^2 \sigma_1 + l_2^2 \sigma_2 + l_3^2 \sigma_3. \tag{32}$$

An important special case of (32) arises with the optically uniaxial group of crystal systems (tetragonal, hexagonal and trigonal). Referring to Table 3, (32) becomes

$$\sigma = \sigma_1(l_1^2 + l_2^2) + \sigma_3 l_3^2 = \sigma_1(1 - l_3^2) + \sigma_3 l_3^2,$$

or
$$\sigma = \sigma_1 \sin^2\theta + \sigma_3 \cos^2\theta,$$

where θ is the angle between l_i and the principal symmetry axis Ox_3.

(ii) *Referred to general axes.* To find a corresponding expression to (32) referred to general axes we proceed as follows. Let l_i now be the direction cosines of $\mathbf{E}$ referred to general axes, so that $E_i = El_i$. The component of $\mathbf{j}$ parallel to $\mathbf{E}$ is $(\mathbf{j} \cdot \mathbf{E})/E$, or in suffix notation $(j_i E_i)/E$.† Therefore the conductivity in the direction l_i is

$$\sigma = \frac{j_i E_i}{E^2} = \frac{\sigma_{ij} E_j E_i}{E^2}$$

or,
$$\sigma = \sigma_{ij} l_i l_j. \tag{33}$$

This is the general expression for σ, of which (32) is a specialization.

It is sometimes convenient to choose new axes of reference x_i' and to take one of them, say x_1', in the direction l_i. Then, referring to the direction cosine table (11), we see that $l_i = a_{1i}$. Thus (33) becomes

$$\sigma = \sigma_{ij} a_{1i} a_{1j} = \sigma_{11}', \tag{34}$$

from (22). Hence, *the magnitude of σ in the direction x_1' is σ_{11}'*. This result is also evident from the definition of σ_{11}'.

Similar expressions to (32), (33) and (34) hold for any property represented by a symmetrical second-rank tensor; (33) and (34) also hold when the tensor is unsymmetrical.

7. Geometrical properties of the representation quadric

7.1. Length of the radius vector. We may now find a geometrical interpretation of σ. Let P be a general point on the ellipsoid

$$\sigma_{ij} x_i x_j = 1, \tag{35}$$

shown in Fig. 1.8. If the direction cosines of OP are l_i, we have

$$x_i = rl_i,$$

where $OP = r$. Therefore, substituting in (35),

$$r^2 \sigma_{ij} l_i l_j = 1;$$

or, from (33),
$$\sigma = 1/r^2, \qquad r = 1/\sqrt{\sigma}. \tag{36}$$

† See Appendix A for scalar product notation.

We have already met special cases of this property, in that the radius vectors in the directions of the semi-axes are of lengths $1/\sqrt{\sigma_1}$, $1/\sqrt{\sigma_2}$, $1/\sqrt{\sigma_3}$.

The same considerations apply to any second-rank symmetrical tensor property S_{ij}. In general

$$S = 1/r^2, \qquad r = 1/\sqrt{S}; \tag{37}$$

that is to say, *the length r of any radius vector of the representation quadric*

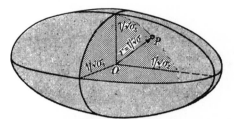

FIG. 1.8. Illustrating the properties of the representation ellipsoid for conductivity:
$$\sigma_{ij} x_i x_j = 1.$$

is equal to the reciprocal of the square root of the magnitude S of the property in that direction.

Case when $S_{ij} x_i x_j = 1$ is a hyperboloid. If the quadric $S_{ij} x_i x_j = 1$ is a hyperboloid the above analysis still holds, but the length of some of the radius vectors is imaginary. When this occurs it is more convenient to consider, instead, the quadric

$$S_{ij} x_i x_j = -1,$$

which the radius vector will meet in a real point.† The radius vector of this quadric is evidently such that

$$S = -1/r^2, \qquad r = 1/\sqrt{(-S)}. \tag{38}$$

Case when $S_{ij} x_i x_j = 1$ is an imaginary ellipsoid. In this case it is evidently more appropriate to consider the quadric

$$S_{ij} x_i x_j = -1,$$

which is a real ellipsoid.

Thus, in general, we use the real branches of the quadrics

$$S_{ij} x_i x_j = \pm 1, \tag{39}$$

and we use equations (37) or (38) respectively according to whether the radius vector meets a real branch of

$$S_{ij} x_i x_j = +1, \quad \text{or} \quad S_{ij} x_i x_j = -1.$$

† If $S_{ij} x_i x_j = 1$ is a hyperboloid of two sheets, $S_{ij} x_i x_j = -1$ will be a hyperboloid of one sheet and vice versa.

7.2. The radius-normal property. As in § 6.2 (i) we use the example of conductivity and take Ox_i as the principal axes of σ_{ij}. As before, let $\mathbf{E} = [l_1 E,\; l_2 E,\; l_3 E]$; then

$$\mathbf{j} = [\sigma_1 l_1 E,\; \sigma_2 l_2 E,\; \sigma_3 l_3 E].$$

Therefore, the direction cosines of $\mathbf{j}$ are proportional to

$$\sigma_1 l_1,\quad \sigma_2 l_2,\quad \sigma_3 l_3.$$

If P is a point on $\sigma_1 x_1^2 + \sigma_2 x_2^2 + \sigma_3 x_3^2 = 1,$

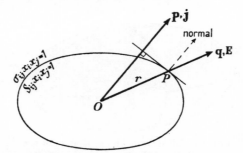

FIG. 1.9. The radius-normal property of the representation ellipsoid. The figure shows the central section of the ellipsoid which contains the radius vector OP and the normal from O on to the tangent plane at P. The tangent plane is thus seen on edge.

(Fig. 1.9) such that OP is parallel to $\mathbf{E}$, then $P = (rl_1, rl_2, rl_3)$, where $OP = r$. The tangent plane at P is

$$rl_1 \sigma_1 x_1 + rl_2 \sigma_2 x_2 + rl_3 \sigma_3 x_3 = 1.$$

Therefore the normal at P has direction cosines proportional to

$$l_1 \sigma_1,\quad l_2 \sigma_2,\quad l_3 \sigma_3.$$

Hence, the normal at P is parallel to $\mathbf{j}$.

This result is evidently general. *If $p_i = S_{ij} q_j$, the direction of $\mathbf{p}$ for a given $\mathbf{q}$ may be found by first drawing, parallel to $\mathbf{q}$, a radius vector OP of the representation quadric, and then taking the normal to the quadric at P.*

Case when $S_{ij} x_i x_j = 1$ is a hyperboloid. If $S_{ij} x_i x_j = 1$ is a hyperboloid, a vector parallel to $\mathbf{q}$ through O may meet the surface in an imaginary point. In this case it is better, as before, to consider the quadric $S_{ij} x_i x_j = -1$. If the radius vector parallel to $\mathbf{q}$ meets this quadric at P, the normal at P is parallel to $\mathbf{p}$, exactly as before.

Fig. 1.10 shows a principal hyperbolic section through a hyperboloid $S_{ij} x_i x_j = 1$. The real branch of the hyperboloid $S_{ij} x_i x_j = -1$ is indicated by a broken line. The construction for the direction of $\mathbf{p}$ has been performed for two different directions of $\mathbf{q}$, OP and OP' respectively. P lies on the real branch of $S_{ij} x_i x_j = 1$

and the construction for the direction of **p** is straightforward. P' lies on the real branch of $S_{ij}x_i x_j = -1$. S for this direction is negative, by equation (38). The component of **p** parallel to **q** is therefore negative, and the sense of **p** is therefore as shown. As **q** rotates anti-clockwise from the direction OA it will be seen from the figure that **p** will rotate clockwise. When **q** reaches the asymptote, **p** will be perpendicular to **q**. After **q** crosses the asymptote **p** continues to rotate anti-clockwise until, when the radius vector reaches OB, **p** and **q** are anti-parallel, corresponding to a negative value of one of the principal components of S_{ij}.

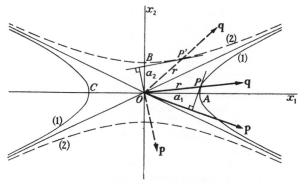

Fig. 1.10. The properties of the representation quadric when it is a hyperboloid. Sections are shown of (1) the hyperboloid

$$S_{ij}x_i x_j = \frac{x_1^2}{a_1^2} - \frac{x_2^2}{a_2^2} + \frac{x_3^2}{a_3^2} = 1,$$

and (2) the hyperboloid

$$S_{ij}x_i x_j = \frac{x_1^2}{a_1^2} - \frac{x_2^2}{a_2^2} + \frac{x_3^2}{a_3^2} = -1.$$

As **q** continues to rotate anti-clockwise **p** moves clockwise again, and eventually, when **q** reaches OC, the two vectors become parallel; this corresponds to a positive value of one of the principal components of S_{ij}.

Case when $S_{ij}x_i x_j = 1$ is an imaginary ellipsoid. We consider the (real) ellipsoid $S_{ij}x_i x_j = -1$, and the construction for the direction of **p** is exactly as before, except that **p** will be in the direction of the inward rather than the outward normal.

7.3. Summary of geometrical properties. We may summarize these two properties of the representation quadric as follows (Figs. 1.8, 1.9 and 1.10). Let **p** and **q** be two vectors connected by the symmetrical second-rank tensor S_{ij} by the relation $p_i = S_{ij}q_j$. Draw the real branches of the quadrics $S_{ij}x_i x_j = \pm 1$. Draw a radius, of length r, parallel to **q**. Then:

(1) The component of **p** parallel to **q** is given by $p_{\parallel} = Sq$, where $S = \pm 1/r^2$, the positive sign to be taken when the radius is to $S_{ij}x_i x_j = 1$, and the negative when it is to $S_{ij}x_i x_j = -1$.

(2) The direction of **p** is the same as that of the normal at the end of the radius vector: the outward normal for $S_{ij} x_i x_j = 1$ and the inward normal for $S_{ij} x_i x_j = -1$.

These two results are obviously true when **q** is parallel to one of the principal axes.

SUMMARY

1. **Transformation of axes** (§§ 2.1, 2.3). A transformation from one orthogonal set of axes to another is given, in the dummy suffix notation, by

$$x_i' = a_{ij} x_j.$$

2. **Definition of a tensor** (§ 3). A tensor is a physical quantity whose components transform according to the following laws.

3. **Transformation laws for tensors** (Table 2, p. 13).

Zero-rank tensor (scalar): $\phi' = \phi$;

First-rank tensor (vector): $p_i' = a_{ij} p_j$;

Second-rank tensor: $T_{ij}' = a_{ik} a_{jl} T_{kl}$;

and so on for higher ranks.

The transformation law for a vector is the same as that for the coordinates. The transformation law for a second-rank tensor is the same as that for products of coordinates (§ 2.5).

4. **Vector-vector relation** (§§ 1, 1.1). If the vectors p_i and q_i are connected by the linear relation

$$p_i = T_{ij} q_j,$$

the coefficients T_{ij} form a second-rank tensor (§ 3).

SECOND-RANK TENSORS

5. **Symmetrical and antisymmetrical tensors** (§ 3.2). $[T_{ij}]$ is symmetrical if $T_{ij} = T_{ji}$, and antisymmetrical if $T_{ij} = -T_{ji}$.

6. **Principal axes** (§ 4.1). A symmetrical tensor $[S_{ij}]$ has principal axes; when referred to them it takes the form

$$\begin{bmatrix} S_1 & 0 & 0 \\ 0 & S_2 & 0 \\ 0 & 0 & S_3 \end{bmatrix},$$

and linear relations $p_i = S_{ij} q_j$ are simplified (§ 4.2) to

$$p_1 = S_1 q_1, \qquad p_2 = S_2 q_2, \qquad p_3 = S_3 q_3.$$

7. **Magnitude of a property** (§ 6.1). If $p_i = S_{ij} q_j$, the magnitude S of the property $[S_{ij}]$ in a certain direction is given by applying **q** in that direction and measuring $p_\parallel / q$, where $p_\parallel$ is the component of **p** parallel to **q**. S in the direction l_i is given by (§ 6.2) $S = S_{ij} l_i l_j$,

or, referred to principal axes,

$$S = S_1 l_1^2 + S_2 l_2^2 + S_3 l_3^2.$$

If Ox_1' is in the chosen direction,

$$S = S_{11}'.$$

8. Representation quadric (§ 4). The representation quadric for the symmetrical tensor $[S_{ij}]$ is defined as

$$S_{ij}\,x_i\,x_j = 1;$$

or, referred to principal axes (§ 4.1),

$$S_1 x_1^2 + S_2 x_2^2 + S_3 x_3^2 = 1.$$

Radius vector property (§ 7.1). The radius vector r of the representation quadric is connected with the magnitude S of the property in that direction by

$$S = 1/r^2, \qquad r = 1/\sqrt{S}.$$

Radius-normal property (§ 7.2). If $p_i = S_{ij}q_j$, and if a radius vector OP of the representation quadric is drawn parallel to $\mathbf{q}$, the direction of $\mathbf{p}$ is that of the normal at P.

Imaginary radius vectors can be avoided by choosing the quadric

$$S_{ij}\,x_i\,x_j = -1,$$

and making appropriate changes of sign (§ 7.3).

9. Effect of crystal symmetry (§§ 5, 5.1). The representation quadric for a second-rank tensor property of a crystal has a shape and orientation determined by the crystal symmetry in accordance with Neumann's Principle (see Table 3, p. 23).

———

EXERCISE 1.3. Readers who wish to test their understanding of this chapter by a simple graphical exercise may try the following.

[1] The electrical conductivity tensor of a certain crystal has the following components referred to axes x_1, x_2, x_3:

$$[\sigma_{ij}] = \begin{bmatrix} 25 \times 10^7 & 0 & 0 \\ 0 & 7 \times 10^7 & -(3\sqrt{3}) \times 10^7 \\ 0 & -(3\sqrt{3}) \times 10^7 & 13 \times 10^7 \end{bmatrix}$$

in m.k.s. units (ohm^{-1} m^{-1}). The axes are now transformed to a new set x_1', x_2', x_3' given by the following angles:

$$x_1' O x_1 = 0°, \qquad x_2' O x_2 = 30°, \qquad x_2' O x_3 = 60°, \qquad x_3' O x_3 = 30°.$$

Draw up a table of the form (11) (p. 9) for this transformation, and check that the sum of the squares of the a_{ij} in each row and column is 1.

[2] Determine the values of the components σ_{ij}', and comment on the result obtained.

[3] Draw on the new axes x_2', x_3' a section of the conductivity ellipsoid (representation quadric) in the plane $x_1' = 0$, and notice that this is a principal section. Insert the old axes x_2, x_3 on the drawing.

[4] Draw a radius vector OP in the direction whose cosines referred to the old axes are $(0, \frac{1}{2}, \sqrt{3}/2)$. Measure the length of this radius vector and so find the electrical conductivity in this direction.

[5] Check the last result by using an analytical expression.

[6] Assume an electric field of 1 volt/m to be established in the direction OP. Calculate the components E_i along the x_i axes, and hence calculate the components of current density j_i.

[7] Insert these components j_i to scale on a vector diagram on the axes x_1, x_2, x_3, and hence determine graphically the magnitude and direction of the resultant current density.

[8] Assuming the same electric field as in [6], repeat the calculation [6] and the construction [7] using the x'_i axes instead of the x_i axes, and using the values of the σ'_{ij} found in [2]. Compare the result with that of [7].

[9] Compare the direction of the resultant current with that of the normal to the conductivity ellipsoid at the point P.

[10] Find graphically the component along OP of the resultant current density and hence find σ in this direction. Compare the value with those found in [4] and [5].

TRANSFORMATIONS AND SECOND-RANK TENSORS: FURTHER DEVELOPMENTS

THIS chapter constitutes a mathematical addendum to Chapter I, and those who wish to go straight on to the physical applications may omit it and pass on to Chapter III. Here we take up again some of the topics of Chapter I and carry them a stage farther.

1. Transformation of axes

1.1. Relations between the direction cosines a_{ij}. We remarked in § 2.1 of Chapter I that, for a transformation from one orthogonal set of axes Ox_i to another set Ox_i' given by

$$
\begin{array}{c|ccc}
 & x_1 & x_2 & x_3 \\
\hline
x_1' & a_{11} & a_{12} & a_{13} \\
x_2' & a_{21} & a_{22} & a_{23} \\
x_3' & a_{31} & a_{32} & a_{33}
\end{array}
\qquad (1)
$$

the nine coefficients a_{ij} are not independent. This is readily seen by considering the number of degrees of freedom of the transformation. If Ox_1, Ox_2, Ox_3 are given, two angles are necessary to specify the direction of Ox_1', the latitude and longitude for example; the new axes may still rotate about Ox_1', and so one further angle, an angle of rotation about Ox_1', is needed to fix them completely. Only three independent quantities are therefore needed to define the transformation, and accordingly we may expect to find six independent relations between the nine coefficients a_{ij}.

Since each row of the array (1) represents the three direction cosines of a straight line with respect to the orthogonal axes Ox_1, Ox_2, Ox_3, we have

$$
\left.
\begin{aligned}
a_{11}^2 + a_{12}^2 + a_{13}^2 &= 1, \quad \text{i.e.} \quad a_{1k}a_{1k} = 1 \\
a_{21}^2 + a_{22}^2 + a_{23}^2 &= 1, \quad \text{i.e.} \quad a_{2k}a_{2k} = 1 \\
a_{31}^2 + a_{32}^2 + a_{33}^2 &= 1, \quad \text{i.e.} \quad a_{3k}a_{3k} = 1
\end{aligned}
\right\}.
\qquad
\begin{aligned}
&(2.1) \\
&(2.2) \\
&(2.3)
\end{aligned}
$$

Hence,
$$a_{ik}a_{jk} = 1, \quad \text{if} \quad i = j. \qquad (3)$$

Again, since each pair of rows of (1) represents the direction cosines of two lines at right angles, we have (compare the scalar product of two vectors at right angles)

$$a_{21}a_{31}+a_{22}a_{32}+a_{23}a_{33} = 0, \quad \text{i.e.} \quad a_{2k}a_{3k} = 0 \tag{4.1}$$

$$a_{31}a_{11}+a_{32}a_{12}+a_{33}a_{13} = 0, \quad \text{i.e.} \quad a_{3k}a_{1k} = 0 \tag{4.2}$$

$$a_{11}a_{21}+a_{12}a_{22}+a_{13}a_{23} = 0, \quad \text{i.e.} \quad a_{1k}a_{2k} = 0 \tag{4.3}$$

Hence, $$a_{ik}a_{jk} = 0, \quad \text{if} \quad i \neq j. \tag{5}$$

Equations (3) and (5), which are called the *orthogonality relations*, can be combined into a single equation

$$a_{ik}a_{jk} = \delta_{ij} \tag{6}$$

by introducing the new symbol δ_{ij}, the *Kronecker delta*, defined by

$$\delta_{ij} = \begin{cases} 1 & (i = j), \\ 0 & (i \neq j). \end{cases} \tag{7}$$

By noticing that each *column* of (1) also represents a set of direction cosines, this time with respect to axes Ox'_i, of three mutually perpendicular lines, we obtain in a similar way the six orthogonality relations for the reverse transformation:

$$a_{ki}a_{kj} = \delta_{ij}. \tag{8}$$

It may be remarked, however, that these relations are not independent of relations (6), for they do not contain any essentially new information.

Note on matrices. When written out in rows and columns according to the values of i and j the δ_{ij} form the *unit matrix*, thus:

$$(\delta_{ij}) = \begin{pmatrix} \delta_{11} & \delta_{12} & \delta_{13} \\ \delta_{21} & \delta_{22} & \delta_{23} \\ \delta_{31} & \delta_{32} & \delta_{33} \end{pmatrix} = \begin{pmatrix} 1 & 0 & 0 \\ 0 & 1 & 0 \\ 0 & 0 & 1 \end{pmatrix}. \tag{9}$$

We enclose matrices within round brackets to distinguish them from the tensor arrays. Just as $[T_{ij}]$ stands for the whole array of T_{ij} components so (δ_{ij}) stands for the whole array of δ_{ij} components.

We have now introduced two arrays, (a_{ij}) and (δ_{ij}), and called them matrices. It is as well at this point to make clear the difference between matrices and tensors. A matrix is essentially a mathematical idea. Any array of numbers arranged in a rectangular table is a matrix. (It happens that the two matrices we have met so far have both been square, but later we shall meet others that are rectangular.) A tensor, on the other hand, is essentially a physical concept, for, as we have seen (p. 14), a tensor is a physical quantity (such as electrical conductivity). The formal similarity between matrices and tensors arises because in order to represent a tensor we make use of matrix arrays of coefficients. We have seen how, for a given second-rank tensor, the matrix of coefficients takes different forms according to the choice of reference axes. All tensors can be represented by matrices (and for second-rank tensors this method of representation is particularly natural); but all matrices do not necessarily represent tensors.

Substitution property of (δ_{ij}). If p_i is the ith component of a vector and we form the expression $\delta_{ij} p_i$, it will be seen, by performing the summation with respect to i and substituting the values of the δ_{ij}, that

$$\delta_{ij} p_i = p_j.$$

Thus, the effect of multiplying p_i by δ_{ij} is to substitute j for i. Similarly,

$$\delta_{ij} p_j = p_i;$$

and, in the same way, for a tensor component T_{jl}

$$\delta_{ij} T_{jl} = T_{il} \quad \text{and} \quad \delta_{il} T_{jl} = T_{ji}.$$

Because of this convenient property (δ_{ij}) is sometimes called the *substitution matrix*.

EXERCISE 2.1. Prove, using the dummy suffix notation, that the square of the length of a vector $[p_i]$, defined as $p_i p_i$, is unchanged by a transformation of axes. (Such a quantity is called an *invariant*.)

1.2. The value of $|a_{ij}|$. Up to this point it has not been necessary to specify whether right-handed or left-handed axes were being used.

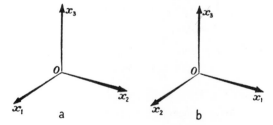

FIG. 2.1. (a) Right-handed axes; (b) left-handed axes.

Now we have to make a distinction. Fig. 2.1 a shows a right-handed, and Fig. 2.1 b a left-handed, set. It will be seen that, in the right-handed set, a right-handed screw motion through 90° along the positive x_1 direction rotates x_2 to x_3; similarly, a right-handed motion along positive x_2 rotates x_3 to x_1 (cyclic order), and a right-handed motion along positive x_3 rotates x_1 to x_2. For the left-handed set the same description applies with 'left' substituted for 'right'.

The introduction of left-handed axes is an unfortunate necessity forced on us in this subject because we shall have to consider symmetry operations, such as reflection in a plane and inversion through a centre, which change the hand of the axes. We now show that the determinant† of the transformation matrix (a_{ij}), which we write as $|a_{ij}|$, is equal to -1 or $+1$, respectively, according to whether the hand of the axes is changed or unchanged by the transformation (Jeffreys 1931).

We have

$$|a_{ij}| = \begin{vmatrix} a_{11} & a_{12} & a_{13} \\ a_{21} & a_{22} & a_{23} \\ a_{31} & a_{32} & a_{33} \end{vmatrix}$$

$$= a_{11}(a_{22}a_{33}-a_{23}a_{32})+a_{12}(a_{23}a_{31}-a_{21}a_{33})+$$
$$+a_{13}(a_{21}a_{32}-a_{22}a_{31}). \quad (10)$$

But, by solving (4.2) and (4.3) for the ratios of a_{11}, a_{12}, a_{13}, we obtain

$$\frac{a_{11}}{a_{32}a_{23}-a_{33}a_{22}} = \frac{a_{12}}{a_{33}a_{21}-a_{31}a_{23}} = \frac{a_{13}}{a_{31}a_{22}-a_{32}a_{21}} = -\frac{1}{k} \quad \text{say.} \quad (11)$$

Hence, by substituting the values of $(a_{22}a_{33}-a_{23}a_{32})$ etc. in (10), we have

$$|a_{ij}| = k(a_{11}^2+a_{12}^2+a_{13}^2) = k, \quad \text{from (2.1).}$$

Now (11) gives

$$\frac{a_{11}^2+a_{12}^2+a_{13}^2}{(a_{32}a_{23}-a_{33}a_{22})^2+(a_{33}a_{21}-a_{31}a_{23})^2+(a_{31}a_{22}-a_{32}a_{21})^2} = \frac{1}{k^2}.$$

Hence, from (2.1),

$$(a_{32}a_{23}-a_{33}a_{22})^2+(a_{33}a_{21}-a_{31}a_{23})^2+(a_{31}a_{22}-a_{32}a_{21})^2 = k^2.$$

By using the general identity,

$$(bc'-cb')^2+(ca'-ac')^2+(ab'-ba')^2$$
$$= (a^2+b^2+c^2)(a'^2+b'^2+c'^2)-(aa'+bb'+cc')^2,$$

this may be expressed as

$$(a_{31}^2+a_{32}^2+a_{33}^2)(a_{21}^2+a_{22}^2+a_{23}^2)-(a_{31}a_{21}+a_{32}a_{22}+a_{33}a_{23})^2 = k^2.$$

† The determinant of a 2×2 matrix, known as a *determinant of the second order*, is defined as follows. The determinant $\begin{vmatrix} a & b \\ c & d \end{vmatrix}$ stands for the expression $(ad-bc)$. The determinant of a 3×3 matrix, known as a *determinant of the third order*, may be defined in terms of determinants of the second order as follows. The determinant

$$\begin{vmatrix} a & b & c \\ d & e & f \\ g & h & k \end{vmatrix}$$

stands for $a\begin{vmatrix} e & f \\ h & k \end{vmatrix} -b\begin{vmatrix} d & f \\ g & k \end{vmatrix} +c\begin{vmatrix} d & e \\ g & h \end{vmatrix}.$

Definitions of determinants of higher order may be given in a similar way. A general definition appears in the footnote on p. 156. For further information on determinants see Ferrar (1941) or Aitken (1948).

But from (2.3), (2.2) and (4.1) the left-hand side is equal to unity. Hence
$$k = |a_{ij}| = \pm 1. \tag{12}$$

If two rows of $|a_{ij}|$ are interchanged, the sign changes. The sign of $|a_{ij}|$ therefore depends on how we choose to number the axes. Consider first the transformation that leaves the axes unchanged, called the *identical transformation*,

$$(a_{ij}) = \begin{pmatrix} 1 & 0 & 0 \\ 0 & 1 & 0 \\ 0 & 0 & 1 \end{pmatrix} = (\delta_{ij}),$$

by equation (9). In this case evidently $|a_{ij}| = +1$. Now imagine the new set of axes to rotate relative to the old set. Since all the a_{ij} change continuously, the value of $|a_{ij}|$ cannot change discontinuously from $+1$ to -1, and so we conclude that, for any transformation that corresponds simply to a rotation of the axes, $|a_{ij}| = +1$. Any set of right-handed axes can be brought into coincidence with any other right-handed set by a suitable rotation, and the same holds for any two left-handed sets. Thus *for a transformation that leaves the hand of the axes unchanged* $|a_{ij}| = +1$.

Consider now the transformation

$$(a_{ij}) = \begin{pmatrix} -1 & 0 & 0 \\ 0 & 1 & 0 \\ 0 & 0 & 1 \end{pmatrix},$$

which gives $x_1' = -x_1$, $x_2' = x_2$, $x_3' = x_3$ and $|a_{ij}| = -1$. An inspection of Figs. 2.1 *a* and 2.1 *b* shows that this transformation changes the hand of the axes. Now imagine the new axes to rotate. Again, $|a_{ij}|$ cannot change discontinuously and so $|a_{ij}|$ remains equal to -1. If the old axes are right-handed, such a rotation will bring the new set into any desired left-handed orientation; and a similar statement holds if the old axes are left-handed and the new ones right-handed. It follows that *for a transformation that changes the hand of the axes* $|a_{ij}| = -1$.†

Putting $k = \pm 1$ in (11) gives another useful property of the a_{ij}:

$$\left. \begin{aligned} a_{11} &= \pm(a_{22}a_{33} - a_{23}a_{32}) \\ a_{12} &= \pm(a_{23}a_{31} - a_{21}a_{33}) \\ a_{13} &= \pm(a_{21}a_{32} - a_{22}a_{31}) \end{aligned} \right\}, \tag{13}$$

---

† $|a_{ij}|$ is equal to the triple product $(\mathbf{a}' \, \mathbf{b}' \, \mathbf{c}')$, defined as $(\mathbf{a}' \wedge \mathbf{b}') . \mathbf{c}'$, where $\mathbf{a}'$, $\mathbf{b}'$, $\mathbf{c}'$ are unit vectors parallel to the Ox_i' axes. However, a proof that $|a_{ij}| = \pm 1$ on these lines requires a definition of the sign of a vector product, and we postpone this question until § 2.

the sign outside the brackets being $+$ for transformations that do not change the hand and $-$ for those that do. In general, each a_{ij} is equal to plus or minus its cofactor (signed minor) in $|a_{ij}|$.

2. Vector product. Polar and axial vectors

In vector analysis it is customary, given two vectors $\mathbf{p}$ and $\mathbf{q}$, to define a vector called the vector product $\mathbf{p} \wedge \mathbf{q}$. There are certain difficulties and distinctions in connexion with this idea that we must now discuss.

In this book, if p_i and q_i are vectors, we define the vector product of p_i and q_i as the quantities

$$V_{ij} = -p_i q_j + p_j q_i. \tag{14}$$

It is easy to show that the V_{ij} obey the transformation law of a second-rank tensor [Ch. I, equation (22)]. We also see that

$$V_{ij} = -(-p_j q_i + p_i q_j) = -V_{ji}.$$

Hence $[V_{ij}]$ is an antisymmetrical second-rank tensor. Written out in full, we have

$$[V_{ij}] = \begin{bmatrix} 0 & -(p_1 q_2 - p_2 q_1) & (p_3 q_1 - p_1 q_3) \\ (p_1 q_2 - p_2 q_1) & 0 & -(p_2 q_3 - p_3 q_2) \\ -(p_3 q_1 - p_1 q_3) & (p_2 q_3 - p_3 q_2) & 0 \end{bmatrix}, \tag{15}$$

which may be written as

$$\begin{bmatrix} 0 & -r_3 & r_2 \\ r_3 & 0 & -r_1 \\ -r_2 & r_1 & 0 \end{bmatrix},$$

where

$$r_1 = p_2 q_3 - p_3 q_2, \quad r_2 = p_3 q_1 - p_1 q_3, \quad r_3 = p_1 q_2 - p_2 q_1. \tag{16}$$

We have given negative signs to the terms one place to the right of the leading diagonal in order to preserve the cyclic order in equations (16).

In vector analysis r_1, r_2, r_3 are taken as the components of a vector $\mathbf{r}$, but we shall now show that they do *not* form a vector in the sense we have defined (p. 14).

When the axes are changed from x_i to x_i', p_i and q_i become p_i' and q_i', and if the definition (16) of r_i is to continue to hold for the new axes

$$r_1' = p_2' q_3' - p_3' q_2',$$
$$= a_{2i} p_i a_{3j} q_j - a_{3j} p_j a_{2i} q_i, \quad \text{from Chapter I, equation (13)},$$
$$= a_{2i} a_{3j} (p_i q_j - p_j q_i).$$

When this is expanded, each term in the bracket appears twice, with opposite signs. Hence,

$$r'_1 = (a_{22}a_{33} - a_{23}a_{32})(p_2 q_3 - p_3 q_2) + (a_{23}a_{31} - a_{21}a_{33})(p_3 q_1 - p_1 q_3) +$$
$$+ (a_{21}a_{32} - a_{22}a_{31})(p_1 q_2 - p_2 q_1).$$

But, with equations (13) and (16), this gives

$$r'_1 = \pm a_{11} r_1 \pm a_{12} r_2 \pm a_{13} r_3, \qquad (17)$$

where the positive signs are to be taken for a transformation that leaves the hand of the axes unchanged ($|a_{ij}| = +1$), and the negative signs for a transformation that changes the hand ($|a_{ij}| = -1$). There are two similar equations for r'_2 and r'_3. We therefore have

$$r'_i = \pm a_{ij} r_j. \qquad (18)$$

If it were not for the possible negative signs in equations (18), r_1, r_2, r_3 would transform like, and therefore be, the components of a vector. As it is, however, a change from right-handed to left-handed axes results in changes in the components r_1, r_2, r_3 which are not those given by the vector transformation law

$$p'_i = a_{ij} p_j. \qquad (19)$$

Quantities which transform according to (18) are called *axial vectors*. The true vectors that transform according to (19) may be called *polar vectors* when it is necessary to make an explicit distinction.

Our conclusions up to this point may be summarized as follows. The vector product of two polar vectors is not a true (polar) vector but an antisymmetrical tensor. The suffix notation in equation (14) leads us to expect this.† The three independent components of the antisymmetrical tensor, when denoted by r_i, transform according to the law (18), and are said to form an axial vector.

We can represent a polar vector without ambiguity by an arrow pointing in a certain direction (Fig. 2.2a). Axial vectors can be recognized by the fact that, before they can be represented in this way, the notion of a right-handed or left-handed screw motion has to be introduced. Consider, for example, how the vector product of two vectors is given a definite direction in vector analysis. The vector product $\mathbf{p} \wedge \mathbf{q}$ is defined as $pq \sin\theta . \mathbf{1}$, where θ is the angle between $\mathbf{p}$ and $\mathbf{q}$, and

---

† It happens, as a coincidence, that in three dimensions an antisymmetrical tensor of the second rank has the same number of independent components as a vector, but this is not true in any other number of dimensions. For instance, in four dimensions an antisymmetrical tensor of the second rank is a 4×4 array with 6 independent components; a vector, on the other hand, has only 4 components.

where l is a unit vector perpendicular to **p** and **q**, such that **p**, **q**, l form a *right-handed* set. Or, alternatively, the vector product of **p** and **q** is defined as the vector whose components referred to *right-handed* axes are $[p_2 q_3 - p_3 q_2, p_3 q_1 - p_1 q_3, p_1 q_2 - p_2 q_1]$. Angular velocity, angular momentum, mechanical couple, and the curl of a polar vector, are other examples of axial vectors. A method of representing axial vectors that does not depend on any convention about a positive screw motion is shown in Fig. 2.2 b. The symbol consists of a line of definite length and orientation, with a definite sense of rotation attached to it.

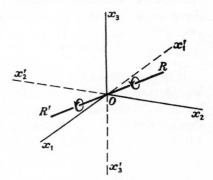

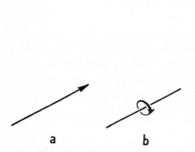

FIG. 2.2. Symbolic representation of (a) a polar vector, and (b) an axial vector.

FIG. 2.3.

To see that there is a difference between quantities symbolized by Fig. 2.2 a and those symbolized by Fig. 2.2 b, simply reflect each symbol in a plane perpendicular to its length. This evidently reverses quantities which are polar vectors, but leaves unchanged quantities which are axial vectors. On the other hand, reflection of each symbol in a plane parallel to its length has the opposite effect.

The prescription for constructing the symbol of Fig. 2.2 b from known components is as follows. To represent the axial vector whose components referred to a certain set of axes are r_1, r_2, r_3, draw out from the origin a line whose projections on the axes are r_1, r_2, r_3. If the axes are right-handed, attach to the line a sense of rotation given by a right-handed screw motion out from the origin. If the axes are left-handed, attach a rotation given by a left-handed screw motion. To illustrate this procedure let us imagine (Fig. 2.3) that we have constructed the symbol OR for an axial vector whose .components on a certain set of *right-handed* axes Ox_1, Ox_2, Ox_3 are r_1, r_2, r_3. Now suppose we transform to a new, *left-handed* set of axes, given by $a_{ij} = -\delta_{ij}$

(i.e all the axes are reversed in direction). The new components are, by (18),

$$r'_i = -a_{ij} r_j = \delta_{ij} r_j = r_i;$$

they are therefore unchanged. The line we draw out from the origin to represent the vector according to the new components, namely OR', is in the opposite direction to OR but of the same length. Since the axes are now left-handed we attach to OR' the sense of rotation given by a left-handed screw. Note that the symbol so constructed is the same as before (except for a translation which does not concern us). It remains, in fact, unchanged by any transformation of axes—which is clearly the necessary property of such a symbol.

3. The principal axes of a tensor

We remarked without proof in Chapter I, § 4.1 that the equation of the representation quadric

$$S_{ij} x_i x_j = 1 \quad (S_{ij} = S_{ji}) \tag{20}$$

could be simplified by change of axes to the form

$$S_1 x_1^2 + S_2 x_2^2 + S_3 x_3^2 = 1.$$

This question merits further discussion.

For convenience, we take as the defining property of principal axes that, at their intersections with the quadric, the normal to the quadric is parallel to the radius vector. Let P be a point on the quadric (20) and let the vector **OP** be X_i. Then, if we form the vector $S_{ij} X_j$, the radius-normal property of the representation quadric shows that it is parallel to the normal at P. The condition that the radius vector and normal should be parallel is that corresponding components should be proportional, thus

$$S_{ij} X_j = \lambda X_i, \tag{21}$$

where λ is a constant. (21) represents three homogeneous linear equations in the variables X_i. The condition that they have a solution other than $X_i = 0$ is that the determinant of the coefficients, which we denote by $F(\lambda)$, is zero. That is

$$F(\lambda) \equiv \begin{vmatrix} S_{11}-\lambda & S_{12} & S_{31} \\ S_{12} & S_{22}-\lambda & S_{23} \\ S_{31} & S_{23} & S_{33}-\lambda \end{vmatrix} = 0, \tag{22}$$

which may be written shortly as

$$F(\lambda) \equiv |S_{ij} - \lambda \delta_{ij}| = 0. \tag{23}$$

This cubic equation in λ is called the *secular equation*. The three roots, λ', λ'', λ''', say, give the three possible values of λ that ensure that equations (21) have a non-zero solution.† Each of the roots defines a direction in which the radius vector of the quadric is parallel to the normal, that is, the direction of one of the principal axes.

We may show that these three directions are mutually perpendicular. For consider any two of them, defined by λ' and λ'', say. Let the two corresponding radius vectors be denoted by X_i' and X_i''. Then

$$S_{ij} X_j' = \lambda' X_i' \Big\rbrace \tag{24}$$
$$S_{ij} X_j'' = \lambda'' X_i'' \Big\rbrace . \tag{25}$$

Multiplying (24) by X_i'' and (25) by X' and subtracting, we have

$$S_{ij}(X_j' X_i'' - X_j'' X_i') = (\lambda' - \lambda'') X_i' X_i''.$$

Since $S_{ij} = S_{ji}$, the left-hand side is zero. Hence

$$(\lambda' - \lambda'') X_i' X_i'' = 0.$$

Therefore, if $\lambda' \neq \lambda''$, the scalar product of X_i' and X_i'' is zero, and so the two vectors are perpendicular.

Directions of the principal axes. The directions of the three principal axes X_i', X_i'', X_i''' could be found in a particular case as follows. First solve equation (22) for λ. Then, with one of the values of λ so found, say λ', form the three equations (24) and solve for the ratios $X_1' : X_2' : X_3'$. Repeat the procedure for another root, say λ'', and so find $X_1'' : X_2'' : X_3''$. The third direction $X_1''' : X_2''' : X_3'''$ may then be found as the one perpendicular to the other two.

This method of finding the directions of the principal axes involves the solution of a cubic equation, and is not well suited for use with numerical data. To obtain numerical results it is better, in general, to use a method of successive approximation, such as the one we describe and illustrate in Chapter IX. For two-dimensional cases the problem is simplified very considerably; we deal with this in § 4 of the present chapter.

Lengths of the principal axes. Multiply both sides of equation (24) by X_i'. Then
$$S_{ij} X_j' X_i' = \lambda' X_i' X_i' = 1,$$

by (20), since X_i' is a radius vector. The length of the radius vector corresponding to λ' is therefore

$$\sqrt{(X_i' X_i')} = 1/\sqrt{\lambda'}.$$

---

† It may be proved (Eisenhart 1939, p. 259) that, since the S_{ij} are all real, by definition, the roots λ', λ'', λ''' are all real numbers.

Denoting the lengths of the principal axes, as before, by $1/\sqrt{S_1}$, $1/\sqrt{S_2}$, $1/\sqrt{S_3}$, and associating λ' with S_1 etc., we see that

$$\lambda' = S_1, \quad \lambda'' = S_2, \quad \lambda''' = S_3. \tag{26}$$

The three roots of (22) are therefore the three principal coefficients S_1, S_2, S_3.

As an aid to memory it may be noticed that, if the quadric is already referred to its principal axes, the secular equation is simply

$$\begin{vmatrix} S_1-\lambda & 0 & 0 \\ 0 & S_2-\lambda & 0 \\ 0 & 0 & S_3-\lambda \end{vmatrix} = 0,$$

i.e. $\qquad\qquad (S_1-\lambda)(S_2-\lambda)(S_3-\lambda) = 0,$

and the roots are evidently S_1, S_2, S_3.

4. The Mohr circle construction

The following helpful construction (due to Otto Mohr, 1835–1918) is used by engineers in the analysis of the stress and strain tensors (Chs. V and VI); it is, however, equally useful in dealing with any other symmetrical second-rank tensor.

4.1. Rotation about a principal axis. It often happens that we want to transform the components of a symmetrical second-rank tensor from one set of axes to another set which is obtained from the first set by a simple rotation about one of the axes of reference. The transformation equations are then much simplified.

Suppose, as in Fig. 2.4 a, that the axes Ox_i' are obtained from Ox_i by rotation about Ox_3 through an angle θ measured from Ox_1 towards Ox_2. Then, from the meaning of the a_{ij} (Ch. I, § 2.1),

$$(a_{ij}) = \begin{pmatrix} a_{11} & a_{12} & a_{13} \\ a_{21} & a_{22} & a_{23} \\ a_{31} & a_{32} & a_{33} \end{pmatrix} = \begin{pmatrix} \cos\theta & \sin\theta & 0 \\ -\sin\theta & \cos\theta & 0 \\ 0 & 0 & 1 \end{pmatrix}. \tag{27}$$

To simplify the argument we shall first deal with the case when Ox_1, Ox_2, Ox_3 are principal axes of the tensor $[S_{ij}]$. Thus,

$$[S_{ij}] = \begin{bmatrix} S_1 & 0 & 0 \\ 0 & S_2 & 0 \\ 0 & 0 & S_3 \end{bmatrix}.$$

The components referred to Ox_i' are, from Chapter I, equation (22),

$$S_{ij}' = a_{ik} a_{jl} S_{kl} \qquad (S_{ij}' = S_{ji}'). \tag{28}$$

Using the values of the various a_{ij} and S_{ij}, the transformed tensor is calculated as

$$\begin{bmatrix} S_{11}' & S_{12}' & 0 \\ S_{12}' & S_{22}' & 0 \\ 0 & 0 & S_3 \end{bmatrix},$$

where

$$S'_{11} = S_1\cos^2\theta + S_2\sin^2\theta$$
$$S'_{22} = S_1\sin^2\theta + S_2\cos^2\theta$$
$$S'_{12} = -S_1\sin\theta\cos\theta + S_2\sin\theta\cos\theta$$

(29)

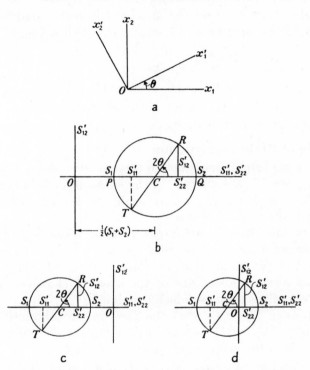

FIG. 2.4. The Mohr circle construction. Showing (a) the rotation of axes, (b) the case when S_1 and S_2 are both positive, (c) the case when S_1 and S_2 are both negative, and (d) the case when S_1 and S_2 are opposite in sign.

Note that only the four components in the upper left-hand corner of the tensor change. Equations (29) may be written

$$S'_{11} = \tfrac{1}{2}(S_1+S_2) - \tfrac{1}{2}(S_2-S_1)\cos 2\theta$$
$$S'_{22} = \tfrac{1}{2}(S_1+S_2) + \tfrac{1}{2}(S_2-S_1)\cos 2\theta$$
$$S'_{12} = \qquad\qquad \tfrac{1}{2}(S_2-S_1)\sin 2\theta$$

(30)

This result may be expressed graphically in a plane diagram (Fig. 2.4 b). Suppose $S_1 < S_2$. On the horizontal axis two points P, Q are taken at distances S_1 and S_2 from O. A circle, centre C, is then drawn on PQ as diameter. A radius CR is drawn so that CR makes an angle of 2θ with

CQ, measured anti-clockwise (the same sense as in Fig. 2.4 a). Then, since $OC = \frac{1}{2}(S_1+S_2)$ and $CR = \frac{1}{2}(S_2-S_1)$, equations (30) show that the coordinates of R, relative to the axes shown in the diagram, are S'_{22} and S'_{12}. If RC is produced to cut the circle in T, the abscissa of T is S'_{11}. This construction, called the *Mohr circle construction*, gives a visual picture of the way S'_{11}, S'_{22} and S'_{12} change as the axes of reference are rotated.

As Ox'_1 in Fig. 2.4 a rotates from Qx_1 to Ox_2, θ changes from 0 to $\frac{1}{2}\pi$, and R moves round the upper half of the circle in Fig. 2.4 b from Q to P. Clearly, S'_{11} and S'_{22} are always intermediate in value between S_1 and S_2 and reach their extreme values when R is at P and Q, that is, when the axes Ox'_i coincide with the principal axes. We see too that $(S'_{11}+S'_{22})$ has the same value for all positions of the axes; it is an *invariant* for this transformation:
$$S'_{11}+S'_{22} = S_1+S_2.$$
S'_{12} reaches its greatest value, $\frac{1}{2}(S_2-S_1)$, when $2\theta = \frac{1}{2}\pi$ or $\theta = \frac{1}{4}\pi$. If Ox'_1 rotates in the opposite direction, so does R. The values of S'_{12} are then negative when R is below Q.

S_1 and S_2 are taken as positive in Fig. 2.4 b, but the construction works equally well if they are both negative as in Fig. 2.4 c, or if they have opposite signs as in Fig. 2.4 d. The values of S'_{22} and S'_{12} are still the coordinates of R, and the value of S'_{11} is the abscissa of T, with due regard to sign throughout.

If we were given the values of S'_{11}, S'_{12}, S'_{22}, we could use the Mohr circle to find the principal components S_1 and S_2. R and T could be plotted as $R = (S'_{22}, S'_{12})$, $T = (S'_{11}, -S'_{12})$; R and T would define the circle, and the points P, Q in which it cuts the horizontal axis would give S_1 and S_2. In practice this construction is mainly useful as a quick way of deriving formulae. For instance, if S'_{11}, S'_{12}, S'_{22} are given, it is evident from the geometry of Fig. 2.4 b that

$$\left.\begin{array}{l} S_1 = OC-CP = \frac{1}{2}(S'_{11}+S'_{22})-r \\ S_2 = OC+CQ = \frac{1}{2}(S'_{11}+S'_{22})+r \end{array}\right\}, \tag{31}$$

and
$$\tan 2\theta = 2S'_{12}/(S'_{22}-S'_{11}), \tag{32}$$

where
$$r = CR = \sqrt{\{\tfrac{1}{4}(S'_{22}-S'_{11})^2+S'^2_{12}\}}.$$

If two sets of axes, say Ox'_i and Ox''_i, have to be considered, neither of which is the principal set, the Mohr circle gives a means of writing down formulae connecting the respective components. Again, we sometimes want to know how rapidly a certain component varies with respect to rotation of the axes. For example, it is clear from Fig. 2.4 b that, when $\theta = \frac{1}{4}\pi$, S'_{12} has a stationary value; it is varying most rapidly when $\theta = 0$.

In drawing Figs. 2.4 a, b, c, d we have assumed that the rotation is about Ox_3 and that $S_1 < S_2$. In practice one often has to apply the construction to cases where the labelling of the axes does not correspond to this scheme. The following general procedure is suggested. Choose the greater of the two principal components that are concerned in the transformation and draw the corresponding axis vertically upwards in Fig. 2.4 a. Draw the axis for the lesser principal component out to the right. Label the axes in whatever way is convenient for the problem and then label the diagram in Fig. 2.4 b, c or d to correspond.

EXERCISE 2.2. Show from the Mohr circle construction that

$$\begin{vmatrix} S_{11} & S_{12} \\ S_{12} & S_{22} \end{vmatrix}$$

is invariant for rotations about x_3. Interpret this geometrically.

EXERCISE 2.3. Carry out the relevant parts of steps [1] to [5] in Exercise 1.3 on p. 31 by means of the Mohr circle construction.

4.2. Rotation about an arbitrary axis. The above application of the Mohr circle construction (§ 4.1) is the most useful one in practice, but the construction still works when Ox_3, the axis of rotation, is no longer a principal axis of the tensor. Thinking for a moment of the representation quadric (Ch. I, § 4) we see that the construction just given is closely related to finding the principal axes of one of the principal sections. We could equally well consider an arbitrary central section, which would be a conic, and then apply the construction to finding its principal axes.

In this case we still have the transformation (27), but since Ox_i are no longer principal axes equations (28) contain more terms. Referred to the two principal axes, Ox_1 and Ox_2, of the section in question, and the normal Ox_3, the tensor has the form

$$\begin{bmatrix} S_{11} & 0 & S_{31} \\ 0 & S_{22} & S_{23} \\ S_{31} & S_{23} & S_{33} \end{bmatrix}.$$

The tensor components referred to Ox_i' are then

$$\left. \begin{aligned} S_{11}' &= S_{11}\cos^2\theta + S_{22}\sin^2\theta \\ S_{22}' &= S_{11}\sin^2\theta + S_{22}\cos^2\theta \\ S_{33}' &= S_{33} \\ S_{12}' &= -S_{11}\sin\theta\cos\theta + S_{22}\sin\theta\cos\theta \\ S_{23}' &= -S_{13}\sin\theta + S_{23}\cos\theta \\ S_{31}' &= S_{13}\cos\theta + S_{23}\sin\theta \end{aligned} \right\} . \tag{33}$$

So far as the components S'_{11}, S'_{22}, S'_{12} are concerned, therefore, it makes no difference whether Ox_3 is a principal axis or not, and the Mohr circle construction applies equally well to both cases.

EXERCISE 2.4. Transform the following tensors to their principal axes, using the Mohr circle construction

$$(a) \begin{bmatrix} 11 \cdot 06 & 3 \cdot 08 & 0 \\ 3 \cdot 08 & 18 \cdot 94 & 0 \\ 0 & 0 & 43 \end{bmatrix}, \qquad (b) \begin{bmatrix} -6 & -3\sqrt{3} & 0 \\ -3\sqrt{3} & 0 & 0 \\ 0 & 0 & 10 \end{bmatrix},$$

$$(c) \begin{bmatrix} 2 & 2 & 0 \\ 2 & 2 & 0 \\ 0 & 0 & 9 \end{bmatrix}, \qquad (d) \begin{bmatrix} 8 & 0 & -4 \\ 0 & 12 & 0 \\ -4 & 0 & 2 \end{bmatrix}.$$

EXERCISE 2.5. It is required to rotate the tensor

$$\begin{bmatrix} -1 & 3 & 8 \\ 3 & 10 & 6 \\ 8 & 6 & 2 \end{bmatrix}$$

about x_3 so that the 22 component becomes zero. Find the two possible angles of rotation which are less than 90°, and illustrate by the Mohr circle construction.

5. The magnitude ellipsoid

We have discussed the representation quadric (Ch. I, § 4) as giving a geometrical picture of a symmetrical second-rank tensor $[S_{ij}]$. There is another surface connected with $[S_{ij}]$ that is of interest, namely the surface whose equation when referred to the principal axes of $[S_{ij}]$ is

$$\frac{x_1^2}{S_1^2} + \frac{x_2^2}{S_2^2} + \frac{x_3^2}{S_3^2} = 1. \qquad (34)$$

This is an ellipsoid (note, never a hyperboloid), whose semi-axes are $|S_1|$, $|S_2|$, $|S_3|$.

Using the example of electrical conductivity (p. 4), we know that, if a field $\mathbf{E}$ of unit strength is applied in a certain direction (Fig. 2.5), $\mathbf{j}$ is determined. If the vector $\mathbf{E}$ rotates about O but remains of the same length, $\mathbf{j}$ will both rotate and change in length. We now prove that the end of $\mathbf{j}$ traces out the surface (34), constructed for the conductivity tensor $[\sigma_{ij}]$, denoted (1) in Fig. 2.5.

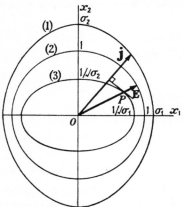

FIG. 2.5. A principal section through (1) the magnitude ellipsoid for conductivity, (2) a unit sphere, and (3) the representation quadric for conductivity.

Let $\mathbf{E} = [E_1, E_2, E_3]$. Then $\mathbf{j} = [j_1, j_2, j_3] = [\sigma_1 E_1, \sigma_2 E_2, \sigma_3 E_3]$. Therefore, since $E_1^2 + E_2^2 + E_3^2 = 1$,

$$\frac{j_1^2}{\sigma_1^2} + \frac{j_2^2}{\sigma_2^2} + \frac{j_3^2}{\sigma_3^2} = 1.$$

Hence, the extremity of the vector $\mathbf{j}$ lies on the ellipsoid

$$\frac{x_1^2}{\sigma_1^2} + \frac{x_2^2}{\sigma_2^2} + \frac{x_3^2}{\sigma_3^2} = 1.$$

The property of this surface just described obviously holds along the principal axes and this is the best way of remembering it. It gives essentially the *length* of $\mathbf{j}$ once the direction is known; to find the *direction* of $\mathbf{j}$ we can use the radius-normal property of the representation quadric for conductivity, denoted by (3) in the figure.

EXERCISE 2.6. Prove that the surface whose radius vector in a given direction is directly proportional to the magnitude of the property S in that direction is the ovaloid, $\quad (S_1 x_1^2 + S_2 x_2^2 + S_3 x_3^2)^2 = (x_1^2 + x_2^2 + x_3^2)^3.$

(Strictly, this is the equation of a surface surrounding the origin, together with the origin itself.)

SUMMARY

Transformation of axes (§ 1). If Ox_i, Ox_i' are two sets of mutually orthogonal axes and the transformation equation is

$$x_i' = a_{ij} x_j,$$

then the following relations, known as the *orthogonality relations*, hold between the direction cosines a_{ij},
$$\left. \begin{matrix} a_{ik} a_{jk} = \delta_{ij} \\ a_{ki} a_{kj} = \delta_{ij} \end{matrix} \right\},$$

where
$$\delta_{ij} = \begin{cases} 1 & (i = j), \\ 0 & (i \neq j). \end{cases}$$

We also have $\qquad |a_{ij}| = +1$

if the hand of the axes is unchanged by the transformation, and

$$|a_{ij}| = -1$$

if the hand is changed by the transformation.

Definition of vector product (§ 2). If p_i and q_i are (polar) vectors, the vector product of p_i and q_i is the antisymmetrical tensor V_{ij}, whose components are given by $\qquad V_{ij} = -p_i q_j + p_j q_i.$

If the three independent components V_{ij} are denoted by r_i, where

$$r_1 = p_2 q_3 - p_3 q_2, \qquad r_2 = p_3 q_1 - p_1 q_3, \qquad r_3 = p_1 q_2 - p_2 q_1,$$

the r_i transform according to the law

$$r_i' = \pm a_{ij} r_j \tag{18}$$

($+$ for transformations that do not change the hand of the axes and $-$ for those that do).

Axial vectors (§ 2). Quantities transforming by the law (18) are said to be the components of an *axial vector*. A true (polar) vector can be visualized as an arrow of definite length and direction (Fig. 2.2 *a*). An axial vector can be visualized as a length with a definite orientation and with a definite sense of rotation attached to it (Fig. 2.2 *b*).

Principal axes of a tensor (§ 3). The principal components of the symmetrical tensor S_{ij} are the three roots of the cubic equation in λ

$$|S_{ij} - \lambda \delta_{ij}| = 0,$$

known as the *secular equation*. If λ' is a root, the direction of the corresponding principal axis may be found by solving the equations

$$S_{ij} X'_j = \lambda' X'_i$$

for the ratios $X'_1 : X'_2 : X'_3$.

Mohr circle construction (§ 4). If the axes of reference Ox'_i of a symmetrical second-rank tensor S_{ij} are rotated about one of them, Ox'_3 (Fig. 2.4 *a*), the transformation of the components S'_{11}, S'_{22}, S'_{12} follows the construction shown in Figs. 2.4 *b, c, d*.

$$R = (S'_{22}, S'_{12}); \qquad T = (S'_{11}, -S'_{12}); \qquad \text{angle } RCQ = 2\theta.$$

The magnitude ellipsoid (§ 5). Let two vectors p_i and q_i be connected by the linear relations

$$p_i = S_{ij} q_j \quad (S_{ij} = S_{ji}).$$

Then, if q_i is of unit length and is allowed to rotate, the extremity of p_i traces out the ellipsoid

$$\frac{x_1^2}{S_1^2} + \frac{x_2^2}{S_2^2} + \frac{x_3^2}{S_3^2} = 1.$$

PART 2

EQUILIBRIUM PROPERTIES

Chapters III to IX discuss a number of crystal
properties in the light of the principles developed
in the two preceding chapters. All the properties
may be described by reference to equilibrium
states and to thermodynamically reversible
changes. The significance of this appears in
Chapter X where a unified treatment of the properties is given.

III

PARAMAGNETIC AND DIAMAGNETIC SUSCEPTIBILITY

THE magnetic susceptibility of paramagnetic and diamagnetic crystals is a typical anisotropic crystal property represented by a second-rank tensor. In this chapter we first formulate the property, and then show how the formulation leads to expressions which govern the behaviour of a crystal in a magnetic field. We shall not be concerned with ferromagnetism; this is an anisotropic crystal property of great interest both from a theoretical and a practical point of view, but it is not suitable for treatment within the tensor framework set up in this book.

1. General relations

The reader is assumed to be familiar with the elementary theory of magnetization in isotropic bodies. We are concerned with the following three vectors:

H, the *magnetic field intensity* or *field strength*;

I, the *intensity of magnetization*, that is, the magnetic moment per unit volume of the crystal;

B, the *magnetic induction* or *flux density*.

The vectors are connected by the relation

$$\mathbf{B} = \mu_0\mathbf{H} + \mathbf{I}, \tag{1}$$

where μ_0 is a scalar constant, the permeability of a vacuum, with the value in rationalized m.k.s. units† of $4\pi/10^7 = 1\cdot257 \times 10^{-6}$. For the definitions of **H**, **I** and **B** reference may be made to standard textbooks (for example, Abraham and Becker 1937, Sears 1946, or Slater and Frank 1947).

In many isotropic substances the intensity of magnetization is directly proportional to the field strength, except for very high fields, and we write

$$\mathbf{I} = \mu_0\psi\mathbf{H}, \tag{2}$$

where **H** is the field strength within the substance and ψ is a constant

---

† The m.k.s. system of rationalized units is used throughout this book. An account of the system, with conversion tables, is given by Stratton (1941), pp. 16–23. There is a useful discussion of units in Slater and Frank (1947), a small book on m.k.s. units by Sas and Pidduck (1947), and a fuller account in a book by McGreevy (1953).

called the *magnetic susceptibility*. From equations (1), $\mathbf{I}$ and $\mu_0\mathbf{H}$ have the same dimensions, and ψ is therefore a dimensionless ratio. This is the reason for introducing μ_0 into equation (2).† Since $\mathbf{I}$ refers to unit volume, ψ in equation (2), although dimensionless, is called the *volume susceptibility*. When ψ is positive the substance is said to be *paramagnetic*; when ψ is negative it is *diamagnetic*.

Although ψ is dimensionless, as defined here, the numerical value of the volume susceptibility for a given substance depends upon whether rationalized or unrationalized units are being used. Values of volume susceptibility in unrationalized units have to be multiplied by 4π to obtain the value of ψ as defined here in rationalized units. Other quantities frequently used in quoting numerical data are: the *specific susceptibility* or *susceptibility per unit mass*, ψ/ρ, and the *atomic*, or *molar susceptibility*, $A\psi/\rho$, where ρ is the density and A is the atomic or molecular weight. Note that these are the definitions for rationalized units.

Combining (1) and (2) we have

$$\mathbf{B} = (1+\psi)\mu_0\mathbf{H} = \mu\mathbf{H}, \tag{3}$$

where
$$\mu = \mu_0(1+\psi). \tag{4}$$

μ is the *permeability* of the substance. It is sometimes useful to introduce the dimensionless ratio,

$$M = \mu/\mu_0 = 1+\psi,$$

known as the *relative permeability*.

In a crystal $\mathbf{I}$ is, in general, not parallel to $\mathbf{H}$ and (2) is replaced by

$$I_i = \mu_0\psi_{ij}H_j, \tag{5}$$

where the ψ_{ij} are components of the *magnetic susceptibility tensor*.‡ Now

---

† There is no general agreement among authors on symbols or terminology in this subject. Some replace (1) by $\mathbf{B} = \mu_0(\mathbf{H}+\mathbf{I})$ and omit μ_0 from (2); this formulation gives the same meaning to ψ but a different meaning to $\mathbf{I}$. Other authors simply omit the μ_0 from (2); in this case $\mathbf{I}$ has the meaning we have given it but ψ has not, its dimensions being the same as those of μ_0.

‡ Since electric current density $\mathbf{j}$ is a polar vector, it follows from the electromagnetic field equation, curl $\mathbf{H} = \mathbf{j}+\dot{\mathbf{D}}$, that $\mathbf{H}$ is an axial vector (see Ch. II, § 2, for the distinction between polar and axial vectors). As confirmation, note the need for the right-hand rule in defining the direction of $\mathbf{H}$ relative to the current. $\mathbf{I}$ and $\mathbf{B}$ are also axial vectors, by equation (1) above. The reader may easily prove that, since both I_i and H_j in (5) transform as the components of axial vectors, the ψ_{ij} transform as the components of an ordinary (polar) second-rank tensor with no choice of signs in the transformation law.

There is, nevertheless, an element of arbitrariness in concluding that $\mathbf{H}$ is axial. The conclusion depends on the *assumption* that $\mathbf{j}$ is polar. The above field equation is satisfied equally consistently if $\mathbf{j}$ is axial and $\mathbf{H}$ is polar; and, in general, the electromagnetic field equations are satisfied either by having $\mathbf{E}, \mathbf{D}, \mathbf{P}, \mathbf{j}$ polar and $\mathbf{H}, \mathbf{B}, \mathbf{I}$ axial or vice versa. We have decided between the two alternatives by taking charge density

(1) is a general relation that holds whether $\mathbf{H}$ and $\mathbf{I}$ are parallel or not (Fig. 3.1). It may be written in suffix notation as

$$B_i = \mu_0 H_i + I_i. \tag{6}$$

Hence,
$$B_i = \mu_0(H_i + \psi_{ij} H_j)$$
$$= \mu_0(\delta_{ij} + \psi_{ij}) H_j,$$

using the substitution property of δ_{ij} (p. 35). This equation may be written

$$B_i = \mu_{ij} H_j, \tag{7}$$

where
$$\mu_{ij} = \mu_0(\delta_{ij} + \psi_{ij}). \tag{8}$$

Since $\mu_0 \delta_{ij}$ and $\mu_0 \psi_{ij}$ are second-rank tensors, μ_{ij} is also a second-rank tensor, the *permeability tensor*.† Equation (8) is the generalization for anisotropic substances of equation (4). Written out in full we have

$$\begin{bmatrix} \mu_{11} & \mu_{12} & \mu_{13} \\ \mu_{21} & \mu_{22} & \mu_{23} \\ \mu_{31} & \mu_{32} & \mu_{33} \end{bmatrix} = \begin{bmatrix} \mu_0(1+\psi_{11}) & \mu_0 \psi_{12} & \mu_0 \psi_{13} \\ \mu_0 \psi_{21} & \mu_0(1+\psi_{22}) & \mu_0 \psi_{23} \\ \mu_0 \psi_{31} & \mu_0 \psi_{32} & \mu_0(1+\psi_{33}) \end{bmatrix}.$$

The *relative permeability tensor* of a crystal may be defined as

$$M_{ij} = \mu_{ij}/\mu_0 = \delta_{ij} + \psi_{ij}.$$

In § 2 it is proved, by considering the energy of a magnetized crystal, that the μ_{ij} form a symmetrical tensor, that is,

$$\mu_{ij} = \mu_{ji}. \tag{9}$$

It follows that $[\psi_{ij}]$ is also symmetrical, since, from (8),

$$\psi_{ij} = \mu_{ij}/\mu_0 - \delta_{ij} = \mu_{ji}/\mu_0 - \delta_{ji} = \psi_{ji}.$$

Thus, both $[\mu_{ij}]$ and $[\psi_{ij}]$ may be conveniently referred to their common principal axes. *If* $\mathbf{H}$ *is directed along any one of these three mutually perpendicular directions,* $\mathbf{H}$, $\mathbf{I}$ *and* $\mathbf{B}$ *are all parallel, as they are for isotropic bodies.* If $\mathbf{H}$ points along a principal axis denoted by Ox_1, for example, we have

$$\mathbf{B} = \mu_1 \mathbf{H}, \qquad \mathbf{I} = \mu_0 \psi_1 \mathbf{H},$$

$$\mu_1 = \mu_0(1+\psi_1).$$

The susceptibility of a crystal is therefore completely defined by the magnitudes and directions of its principal susceptibilities ψ_1, ψ_2, ψ_3.

as a true scalar and therefore $\mathbf{j}$ as a polar vector. It would have been permissible, however, to take magnetic pole strength as a true scalar and so to have reached the conclusion that $\mathbf{H}$ (as the force on unit pole) was polar.

We may note that both alternatives give the same transformation law, and therefore the same 'tensor character', to $[\psi_{ij}]$.

† It is easy to show that the sum (or difference) of any two tensors of equal rank is another tensor.

These are, of course, subject to any restrictions that the symmetry of the crystal may impose (see Table 3, p. 23).

A crystal is said to be paramagnetic or diamagnetic along a particular one of the principal axes if ψ for that direction is respectively positive or negative. A few crystals are paramagnetic along one axis and diamagnetic along another. The principal susceptibilities of diamagnetic and paramagnetic crystals are usually in the neighbourhood of -10^{-5} and $+10^{-5}$ respectively; a few values are given in Table 4.

TABLE 4

Susceptibilities of paramagnetic and diamagnetic crystals

When multiplied by 10^{-5} the numbers in the table are the volume susceptibilities in rationalized units.

Crystal	System	ψ_1	ψ_2	ψ_3	Ref.
Aragonite .	orthorhombic	$-1\cdot44$	$-1\cdot42$	$-1\cdot63$	1
Quartz . .	trigonal		$-1\cdot51$	$-1\cdot52$	2
Calcite . .	,,		$-1\cdot24$	$-1\cdot38$	1
Cadmium. .	hexagonal		$-1\cdot74$	$-2\cdot85$	3
Beryl . .	,,		$2\cdot76$	$1\cdot29$	1
Rutile . .	tetragonal		$10\cdot5$	$11\cdot2$	1
Fluorite . .	cubic		$-1\cdot14$		1
Sodium chloride	,,		$-1\cdot36$		1

References. (1) Computed from the specific susceptibilities given in *International Critical Tables* (1929), 6, 364. (2) The same, p. 341. (3) Computed from atomic susceptibilities given by Bates (1937).

Since the ψ_{ij} in equation (3) are small compared with 1, the components of μ_{ij} differ only slightly from those of $\mu_0 \delta_{ij}$. The smaller the values of the ψ_{ij} the smaller is I, and the more nearly is **B** parallel to **H** (Fig. 3.1).

The smallness of the ψ_{ij} allows an important simplification to be made, which we must now describe.

When a crystal is placed in a field and becomes magnetized it sets up a magnetic field of its own, which depends on the susceptibility, the shape, and the size of the crystal. The actual magnetic field acting at any point of space, either inside or outside the crystal, is therefore not the same as the applied field. Let us denote the field due to the crystal by $\mathbf{H}_o$ and the applied field by $\mathbf{H}_a$. $\mathbf{H}_a$ is the field produced by sources external to the crystal; it is the field that would exist in the absence of the crystal. The actual field, or total field, $\mathbf{H}_t$, acting at any point is the sum of the fields produced by the external sources and by the crystal itself (Fig. 3.2):

$$\mathbf{H}_t = \mathbf{H}_a + \mathbf{H}_c. \tag{10}$$

$\mathbf{H}_a$ and $\mathbf{H}_t$ will, in general, vary from point to point, and they will differ from each other both inside and outside the crystal.

The field that enters equations (2) and (5) is $\mathbf{H}_t$. However, since the crystal field H_c is proportional to I/μ_0, the disturbing effect on $\mathbf{H}_a$ is small and can usually be neglected. When, therefore, a crystal is introduced into a uniform field, $\mathbf{H}_t$ is nearly the same as $\mathbf{H}_a$ and the field is

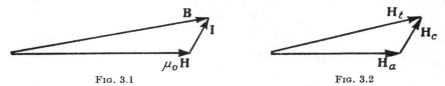

FIG. 3.1   FIG. 3.2

practically the same both inside and outside the crystal. Paramagnetic and diamagnetic effects *in a uniform field* are therefore almost independent of the crystal shape. The reason for the qualification 'in a uniform field' will appear in § 3.3. We emphasize that our discussion applies to para- and diamagnetics only and that this last result would not be true for a ferromagnetic.

2. The energy associated with a magnetized crystal

The energy stored in a body by magnetization is found by considering how much work has to be done to magnetize the body. This, in turn, is obtained by first considering how much work is done in causing a small change in the magnetization. Now the magnetization of a crystal can be changed in various ways. For instance, it may be altered by moving permanent magnets, or it may be altered by changing the current in a solenoid. It is important to realize that the work needed to produce a given change in the magnetization of a body can depend on how this change is produced. We shall show below (for a special case) that if the change is brought about while the field is maintained by the current in a solenoid, and the field is confined to the crystal, the total work done is $dW = v\mathbf{H}.d\mathbf{B}$, where v is the volume of the crystal. If, on the other hand, the change is brought about by moving permanent magnets, the work done can be shown to be $dW = -v\mathbf{I}.d\mathbf{H}$. The reason for the difference is that the work depends not only on the change in the magnetization of the body, but on all the changes that take place in the field both inside and outside the body—and these are different in the two cases. Although the change in magnetization of the body is the same in the two cases, the change in the energy of the whole system is not. We give a fuller discussion of this question in Appendix F. For the purposes of this chapter it is sufficient to state (1) that the expression we are now going to derive for dW is not the only possible one and (2) that our final result, which is the symmetry of $[\mu_{ij}]$, would follow whichever of the various alternative expressions we used.

Consider now the following simple case. A long rod of crystal, of

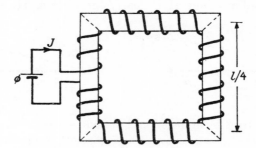

Fig. 3.3. Circuit for finding the energy associated with a magnetized crystal.

length l and cross-section A, is cut into four equal lengths, which are then fitted together so as to form a square 'picture frame', as in Fig. 3.3. The whole square is wound with a solenoid of resistance R, with n turns per unit length. A current J is maintained in the solenoid by a battery of e.m.f. ϕ. If l is made very large compared with A in this arrangement, the field is essentially uniform along each arm and is entirely confined to the crystal; the non-uniformities of field near the corners affect only a very small proportion of the total volume. We seek an expression for the work done when there is a small change in the magnetization of the crystal. $\mathbf{H}$ is determined entirely by J and acts along the axis of each arm of the solenoid:
$$H = nJ.$$

In general, therefore, $\mathbf{B}$ in each arm will be inclined to the axis. Let B_n be the component of $\mathbf{B}$ along the axis of the solenoid in the direction of $\mathbf{H}$. Then, as $\mathbf{B}$ changes, there is an induced e.m.f. in the circuit,
$$Anl\frac{dB_n}{dt}.$$

Therefore, adding the e.m.f.'s round the circuit, we have
$$\phi - Anl\frac{dB_n}{dt} = JR.$$

The work done by the battery in time dt is

$$\phi J \, dt = J^2 R \, dt + AnlJ \, dB_n$$
$$= J^2 R \, dt + AlH \, dB_n$$
$$= J^2 R \, dt + v\mathbf{H}.d\mathbf{B},$$

where v is the volume of the crystal ($v = Al$). The directions of $\mathbf{H}$ and $\mathbf{B}$ in this last equation are to be taken relative to the crystal axes, in order that we may regard $\mathbf{H}$ and $\mathbf{B}$ as the same in all four arms of the square. The first term in the expression for the work represents the Joule heat. The second term,

$$dW = v\mathbf{H}.d\mathbf{B} \quad \text{or} \quad dW = vH_i \, dB_i, \tag{11}$$

we interpret as the work done as a result of the change in magnetization.

We now equate this expression for dW to the increase in energy† $d\Psi$ of the crystal:

$$d\Psi = vH_i \, dB_i. \tag{12}$$

Substituting from equation (7) into equation (12) we have

$$d\Psi = v\mu_{ij} H_i \, dH_j, \tag{13}$$

which, written out in full, is

$$d\Psi = v(\mu_{11} H_1 \, dH_1 + \mu_{12} H_1 \, dH_2 + \mu_{13} H_1 \, dH_3 +$$
$$+ \mu_{21} H_2 \, dH_1 + \mu_{22} H_2 \, dH_2 + \mu_{23} H_2 \, dH_3 +$$
$$+ \mu_{31} H_3 \, dH_1 + \mu_{32} H_3 \, dH_2 + \mu_{33} H_3 \, dH_3). \tag{14}$$

Hence,
$$\frac{1}{v} \frac{\partial \Psi}{\partial H_1} = \mu_{11} H_1 + \mu_{21} H_2 + \mu_{31} H_3, \tag{15}$$

and
$$\frac{1}{v} \frac{\partial \Psi}{\partial H_2} = \mu_{12} H_1 + \mu_{22} H_2 + \mu_{32} H_3. \tag{16}$$

Now, Ψ is a function only of the independent variables H_1, H_2, H_3,‡ and it may be proved (e.g. Woods 1934, p. 78) that for such a function

$$\frac{\partial}{\partial H_1}\left(\frac{\partial \Psi}{\partial H_2}\right) = \frac{\partial}{\partial H_2}\left(\frac{\partial \Psi}{\partial H_1}\right).$$

Hence, differentiating (15) with respect to H_2 and (16) with respect to H_1 we have
$$\mu_{12} = \mu_{21}.$$

---

† Strictly, for an isothermal and reversible change, this is the increase in *free energy* (Appendix F).

‡ Since equation (12) only determines the changes in Ψ, Ψ itself contains an arbitrary constant which may be conveniently fixed by agreeing to call Ψ zero when

$$H_1 = H_2 = H_3 = 0.$$

In a similar way we may prove that

$$\mu_{13} = \mu_{31} \quad \text{and} \quad \mu_{23} = \mu_{32}.$$

Hence, the tensor $[\mu_{ij}]$ is symmetrical.

Integrating equation (14), we find for the energy of a magnetized crystal,†

$$\Psi = v(\tfrac{1}{2}\mu_{11}H_1^2 + \mu_{12}H_1H_2 + \mu_{13}H_1H_3 +$$
$$+ \ \tfrac{1}{2}\mu_{22}H_2^2 + \mu_{23}H_2H_3 +$$
$$+ \ \tfrac{1}{2}\mu_{33}H_3^2),$$

or, in suffix notation, $\Psi = \tfrac{1}{2}v\mu_{ij}H_iH_j.$ (17)

EXERCISE 3.1. Suppose a field of strength H_1 is applied to a crystal along Ox_1 and then a field of strength H_2 is added along Ox_2. Calculate the work done, by using (7) and (12). Now suppose the experiment is repeated, but this time H_2 is applied first along Ox_2 and followed by H_1 along Ox_1. Again calculate the work done. Hence show that the work done in applying the total field $\mathbf{H} = [H_1, H_2, 0]$ is the same for both paths only if $\mu_{12} = \mu_{21}$.

3. Couples and forces

The couple and force on a crystal in a magnetic field provide means of measuring its susceptibility (Bates 1951, Lonsdale 1937). In the general

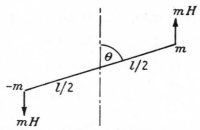

FIG. 3.4. The couple on a dipole in a uniform magnetic field.

case the couple and force depend not only on the susceptibility of the crystal, but also on its shape, orientation and volume, on the strength of the field and on the variation of the field over the specimen. To simplify matters we shall consider the various effects separately.

3.1. The couple on an anisotropic crystal in a uniform magnetic field. Suppose a magnetic dipole of strength $\mathbf{M}$ is situated in a uniform magnetic field of intensity $\mathbf{H}$ so that the direction of the dipole makes an angle θ with the direction of $\mathbf{H}$ (Fig. 3.4). The dipole may be imagined to consist of poles of strength $-m$ and m. If $\mathbf{l}$ denotes the vector drawn from $-m$ to m, we have $\mathbf{M} = m\mathbf{l}$.

† See note ‡ on previous page.

There will be a couple acting on the dipole, consisting of opposite forces $-m\mathbf{H}$ and $m\mathbf{H}$ acting at a distance apart of $l\sin\theta$. The magnitude of the couple is therefore

$$G = mHl\sin\theta = MH\sin\theta,$$

acting so as to turn $\mathbf{M}$ parallel to $\mathbf{H}$. In vector notation

$$\mathbf{G} = \mathbf{M}\wedge\mathbf{H}. \tag{18}$$

We emphasize that $\mathbf{H}$ here refers to the field produced by sources external to the dipole.

Formula (18) holds equally well whether the dipole is a permanent one or whether it is induced by the field itself. We have seen that if a magnetizable crystal is placed in a uniform field the induced $\mathbf{M}$ is not, in general, parallel to $\mathbf{H}$, and such a crystal will therefore experience a couple. On the other hand, in an isotropic substance or a cubic crystal, if shape effects are neglected, $\mathbf{M}$ is parallel to $\mathbf{H}$ and so the couple is zero.

In suffix notation (18) is written

$$G_{ij} = -M_i H_j + M_j H_i, \tag{19}$$

G_{ij} being an antisymmetrical tensor whose three independent non-zero components form an axial vector (Ch. II, § 2). If v is the volume of the crystal we have

$$M_i = v\mu_0\psi_{ij}H_j. \tag{20}$$

Hence

$$G_{ij} = v\mu_0(-\psi_{ik}H_k H_j + \psi_{jk}H_k H_i). \tag{21}$$

If the axes are chosen as the principal axes of ψ_{ij}, the components of G_{ij} simplify to

$$\begin{bmatrix} 0 & -v\mu_0(\psi_1-\psi_2)H_1 H_2 & v\mu_0(\psi_3-\psi_1)H_3 H_1 \\ v\mu_0(\psi_1-\psi_2)H_1 H_2 & 0 & -v\mu_0(\psi_2-\psi_3)H_2 H_3 \\ -v\mu_0(\psi_3-\psi_1)H_3 H_1 & v\mu_0(\psi_2-\psi_3)H_2 H_3 & 0 \end{bmatrix}.$$

The couple thus depends only on the *differences* between the principal susceptibilities and not on their absolute magnitudes. This shows once again that the couple is essentially a result of the anisotropy.

Special case. If the field acts in the plane of ψ_1 and ψ_2, the only component of the couple is about Ox_3 and is of magnitude

$$v\mu_0(\psi_1-\psi_2)H_1 H_2 = v\mu_0(\psi_1-\psi_2)H^2\sin\phi\cos\phi = \tfrac{1}{2}v\mu_0(\psi_1-\psi_2)H^2\sin 2\phi, \tag{22}$$

where ϕ is the angle between $\mathbf{H}$ and Ox_1.

The small couple due to the effect of shape and orientation in a uniform field. Strictly speaking, the **H** in equation (19) is $\mathbf{H}_a$ while the **H** in equation (20) is $\mathbf{H}_t$. Thus, when the small field $\mathbf{H}_c$ produced by the crystal itself is taken into account, formula (21) is not exact. As a result of $\mathbf{H}_c$ even an isotropic substance will experience a couple in a uniform field, if **H** is not directed along a line of symmetry; for **M**, being parallel to $\mathbf{H}_t$, will not be quite parallel to $\mathbf{H}_a$. The couple is an effect of shape and orientation, because in a given applied field the same volume of an isotropic crystal placed in different orientations or made into different shapes will experience different couples. The size of the couple may be estimated as follows. Since the field due to the crystal is of order I/μ_0, the angle between **M** and $\mathbf{H}_a$ will be of order $I/(\mu_0 H)$, which equals ψ. The couple is therefore of order $vH^2\mu_0\psi^2$. By comparison, the order of magnitude of the couple on an anisotropic crystal is, from (21), $vH^2\mu_0\psi$, if we assume that the differences between the principal susceptibilities are of the same order as the susceptibilities themselves. The ratio of (i) the couple due to the effect of shape and orientation *in a uniform field* to (ii) the couple due to anisotropy, is thus ψ, which, as we have seen, is usually a small number of order 10^{-5}.

3.2. The force on a crystal in a non-uniform magnetic field. Suppose a magnetic dipole of strength **M** is placed in a non-uniform field. As before, we may imagine the dipole to consist of poles of strength $-m$ and m separated by a small distance l and we denote the vector drawn from $-m$ to m by **l**. The ith component of the field acting at the positive pole will exceed the ith component at the negative pole by an amount

$$\frac{\partial H_i}{\partial x_j} l_j,$$

H_i being the field caused by sources external to the dipole. There will thus be a force on the dipole whose ith component is

$$F_i = m \frac{\partial H_i}{\partial x_j} l_j = M_j \frac{\partial H_i}{\partial x_j}. \tag{23}$$

This formula holds whether the dipole is a permanent one or whether it is induced by the field itself. Hence, if a small volume v of a paramagnetic or diamagnetic crystal obeying the law (5) is placed in a non-uniform field, it is acted on by a force

$$F_i = v I_j \frac{\partial H_i}{\partial x_j} \tag{24}$$

$$= v \mu_0 \psi_{jk} H_k \frac{\partial H_i}{\partial x_j} \tag{25}$$

(the field of the crystal itself being neglected); or, since $\operatorname{curl} \mathbf{H} = 0$, and $\psi_{jk} = \psi_{kj}$, we have, finally,

$$F_i = v\mu_0 \psi_{jk} H_k \frac{\partial H_j}{\partial x_i} \tag{26}$$

$$= \tfrac{1}{2} v\mu_0 \psi_{jk} \frac{\partial}{\partial x_i} (H_j H_k). \tag{27}$$

As a special case to illustrate the application of these expressions, suppose the crystal is placed so that the field acts parallel to one of the

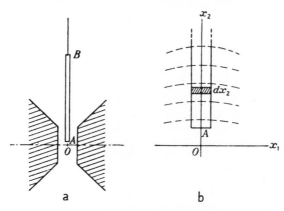

Fig. 3.5 a and b. Arrangement for the measurement of magnetic susceptibility.

principal susceptibility directions, and let this be the Ox_1 direction. Then $H_2 = H_3 = 0$, and the components of force are, from (26),

$$F_1 = v\mu_0 \psi_1 H_1 \frac{\partial H_1}{\partial x_1}, \qquad F_2 = v\mu_0 \psi_1 H_1 \frac{\partial H_1}{\partial x_2}, \qquad F_3 = v\mu_0 \psi_1 H_1 \frac{\partial H_1}{\partial x_3}, \tag{28}$$

or $$\mathbf{F} = \tfrac{1}{2} v\mu_0 \psi_1 \operatorname{grad}(H^2). \tag{29}$$

The direction of the force is in this case the direction of the greatest change in H^2—a result which is evidently true also for isotropic bodies. If ψ_1 is positive, so that the crystal is paramagnetic in the Ox_1 direction, the crystal will be drawn towards the strongest part of the field; conversely, if Ox_1 is a diamagnetic direction, the crystal will tend to move towards the weakest part of the field.

The force in a non-uniform field is much used in the measurement of susceptibility (Bates 1951). A common arrangement is that shown in Fig. 3.5 a. A long, thin rod of crystal is hung with its axis vertical, so that its lower end A is near the mid-point of the field between the two flat pole tips of an electromagnet, while its upper end B is well outside the gap. It is instructive to calculate the forces in this arrangement from equation (26). Taking axes as shown in Fig. 3.5 b, with x_1 parallel to the field, the symmetry of the field is such that on Ox_2,

$$H_2 = H_3 = 0.$$

In equation (26), therefore, we need only consider the terms with $k = 1$. Inspection further shows that on Ox_2 the only non-zero first derivatives of the field are $\partial H_1/\partial x_2 = \partial H_2/\partial x_1$. The components of force on each small volume dv of the crystal are therefore, from (26),

$$dF_1 = dv.\mu_0\psi_{21}H_1(\partial H_2/\partial x_1), \qquad dF_2 = dv.\mu_0\psi_{11}H_1(\partial H_1/\partial x_2), \qquad dF_3 = 0.$$

Putting $dv = \alpha\,dx_2$, where α is the cross-sectional area of the rod, we find for the components of the total force on the rod

$$F_1 = \mu_0\int_A^B \alpha\psi_{21}H_1\left(\frac{\partial H_1}{\partial x_2}\right)dx_2, \qquad F_2 = \mu_0\int_A^B \alpha\psi_{11}H_1\left(\frac{\partial H_1}{\partial x_2}\right)dx_2, \qquad F_3 = 0;$$

or

$$F_1 = \mu_0\int_A^B \alpha\psi_{21}H_1\,dH_1 = -\tfrac{1}{2}\mu_0\alpha\psi_{21}H_0^2, \tag{30}$$

and, similarly,

$$F_2 = -\tfrac{1}{2}\mu_0\alpha\psi_{11}H_0^2, \tag{31}$$

where H_0 is the field at A. The negative sign in the expression for F_2 shows that for ψ_{11} positive (paramagnetic) this component of force is downwards.

The origin of the two components of force may be understood in this way. The field H_1 produces components of magnetization along all three axes. In particular, there is a moment proportional to $\psi_{11}H_1$ along x_1. The force F_2 arises from the action of the field component H_2 on this moment. Although $H_2 = 0$ on the axis Ox_2, $\partial H_2/\partial x_1$ on this axis is not zero (except at O), for H_2 is changing sign. H_2 therefore exerts a different force on the two ends of the moment $\psi_{11}H_1$ and gives a force F_2 proportional to $\psi_{11}H_1(\partial H_2/\partial x_1) = \psi_{11}H_1(\partial H_1/\partial x_2)$. The force F_1 arises from the action of H_1 on the moment proportional to $\psi_{21}H_1$ set up along x_2. The force is proportional to $\psi_{21}H_1(\partial H_1/\partial x_2) = \psi_{21}H_1(\partial H_2/\partial x_1)$.

EXERCISE 3.2. A small paramagnetic crystal is placed in an inhomogeneous field of an electromagnet and supported by a bifilar suspension to form a pendulum that can swing in a plane perpendicular to the horizontal lines of force. The radius of swing is 20 cm. The direction of maximum susceptibility of the crystal is set parallel to the magnetic field, and the deflexion of the crystal observed when the magnet is switched off is 1·20 mm. The value of $H\,\partial H/\partial x$, where $\partial H/\partial x$ is the rate of change of field in the direction of the movement of the crystal, is uniform and equal to 10^{13} m.k.s. units, and the density of the crystal is $2\cdot7\times10^3$ kg/m³. Calculate the greatest principal susceptibility.

EXERCISE 3.3. A single crystal of cadmium (hexagonal) in the form of a long wire of radius 1·00 mm. is suspended from the arm of a balance in such a way that it lies in the transverse and longitudinal planes of symmetry of an electromagnet. The principal axis of the crystal is inclined at an angle θ to the axis of the wire and lies in the longitudinal plane of symmetry of the magnet. When the magnet is switched on the upper end of the wire is in a negligible field and the lower end is in a field of intensity $4\cdot00\times10^5$ m.k.s. units (ampere-turns/m). Under these conditions the magnetic force on the wire parallel to its axis is found to be

$$8\cdot11\times10^{-6} \text{ m.k.s. units (newtons)}.$$

Given that the principal volume susceptibilities of cadmium are (in rationalized units)
$$\psi_1 = \psi_2 = -17{\cdot}4 \times 10^{-6}, \qquad \psi_3 = -28{\cdot}5 \times 10^{-6},$$
determine the angle θ.

EXERCISE 3.4. A small crystal 1 mm.³ in volume, whose principal susceptibilities are (in rationalized units)
$$\psi_1 = 1{\cdot}0 \times 10^{-5}, \qquad \psi_2 = 0{\cdot}6 \times 10^{-5}, \qquad \psi_3 = 2{\cdot}0 \times 10^{-5},$$
is placed in an inhomogeneous static magnetic field. The components of the field and certain of their gradients along the directions of the principal axes are (in m.k.s. units):
$$H_1 = 1{\cdot}0 \times 10^6, \qquad H_2 = 0{\cdot}5 \times 10^6, \qquad H_3 = 2{\cdot}0 \times 10^6;$$
$$\partial H_1/\partial x_1 = 1{\cdot}0 \times 10^8, \qquad \partial H_1/\partial x_2 = 1{\cdot}2 \times 10^8, \qquad \partial H_1/\partial x_3 = 0{\cdot}5 \times 10^8,$$
$$\partial H_2/\partial x_2 = 0{\cdot}8 \times 10^8, \qquad \partial H_2/\partial x_3 = 2{\cdot}0 \times 10^8.$$
Calculate the magnitude and direction of the resultant force and couple acting on the crystal.

3.3. The couple on a crystal in a non-uniform magnetic field.

We have discussed the couple on a crystal in a uniform field and the force on a small volume of a crystal in a non-uniform field. In a non-uniform field there is also a couple, but this is only partly due to the effect of anisotropy discussed in § 3.1. A further contribution to the couple comes from the size and shape of the crystal. This is a first-order effect and quite distinct from the second-order effect of shape mentioned in § 3.1. To see how it arises it is only necessary to consider that the expression (24) for the force applies to every small element of the crystal, and that therefore, owing to the finite size of the crystal, the forces on different parts of the crystal will, in general, be different both in magnitude and direction. They will therefore give rise to both a force and a couple.

As a special instance we may consider (Fig. 3.6) two symmetrically placed elements A and B of a long isotropic para-magnetic rod, free to turn about O. Let the field be caused by a pole of strength m at P. Then the forces on A and B are radial and are proportional to the values of $\psi\,\mathrm{grad}(H^2)$ at A and B. Since $H = m/(4\pi\mu_0 r^2)$, where r is the radial distance, each element experiences a force proportional to $\psi m/r^5$, directed towards P. (We are neglecting the field due to the other parts of the rod itself.) A is closer to P than is B, and so the force on A is greater than that on B. Moreover, it will be seen from the figure that the lever arm about O of the force on A is greater than that for B. Consequently the rod will tend to turn parallel to the field. [The reader may

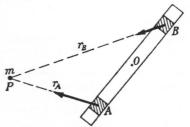

FIG. 3.6. Illustrating the couple on a long isotropic paramagnetic rod in the field of a single magnetic pole.

easily show that the ratio of the moments about O of the forces on A and B respectively is $(r_B/r_A)^6$.] Conversely, if the rod were diamagnetic it would turn perpendicular to the field.

4. The susceptibility of a powder

If a powder consisting of randomly oriented anisotropic crystal grains is placed in a field, each grain will, in general, acquire a moment which is not in the direction of **H**. It is evident, however, on grounds of symmetry, that the resultant moment of the aggregate must be in the direction of **H**; hence, the components of **I** transverse to **H** developed in the individual grains must all cancel on the average. The longitudinal component of **I** in a grain in which **H** has direction cosines (l_1, l_2, l_3) with respect to the principal susceptibility directions is given by

$$\mu_0 \psi H = \mu_0 (\psi_1 l_1^2 + \psi_2 l_2^2 + \psi_3 l_3^2) H,$$

where ψ is the susceptibility in the (l_1, l_2, l_3) direction (Ch. I, § 6). The mean value of I for the powder is the mean value of these longitudinal components taken over all possible values of l_1, l_2, l_3. This must be of the form

$$\mu_0 \alpha (\psi_1 + \psi_2 + \psi_3) H,$$

where α is a numerical factor, since the mean values of l_1^2, l_2^2 and l_3^2 must be equal. But the formula must hold for the isotropic case,

$$\psi_1 = \psi_2 = \psi_3 = \psi,$$

where we know the correct result to be $\mu_0 \psi H$; and so the value of α is seen to be $\frac{1}{3}$.

Thus, the susceptibility of the powder is $\frac{1}{3}(\psi_1 + \psi_2 + \psi_3)$, or, with a general choice of axes, $\frac{1}{3}\psi_{ii}$. This last expression is, of course, an invariant (p. 35).

SUMMARY

(§ 1) In all circumstances, $\qquad B_i = \mu_0 H_i + I_i.$ $\hfill (1)$

In paramagnetic and diamagnetic crystals there is the additional relation

$$I_i = \mu_0 \psi_{ij} H_j \tag{5}$$

(except for very strong fields), where the second-rank tensor $[\psi_{ij}]$ is the (volume) susceptibility. It follows that $\qquad B_i = \mu_{ij} H_j,$ $\hfill (7)$

where $\mu_{ij} = \mu_0(\delta_{ij} + \psi_{ij})$. The second-rank tensor $[\mu_{ij}]$ is the permeability.

(§ 2) The work done in changing the magnetization of a crystal, while the field is maintained by the current in a solenoid and is confined to the crystal, is $v H_i \, dB_i$, where v is the volume of the crystal. With (7) this leads to the expression

$$d\Psi = v \mu_{ij} H_i \, dH_j \tag{13}$$

for the increase in (free) energy. The fact that $d\Psi$ is a perfect differential then

establishes the symmetry of $[\mu_{ij}]$ and hence of $[\psi_{ij}]$. If the arbitrary constant in Ψ is suitably fixed,
$$\Psi = \tfrac{1}{2}v\mu_{ij}H_iH_j. \tag{17}$$

Being symmetrical tensors, $[\mu_{ij}]$ and $[\psi_{ij}]$ can be referred to their (common) principal axes (§ 1). A crystal is said to be paramagnetic or diamagnetic along a particular one of its principal susceptibility directions if ψ for that direction is respectively positive or negative.

Since the principal susceptibilities of paramagnetic and diamagnetic crystals are usually considerably smaller than 1 ($\sim 10^{-5}$), the field set up by the magnetized crystal is small compared with the external field, and can usually be neglected.

(§ 3) **Couples and forces.** Each element of volume v of a crystal in a magnetic field is subjected to a couple (§ 3.1)
$$G_{ij} = v\mu_0(-\psi_{ik}H_kH_j+\psi_{jk}H_kH_i), \tag{21}$$

and a force (§ 3.2)
$$F_i = \tfrac{1}{2}v\mu_0\psi_{jk}\frac{\partial}{\partial x_i}(H_jH_k). \tag{27}$$

In a large crystal the variation of the force from point to point can give rise to a further couple (§ 3.3) depending on the shape of the crystal.

(§ 4) The *susceptibility of a powder* is $\tfrac{1}{3}\psi_{ii}$ or, referred to the principal axes,
$$\tfrac{1}{3}(\psi_1+\psi_2+\psi_3).$$

ELECTRIC POLARIZATION

THE polarization of a crystal produced by an electric field is another example of an anisotropic crystal property that is represented by a second-rank tensor. The formal analysis of electric polarization is closely similar to that of magnetization given in the preceding chapter, but there are some differences which we shall have to note.

1. General relations

The following three vectors are analogous to **H**, **I** and **B** respectively:

E, the *electric field intensity* or *field strength*;

P, the *polarization*; that is the electric moment per unit volume (or the polarization charge per unit area taken perpendicular to the direction of polarization);

D, the *electric displacement* or *electric flux density*.

Corresponding to equation (1) of Chapter III, we have

$$\mathbf{D} = \kappa_0 \mathbf{E} + \mathbf{P}, \tag{1}$$

where κ_0 is a scalar constant, the permittivity of a vacuum, with the numerical value in rationalized m.k.s. units of $8 \cdot 854 \times 10^{-12}$.[†] The vector relation (1) is illustrated in Fig. 4.1.

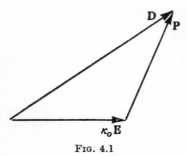

FIG. 4.1

In an *isotropic* substance the counterpart of equation (2), Chapter III, is

$$\mathbf{P} = \kappa_0 \chi \mathbf{E} \tag{2}$$

(except for very large fields), where **E** is the field strength within the substance and χ is the *dielectric susceptibility*.[‡] Analogous to equation (3), Chapter III, is the relation

$$\mathbf{D} = \kappa \mathbf{E}, \tag{3}$$

where $\kappa = \kappa_0(1+\chi)$; κ is the *permittivity*. It is often more convenient to

---

† See the textbooks cited in the footnote on p. 53.

‡ If the crystal shows a spontaneous polarization (§§ 7, 8), **P** must be interpreted as the change in the polarization caused by the field.

express the permittivity in terms of the permittivity of a vacuum. For this purpose we define the dimensionless constant

$$K = \kappa/\kappa_0, \tag{4}$$

known as the *relative permittivity* or the *dielectric constant*.

In an *anisotropic* substance we have, in place of (2),

$$P_i = \kappa_0 \chi_{ij} E_j, \tag{5}$$

where χ_{ij} is the *dielectric susceptibility tensor*; instead of (3) we have

$$D_i = \kappa_{ij} E_j, \tag{6}$$

where κ_{ij} is the *permittivity tensor* given by $\kappa_{ij} = \kappa_0(\delta_{ij} + \chi_{ij})$; and, for the dielectric constant, which is now also a tensor,

$$K_{ij} = \kappa_{ij}/\kappa_0. \tag{7}$$

In § 4 it is proved, from energy considerations, that $\kappa_{ij} = \kappa_{ji}$; hence $K_{ij} = K_{ji}$ and $\chi_{ij} = \chi_{ji}$. Therefore $[\kappa_{ij}]$, $[K_{ij}]$ and $[\chi_{ij}]$ may all be referred to their common principal axes; the relations between their principal components are clearly

$$\kappa_1 = \kappa_0 K_1, \text{ etc. } \text{ and } \chi_1 = K_1 - 1, \text{ etc.}$$

The dielectric properties of a crystal may thus be characterized by the magnitudes and directions of the three principal permittivities, dielectric constants or dielectric susceptibilities. These magnitudes and directions will, in general, depend on the frequency of the electric field, but they must always, of course, conform to any restrictions imposed by crystal symmetry, according to Table 3, p. 23. A few measured values of dielectric constants are given in Table 5.

TABLE 5

Dielectric constants of crystals

Crystal	System	K_1	K_2	K_3	Frequency (cycles/sec.)
Gypsum . .	monoclinic	9·9	5·1	5·0	3×10^8
Aragonite . .	orthorhombic	9·8	7·7	6·6	4×10^8
Quartz . .	trigonal	4·5		4·6	50 to 5×10^6
Calcite . .	,,	8·5		8·0	4×10^8
Rutile . .	tetragonal	89		173	4×10^8
Caesium chloride	cubic	6·3			2×10^5
Sodium chloride	,,	5·6			2×10^5

Values from *International Critical Tables* (1929), **6**, 98–100; *Landolt–Börnstein Physikalisch-Chemische Tabellen* (1923 and 1936); and Cady (1946), p. 414.

2. Differences between electric polarization and magnetization

Up to this stage, and from the point of view we are adopting, the analogy with magnetization is complete. It should be borne in mind, however, that the analogy is no more than a formal one. Compared with electric charges, magnetic poles are an artificial concept, although a convenient one. A physically realistic approach would recognize from the beginning the common origin of both magnetic and electric fields. It would regard magnetic fields and moments as produced by moving charges—the orbital and spin momentum of electrons—rather than by magnetic poles. The lack of complete parallelism between electric polarization and magnetization is emphasized by the fact that, whereas the principal magnetic susceptibilities of substances can be positive or negative, the principal dielectric susceptibilities are always positive. There are also two important practical differences which we now discuss.

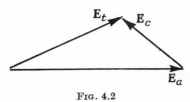

FIG. 4.2

(i) *The depolarizing effect.* If a polarizable crystal is placed in a field $\mathbf{E}_a$, it produces a field of its own, which we may denote by $\mathbf{E}_c$. The total field at any point is then (Fig. 4.2)

$$\mathbf{E}_t = \mathbf{E}_a + \mathbf{E}_c. \tag{8}$$

For paramagnetics and diamagnetics the disturbing effect of the crystal on the field is small, as we have seen, owing to the small values of the susceptibilities. This is not true for ferromagnetics; nor is it true in the electric case, for the χ_{ij} are by no means small compared with 1 (Table 5, p. 69), and, in consequence, $\mathbf{E}_c$ and $\mathbf{E}_a$ are often of the same order of magnitude. $\mathbf{E}_c$ depends on the shape of the crystal. Thus, if two specimens of the same crystal which differ only in shape are placed in identical fields, $\mathbf{E}_c$, and hence $\mathbf{E}_t$, will be different. Since it is $\mathbf{E}_t$ within the crystal that determines $\mathbf{P}$ by equation (5), the two crystals will show different polarizations.

For illustration we may recall here, without proof, the expressions for $\mathbf{E}_c$, $\mathbf{E}_t$ and $\mathbf{P}$ in three simple cases which are treated in detail in the textbooks:

(a) *A long isotropic rod in a uniform field $\mathbf{E}_a$ parallel to its length.*
$$\mathbf{E}_c = 0, \text{ and } \mathbf{E}_t = \mathbf{E}_a. \text{ Hence } \mathbf{P} = \kappa_0 \chi \mathbf{E}_a.$$

(b) *A flat isotropic disk in a uniform field $\mathbf{E}_a$ perpendicular to its faces.*
Inside the disk, $\kappa_0 \mathbf{E}_c = -\mathbf{P}$, and $\mathbf{E}_t = \mathbf{E}_a - \mathbf{P}/\kappa_0$. Hence $\mathbf{P} = \kappa_0 \chi (\mathbf{E}_a - \mathbf{P}/\kappa_0)$,
or
$$\mathbf{P} = \{1/(1+\chi)\}\kappa_0 \chi \mathbf{E}_a.$$

(c) *An isotropic sphere in a uniform field* $\mathbf{E}_a$.

Within the sphere, $\kappa_0 \mathbf{E}_c = -\frac{1}{3}\mathbf{P}$, and $\mathbf{E}_t = \mathbf{E}_a - \mathbf{P}/(3\kappa_0)$. Hence

$$\mathbf{P} = \kappa_0 \chi\{\mathbf{E}_a - \mathbf{P}/(3\kappa_0)\}, \quad \text{or} \quad \mathbf{P} = \{3/(3+\chi)\}\kappa_0 \chi \mathbf{E}_a.$$

In each case any field within the body due to the polarization acts in opposition to the applied field, and the apparent susceptibility is thus lowered. $\mathbf{E}_c$ within the body is accordingly known as the *depolarizing field*.

(ii) *Leakage of charge.* Measurements on dielectrics are complicated by the presence of 'dielectric anomalies', caused by the fact that the dielectrics are not perfectly insulating. Consider the following example.

A flat slab of dielectric is placed between two parallel condenser plates and separated from them by narrow gaps. We show in § 3 that, if the dielectric were perfectly insulating, the dielectric constant would be the ratio of the capacity of the condenser so formed to its capacity with the dielectric removed. Suppose the plates are left connected to a battery. When the crystal is first inserted there will be polarization charges on its surface, but no free charges. In time, however, charges of the *same* sign as the polarization charges will flow to the slab, and will continue to flow until a limiting state is reached in which $\mathbf{E} = 0$ within the slab —for the dielectric may then be thought of as acting like a perfect conductor. The capacity is now simply that of the two narrow gaps. Thus, if the capacity is used to measure the dielectric constant of the slab, the apparent dielectric constant will depend on the time at which the measurement is made and on the conductivity of the specimen.

If the gap between the plates and the crystal were removed, by using sputtered electrodes, for example, there would be, in the ideal case, no migration of free charges. In practice, a crystal is never perfectly homogeneous; it contains small flaws and concentrations of impurities where charges can accumulate. The migration of charge is helped by the fact that these places usually conduct better than the perfect parts of the crystal. Thus even a crystal, such as sulphur or mica, which has a very small gross conductivity, may show a considerable dielectric anomaly by conduction in these small regions.

Migration of charges also takes place when a crystal is suspended in an electric field so that it is free to turn. The crystal takes up a certain orientation when the field is first switched on, but charges leak as time goes on, and the crystal gradually turns to another position. This fact falsified many of the early measurements of permittivity which depended on the couple exerted on a crystal by a static or low frequency field

3. The relations between D, E and P in a parallel plate condenser

If a slab of dielectric is placed between the plates of a parallel plate condenser the capacity of the condenser is increased. It is shown in textbooks on electrostatics that the ratio of the capacities with and without the dielectric gives the dielectric constant of the material:

$$\frac{\text{capacity with dielectric}}{\text{capacity without dielectric}} = \frac{\text{permittivity of dielectric}}{\text{permittivity of vacuum}}$$

$$= \text{dielectric constant.}$$

But, if the dielectric is an anisotropic crystal in an arbitrary orientation, the question arises as to what 'permittivity' and 'dielectric constant' it is that appears in the above equation. In examining this question we have first to find the relations between D, E and P.

Fig. 4.3 a shows the crystal slab in place and Fig. 4.3 b shows the plates with the crystal removed. The plates and the crystal slab are considered to be of large extent compared with the thickness l of the crystal, and to clarify the relations between D and E it is helpful to imagine that narrow gaps exist between the crystal and the condenser plates. The width of these gaps is to be infinitesimally small compared with the thickness of the crystal. Let a constant potential difference ϕ be applied between the plates. The relative magnitudes of the vectors κ_0 E, P and D are roughly indicated in the figures by the spacing of the arrows. Let D_c, E_c and D_v, E_v refer respectively to the crystal and the vacuum gap in Fig. 4.3 a, let D', E' refer to the region between the plates in Fig. 4.3 b, and let σ and σ', respectively, be the charge per unit area of the plates in the two cases. Since the plates are of large extent and the plates themselves are equipotentials, the equipotential surfaces must run parallel to the plates. Therefore E_v, E_c and E' are all perpendicular to the plates. By an application of Gauss's theorem

$$\sigma = \kappa_0 E_v, \quad \text{and} \quad \sigma' = \kappa_0 E',$$

and, since the gap is very narrow,

$$E_c = \phi/l = E' = \sigma'/\kappa_0.$$

Note that the requirement is satisfied that the tangential component of E, which happens to be zero, is continuous across the crystal face. In the gap D is of course parallel to E, but in the crystal D will be at an angle to E. It is a theorem of electrostatics, provable directly from Gauss's theorem, that the normal component of D is continuous across a boundary on which there are no free charges. Therefore, if the normal component of D_c is denoted by $(D_c)_n$,

$$(D_c)_n = D_v = \sigma.$$

The ratio of the capacities in the two cases is therefore

$$\frac{C}{C'} = \frac{\sigma}{\sigma'} = \frac{\text{component of } \mathbf{D}_c \text{ parallel to } \mathbf{E}_c}{\kappa_0 E_c} = \frac{\kappa}{\kappa_0} = K,$$

where κ is the permittivity and K is the dielectric constant *in the direction of the applied field* $\mathbf{E}_c$, in the sense defined in § 6.1 of Chapter I.

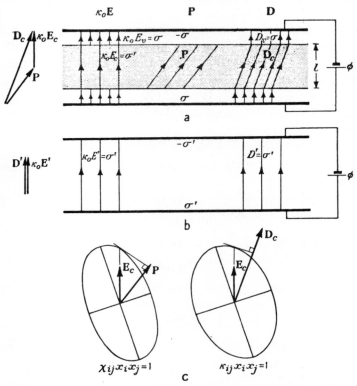

Fig. 4.3. Illustrating the relations between $\mathbf{D}$, $\kappa_0 \mathbf{E}$ and $\mathbf{P}$ in a parallel plate condenser (a) with a crystal slab between the plates and (b) with the crystal removed. In (c) the directions of $\mathbf{E}$ and $\mathbf{P}$ within the crystal are shown in relation to the susceptibility ellipsoid, and the directions of $\mathbf{E}$ and $\mathbf{D}$ within the crystal are shown in relation to the permittivity ellipsoid.

It follows, as a special case, that when one of the principal dielectric constant directions is normal to the plates the ratio of the capacities gives this dielectric constant directly.

Fig. 4.3 c, which is self-explanatory, shows how the vectors $\mathbf{E}$, $\mathbf{P}$ and $\mathbf{E}$, $\mathbf{D}$ within the crystal in Fig. 4.3 a are related in direction by the susceptibility ellipsoid and the permittivity ellipsoid respectively.

4. The energy of a polarized crystal

When the polarization of a crystal is changed it may be proved (Böttcher, 1952) that, if the electric field is entirely confined to the crystal, the work done is given by

$$dW = vE_i\,dD_i, \tag{9}$$

where v is the volume of the crystal. As an example of the derivation of this expression in a special case we may consider the parallel plate condenser of Fig. 4.3 a connected to a battery. If there is a change in the polarization of the crystal, new surface charges, $d\sigma$ and $-d\sigma$ per unit area, will appear on the plates, and work of amount $A\phi\,d\sigma$, where A is the area of the plates, will be done by the battery. This may be expressed in terms of $\mathbf{E}$ and $\mathbf{D}$, and their changes, within the crystal. σ is equal to the normal component of $\mathbf{D}$ in the crystal, and the work is therefore

$$A\phi\,dD_n = A\,El\,dD_n = vE\,dD_n = vE_i\,dD_i,$$

where v is the volume of the crystal, since the electric field is normal to the plates.

The increase in the energy† of the condenser is therefore given by

$$d\Psi = vE_i\,dD_i, \tag{10}$$

and if we use equation (6) this becomes

$$d\Psi = v\kappa_{ij}E_i\,dE_j. \tag{11}$$

Just as in the magnetic case then (p. 59)

$$\kappa_{ij} = \frac{\partial^2\Psi}{\partial E_i\,\partial E_j} = \frac{\partial^2\Psi}{\partial E_j\,\partial E_i} = \kappa_{ji};$$

and, by integrating equation (11), and setting the arbitrary constant equal to zero, we find the expression for the energy

$$\Psi = \tfrac{1}{2}v\kappa_{ij}E_i E_j. \tag{12}$$

5. The force and couple on a crystal in an electric field

The force on a small specimen of dielectric in a non-uniform field is given by an expression analogous to (24), Chapter III, namely

$$F_i = vP_j\frac{\partial E_i}{\partial x_j}, \tag{13}$$

where P_j is the polarization, determined through equation (5) by the

† For an isothermal, reversible change this is the increase in the free energy; see also Appendix F.

total field within the crystal, and E_i is the field produced by sources outside the crystal. Owing to the depolarizing effect the force depends on the shape of the specimen, because, in a given external field, the shape affects **P**.

In the para- and diamagnetic case we were able to calculate the couple acting on a crystal in a uniform field due to anisotropy. A similar couple, due essentially to the non-parallelism between **P** and **E**, acts on each element of an insulating crystal placed in an electric field. Now, however, owing to the depolarizing effect, the expression for the couple depends markedly on the shape of the crystal; what was a second-order effect for para- and diamagnetism becomes here an important part of the couple. As an example of the effect of shape it is easy to show that a disk of *isotropic* dielectric suspended in a uniform field will tend to set so that the axis of the disk is parallel to the field.

The couple on a crystal in a *uniform* field thus arises from two causes: (i), the non-parallelism between **P** and **E** due to anisotropy and (ii), the non-parallelism between **P** and **E** due to shape. In a *non-uniform* field there is a third effect which comes from the finite size of the crystal, as explained in § 3.3, Chapter III, for the corresponding magnetic case.

6.† The electrostatic field in a homogeneous anisotropic dielectric

The electrostatic problem worked out in § 3 was particularly simple because **D** and **E** had uniform values throughout the crystal. In general, an electrostatic field within a crystal may be non-uniform. Let us now consider what general equations it must satisfy. We shall restrict attention to the interior of a homogeneous crystal, and, for generality, we may suppose that the crystal contains a continuous distribution of electric charge.

The two Maxwell equations involving **D** and **E** are

$$\operatorname{div} \mathbf{D} = \rho, \qquad \operatorname{curl} \mathbf{E} = -\dot{\mathbf{B}}, \tag{14}$$

where ρ is the charge density. For a static problem $\dot{\mathbf{B}} = 0$, and we have

$$\operatorname{div} \mathbf{D} = \rho, \qquad \operatorname{curl} \mathbf{E} = 0. \tag{15}$$

Using the principle of vector analysis that, if the curl of a vector is zero, the vector is expressible as the gradient of a scalar potential, we may write

$$\mathbf{E} = -\operatorname{grad} \phi \quad \text{or} \quad E_i = -\partial\phi/\partial x_i. \tag{16}$$

† This section may be omitted on a first reading.

In suffix notation the first of equations (15) is

$$\frac{\partial D_i}{\partial x_i} = \rho;$$

or, using (6), $\qquad\qquad \kappa_{ij}\frac{\partial E_j}{\partial x_i} = \rho,$

since the κ_{ij} are assumed independent of position. Substitution for E_j from (16) then gives

$$\kappa_{ij}\frac{\partial^2\phi}{\partial x_i \, \partial x_j} = -\rho \tag{17}$$

as the equation to be satisfied by the potential ϕ. Referred to the principal axes of $[\kappa_{ij}]$ equation (17) becomes

$$\kappa_1\frac{\partial^2\phi}{\partial x_1^2} + \kappa_2\frac{\partial^2\phi}{\partial x_2^2} + \kappa_3\frac{\partial^2\phi}{\partial x_3^2} = -\rho. \tag{18}$$

In a region of the crystal where there is no volume charge, $\rho = 0$, and (18) reduces to

$$\kappa_1\frac{\partial^2\phi}{\partial x_1^2} + \kappa_2\frac{\partial^2\phi}{\partial x_2^2} + \kappa_3\frac{\partial^2\phi}{\partial x_3^2} = 0. \tag{19}$$

(18) and (19) are the basic equations which must be solved in order to obtain the solution of an electrostatic problem in a crystal. By making the formal substitution,

$$x_1 = \kappa_1^{\frac{1}{2}} X_1, \qquad x_2 = \kappa_2^{\frac{1}{2}} X_2, \qquad x_3 = \kappa_3^{\frac{1}{2}} X_3,$$

in equations (18) and (19) we obtain, respectively, Poisson's equation and Laplace's equation:

$$\frac{\partial^2\phi}{\partial X_1^2} + \frac{\partial^2\phi}{\partial X_2^2} + \frac{\partial^2\phi}{\partial X_3^2} = -\rho \quad \text{and} \quad \frac{\partial^2\phi}{\partial X_1^2} + \frac{\partial^2\phi}{\partial X_2^2} + \frac{\partial^2\phi}{\partial X_3^2} = 0.$$

The known solutions of these two equations may therefore be immediately transformed into solutions of our equations (18) and (19). E_i and D_i then follow from equations (16) and (6) respectively.

We shall meet equations similar to (18) and (19) in studying the steady state conduction of heat in a crystal (Ch. XI, § 3).

Within an isotropic dielectric we have $\mathbf{D} = \kappa\mathbf{E}$; if $\rho = 0$ the electrostatic field equations,

$$\operatorname{div}\mathbf{D} = 0 \quad \text{and} \quad \operatorname{curl}\mathbf{E} = 0,$$

imply at once that, within the medium,

$$\operatorname{div}\mathbf{E} = 0 \quad \text{and} \quad \operatorname{curl}\mathbf{D} = 0. \tag{20}$$

It is worth noticing that equations (20) do not necessarily hold when the medium is anisotropic.

SUMMARY OF §§ 1–6

1. In any material
$$D_i = \kappa_0 E_i + P_i. \tag{1}$$
In an anisotropic dielectric such as a crystal
$$P_i = \kappa_0 \chi_{ij} E_j \tag{5}$$
(except for very strong fields), where $[\chi_{ij}]$ is the dielectric susceptibility tensor. Hence
$$D_i = \kappa_{ij} E_j, \quad \text{where} \quad \kappa_{ij} = \kappa_0(\delta_{ij} + \chi_{ij}). \tag{6}$$
$[\kappa_{ij}]$ is the permittivity tensor. The dielectric constant tensor is defined as
$$K_{ij} = \kappa_{ij}/\kappa_0. \tag{7}$$

2. The work done when the polarization of a crystal is changed, the field being entirely confined to the crystal, is
$$dW = v E_i \, dD_i, \tag{9}$$
where v is the volume of the crystal. Equating this with the increase in energy and putting $D_i = \kappa_{ij} E_j$, we have
$$d\Psi = v \kappa_{ij} E_i \, dE_j. \tag{11}$$
Since $d\Psi$ is a perfect differential it follows that $[\kappa_{ij}]$, and hence $[K_{ij}]$ and $[\chi_{ij}]$, are all symmetrical tensors. Integrating the expression for $d\Psi$ and putting the arbitrary constant zero, we find for the energy of a polarized crystal,
$$\Psi = \tfrac{1}{2} v \kappa_{ij} E_i E_j. \tag{12}$$

3. Although the formal analysis of dielectric polarization is closely similar to that of magnetization of paramagnetics, the depolarizing effect is of a different order of magnitude in the two cases. The leakage of charge found in dielectrics forms another practical difference.

4. The capacity of a parallel plate condenser is a measure of the dielectric constant perpendicular to the plates.

5. The force on a dielectric crystal in a non-uniform electric field is given by
$$F_i = v P_j(\partial E_i/\partial x_j), \tag{13}$$
where P_j is the polarization, determined through (5) by the total field within the crystal, and E_i is the field produced by sources outside the crystal. Owing to the depolarizing effect the force depends on the shape of the crystal.

6. A couple acts on a crystal whenever $\mathbf{P}$ and $\mathbf{E}$ within the crystal are not parallel. This may be caused by shape, through the depolarizing effect, or by crystal anisotropy. A further couple arises in a non-uniform field due to the variation of the force F_i over the specimen.

7. The electrostatic field in an anisotropic crystal obeys the equations
$$\operatorname{div} \mathbf{D} = \rho, \quad\quad \operatorname{curl} \mathbf{E} = 0.$$
It follows that $\mathbf{E}$ may be derived from a scalar potential ϕ, and that in a homogeneous crystal
$$\kappa_{ij} \frac{\partial^2 \phi}{\partial x_i \, \partial x_j} = -\rho, \tag{17}$$
or, referred to the principal axes,
$$\kappa_1 \frac{\partial^2 \phi}{\partial x_1^2} + \kappa_2 \frac{\partial^2 \phi}{\partial x_2^2} + \kappa_3 \frac{\partial^2 \phi}{\partial x_3^2} = -\rho. \tag{18}$$
We note that, in general,
$$\operatorname{div} \mathbf{E} \neq 0 \quad \text{and} \quad \operatorname{curl} \mathbf{D} \neq 0,$$
even in a region free from charge.

7. Pyroelectricity

Certain crystals have the property of developing an electric polarization when their temperature is changed. Alternatively, if a spontaneous polarization is already present, a change of temperature alters it. This phenomenon is called *pyroelectricity*. In practice the electric moment so developed does not persist, for owing to imperfect insulation it becomes neutralized by the migration of charges to the surface.

To observe pyroelectricity we can heat a crystal uniformly and observe the change in polarization. Now, theoretically, this experiment can be done in two different ways: either the shape and size of the crystal can be held fixed during the heating, or alternatively, the crystal may be released so that thermal expansion can occur quite freely. The magnitude of the effect observed in the two experiments would be different. In the first case, with the crystal clamped, the effect observed is called *primary pyroelectricity*. In the second case, with free expansion, which is much easier to achieve experimentally, there is an additional effect called *secondary pyroelectricity*; what is observed in this case is the primary effect plus the secondary effect. An adequate discussion of the relation between the primary and secondary effects, which involves piezoelectricity, necessitates a thermodynamical treatment, and we postpone it until Chapter X. The analysis given in this section applies equally well to both primary and secondary pyroelectricity, and to the sum of the two.†

If there is a small temperature change ΔT, uniform over the crystal, the change in the polarization vector ΔP_i is given by

$$\Delta P_i = p_i \Delta T, \tag{21}$$

where the p_i are the three *pyroelectric coefficients*. The pyroelectric effect in a crystal is thus specified by the vector **p**. This is the first example we have met in this book of a *crystal property* that is represented by a vector.

A polarization may also be produced in some crystals by a hydrostatic pressure; this is a special case of the piezoelectric effect, which we discuss in Chapter VII. Since the polarization is proportional to the pressure, which is a scalar like ΔT above, this property is also represented by a vector. The following discussion of the effect of symmetry on the

---

† For the moments produced by non-uniform heating, known as *tertiary pyroelectricity*, see p. 191. *Tensorial pyroelectricity* refers to the production of quadrupole or higher moments. Its existence is not firmly established experimentally and if it does occur the effect is very minute (Cady, 1946, p. 710).

pyroelectric effect applies equally well to the piezoelectric effect produced by hydrostatic pressure.

By Neumann's Principle (p. 20) $\mathbf{p}$ must conform to the point-group
symmetry of the crystal. It follows immediately that a pyroelectric
effect cannot exist ($\mathbf{p} = 0$) in a crystal possessing a centre of symmetry,
a fact which provides a practical method of testing for the absence of a
centre. A little thought shows that a pyroelectric moment can only lie
along a direction in a crystal which is unique, in the sense that it is not
repeated by any symmetry element (a reversal of direction is regarded
as a repetition here). If there should exist in the point group a unique
direction which is an axis of symmetry (2-, 3-, 4- or 6-fold), this will
necessarily be the direction of $\mathbf{p}$. But the presence of such a unique
symmetry axis (we do not here regard a monad axis as a symmetry axis)
is not essential for the existence of a pyroelectric effect; class m, in which
$\mathbf{p}$ can lie in any direction in the symmetry plane without violating
Neumann's Principle, and class 1 both illustrate this. It may be noted,
in passing, that a unique direction as defined above is not synonymous
with a *polar direction*. A polar direction is any direction of which the
two ends are not related by any symmetry element of the point group.
Thus, a diad axis in class 32 (p. 286) is a polar direction, but it is not
a unique direction. All unique directions are polar, but only some polar
directions are unique.

The direction of the pyroelectric vector $\mathbf{p}$ and the form of its components in the 21 non-centrosymmetrical classes are as follows:

Triclinic
Class 1: no symmetry restriction on the direction of $\mathbf{p}$: (p_1, p_2, p_3).

Monoclinic. x_2 parallel to the diad axis, rotation or inverse, (y).
Class 2: $\mathbf{p}$ parallel to the diad axis: $(0, p, 0)$.
Class m: $\mathbf{p}$ has any direction in the symmetry plane: $(p_1, 0, p_3)$.

Orthorhombic. x_1, x_2, x_3 parallel to crystallographic x, y, z respectively.
Class $mm2$: $\mathbf{p}$ parallel to the diad axis: $(0, 0, p)$.
Class 222: $(0, 0, 0)$.

Tetragonal, trigonal, hexagonal. x_3 parallel to z.
Classes $4, 4mm, 3, 3m, 6, 6mm$: $\mathbf{p}$ parallel to the 4, 3 or 6 axis: $(0, 0, p)$.
Classes $\bar{4}, \bar{4}2m, 422, 32, \bar{6}, \bar{6}m2, 622$: $(0, 0, 0)$.

Cubic
Classes $432, \bar{4}3m, 23$: $(0, 0, 0)$.

Thus, the following 10 classes may theoretically show pyroelectricity under uniform heating or cooling, or a piezoelectric effect under hydro-static pressure:

$$1 \quad 2 \quad 3 \quad 4 \quad 6$$
$$m \quad mm2 \quad 3m \quad 4mm \quad 6mm.$$

They are called the *polar classes*.

Numerical example. To illustrate the order of magnitude of the pyro-electric effect we take the case of tourmaline (trigonal, class *3m*), which is the best-known example of a pyroelectric crystal. p varies somewhat according to the composition of the crystal and the temperature. At room temperature, $p = 1\cdot2$ c.g.s. e.s.u. per ° C is a representative figure for the sum of the primary and secondary effects (Cady, 1946, p. 704). Thus, in rationalized m.k.s. units,

$$p = 1\cdot2(\tfrac{1}{3} \times 10^{-5}) = 4\cdot0 \times 10^{-6} \text{ coulombs m}^{-2} \text{ (° C)}^{-1}.$$

To get a feeling for the size of the effect let us calculate what electric field would produce the same polarization as a change in temperature of 1° C. From the above value of p, the polarization produced by a temperature rise of 1° C is

$$P = 4\cdot0 \times 10^{-6} \text{ m.k.s. units (coulombs/m}^2\text{).}$$

This is directed along the triad axis. The principal dielectric constant of tourmaline parallel to the triad axis is $K_3 = 7\cdot1$. We have for the corresponding susceptibility (p. 69)

$$\chi_3 = K_3 - 1 = 6\cdot1.$$

Hence, the necessary field parallel to x_3 within the crystal is

$$E_3 = \frac{P}{\kappa_0 \chi_3} = \frac{4\cdot0 \times 10^{-6}}{8\cdot85 \times 10^{-12} \times 6\cdot1} = 7\cdot4 \times 10^4 \text{ volts/m.}$$
$$= 740 \text{ volts/cm.}$$

8. Ferroelectricity

Another group of crystals called the *ferroelectrics* are somewhat related to pyroelectric crystals. A ferroelectric, like a pyroelectric, can show a spontaneous polarization, but it has the additional property that the polarization can be reversed by applying a sufficiently large electric field. In a strong alternating field it therefore shows hysteresis.

The electric field can reverse the polarization by causing a small relative shift of the atoms in the crystal, and so turning the crystal into its electric twin. If the atoms make half this shift, or if half the atoms make this shift, the crystal will be in a more symmetrical and non-polar state. Now it commonly happens that this intermediate configuration can be brought about by temperature change; hence most ferroelectrics have a *transition* temperature (Curie point) above which they are normal and non-polar.

When they are above their transition temperatures it happens that the known ferroelectrics are mainly confined to three classes:

>222, e.g. Rochelle salt,
>
>$\bar{4}2m$, e.g. potassium dihydrogen phosphate (KDP),
>
>$m3m$, e.g. barium titanate.

(Rochelle salt is unusual in having a lower as well as an upper transition temperature; it only shows spontaneous polarization between these temperatures.) A crystal in its ferroelectric state, with a spontaneous polarization, must be of lower symmetry than the same crystal in its non-polar state, and must belong to one of the classes that are listed as pyroelectric in § 7. As examples, Rochelle salt in its ferroelectric state belongs to class 2; KDP transforms from class $\bar{4}2m$ to $mm2$; and barium titanate transforms successively from class $m3m$ to the pyroelectric classes $4mm$, $mm2$ and $3m$ as the temperature is lowered.

Ferroelectrics form a group of crystals of great theoretical interest. The spontaneous polarization and hysteresis which characterize them are accompanied by other special properties. For further study see Devonshire (1954) and the references on pp.312, 318, 319.

EXERCISE 4.1. Show that the successive changes of symmetry of barium titanate as the temperature is lowered can be explained by postulating that the cubic crystal first becomes polar along one cube axis, then equally polar along two cube axes together, and finally equally polar along all three cube axes together.

SUMMARY OF §§ 7, 8

Pyroelectricity. The change in the polarization ΔP_i produced in a pyroelectric crystal by a small uniform temperature change ΔT is given by

$$\Delta P_i = p_i \Delta T, \tag{21}$$

where the three p_i are the pyroelectric coefficients. They define a vector which must conform to the crystal symmetry. This requirement makes pyroelectricity under uniform temperature changes impossible in all but ten non-centrosymmetrical classes.

Ferroelectricity. Ferroelectric crystals can show a spontaneous polarization that can be reversed by applying a sufficiently strong electric field. Most ferroelectrics have a transition temperature above which they are normal and non-polar.

V

THE STRESS TENSOR

1. The notion of stress

A BODY which is acted on by external forces, or, more generally, a body in which one part exerts a force on neighbouring parts, is said to be in a state of stress. If we consider a volume element situated within a stressed body, we may recognize two kinds of forces acting upon it.

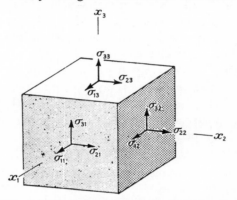

FIG. 5.1. The forces on the faces of a unit cube
in a homogeneously stressed body.

First of all, there are body-forces, such as gravity, which act throughout the body on all its elements and whose magnitudes are proportional to the *volume* of the element. Secondly, there are forces exerted on the surface of the element by the material surrounding it. These forces are proportional to the *area* of the surface of the element, and the force per unit area is called the 'stress'. In the present chapter we discuss how this stress may be exactly specified. A stress is said to be *homogeneous* if the forces acting on the surface of an element of fixed shape and orientation are independent of the position of the element in the body.

1.1. Homogeneous stress. At first we confine the discussion to states in which (1) the stress is homogeneous throughout the body, (2) all parts of the body are in statical equilibrium, and (3) there are no body-forces or body-torques.

Consider a unit cube within the body (Fig. 5.1) with edges parallel

to the axes Ox_1, Ox_2, Ox_3. A force will be transmitted across each face
of the cube, exerted by the material outside the cube upon the material
inside the cube. The force transmitted across each face may be resolved
into three components. Consider first the three faces which are towards
the three positive ends of the axes (those shown in Fig. 5.1). We denote
by σ_{ij} the component of force in the $+Ox_i$ direction transmitted across
that face of the cube which is perpendicular to Ox_j.† Note the sign
convention: σ_{12}, for example, is the force exerted in the $+Ox_1$ direction
on the face normal to Ox_2, *by* the material outside the cube *upon* the
material inside. Since the stress is homogeneous, the forces exerted on
the cube across the three opposite faces must be equal and opposite to
those shown in Fig. 5.1. σ_{11}, σ_{22}, σ_{33} are the *normal components* of stress
and σ_{12}, σ_{21}, σ_{23} etc. are the *shear components*. In § 2 it is proved that
the σ_{ij} thus defined form a second-rank tensor. With the sign conven-
tion we have adopted it will be seen that a positive value of σ_{11}, σ_{22} or
σ_{33} implies a corresponding tensile stress; a negative value implies a
compressive stress. This is the definition normally found in modern
textbooks on elasticity. However, the opposite sign convention (compressive stresses positive) is sometimes used, particularly in applications
to piezoelectricity and photoelasticity, and so care is needed when
consulting numerical data given by different authors.

Our assumption (2), that the unit cube should be in statical equilibrium, imposes conditions on the σ_{ij}. Let us take moments about an
axis parallel to Ox_1 passing through the centre of the cube (Fig. 5.2).
Since the stress is homogeneous the three components of force on any
face all pass through the mid-point of the face. The normal components
and the shear components on the Ox_1 faces therefore give no moment,
and we find as the condition for equilibrium

$$\sigma_{23} = \sigma_{32}.$$

In a similar way $\sigma_{31} = \sigma_{13}$ and $\sigma_{12} = \sigma_{21}$, and so we may write

$$\sigma_{ij} = \sigma_{ji}. \tag{1}$$

1.2. Inhomogeneous stress. The relation (1) continues to hold
even when the stress is inhomogeneous, when the body is not in statical
equilibrium, and when body-forces (but not body-torques) are present.
This may be proved in the following way.

We define components of stress in essentially the same way as before

---

† There is of course no connexion with electrical conductivity which we have also
denoted by σ_{ij}.

but we now have to cover situations where the stress varies from point to point. The notion of the stress at a point is arrived at by a limiting process. We divide the force transmitted across a surface passing through the point by the area of the surface, and then let the area of the surface tend to zero at the point. Specifically, the component of force in the Ox_i direction transmitted across a surface element of area dS perpendicular to Ox_j is defined to be $\sigma_{ij}dS$ as dS tends to zero. The sign convention is the same as for homogeneous stress: namely,

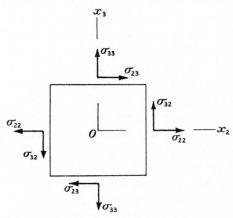

Fig. 5.2. The forces on the faces perpendicular to Ox_2 and Ox_3 of a unit cube in a homogeneously stressed body. The axis Ox_1 is normal to the plane of the figure.

$\sigma_{ij}dS$ denotes the force in the $+Ox_i$ direction exerted *by* the material on the $+Ox_j$ side of the element *upon* the material on the $-Ox_j$ side.

Consider now a small rectangular parallelepiped situated within the stressed body, centred on the origin, and with edges parallel to the axes and of lengths δx_1, δx_2, δx_3. Let σ_{ij} stand for the stresses at the origin. We wiʰ to find the equation of motion of the element in the Ox_1 direction. The average values of the component σ_{11} over the two faces perpendicular to Ox_1 are shown in Fig. 5.3. The forces in the Ox_1 direction on these two faces are

$$-\left(\sigma_{11}-\frac{\partial\sigma_{11}}{\partial x_1}\cdot\tfrac{1}{2}\delta x_1\right)\delta x_2\,\delta x_3 \quad \text{and} \quad \left(\sigma_{11}+\frac{\partial\sigma_{11}}{\partial x_1}\cdot\tfrac{1}{2}\delta x_1\right)\delta x_2\,\delta x_3,$$

so that the resultant is $\dfrac{\partial\sigma_{11}}{\partial x_1}\delta x_1\,\delta x_2\,\delta x_3.$

The forces in the Ox_1 direction on the two faces perpendicular to Ox_2 are

$$-\left(\sigma_{12} - \frac{\partial \sigma_{12}}{\partial x_2} \cdot \tfrac{1}{2}\delta x_2\right)\delta x_3\, \delta x_1 \quad \text{and} \quad \left(\sigma_{12} + \frac{\partial \sigma_{12}}{\partial x_2} \cdot \tfrac{1}{2}\delta x_2\right)\delta x_3\, \delta x_1,$$

with resultant

$$\frac{\partial \sigma_{12}}{\partial x_2}\delta x_1\, \delta x_2\, \delta x_3.$$

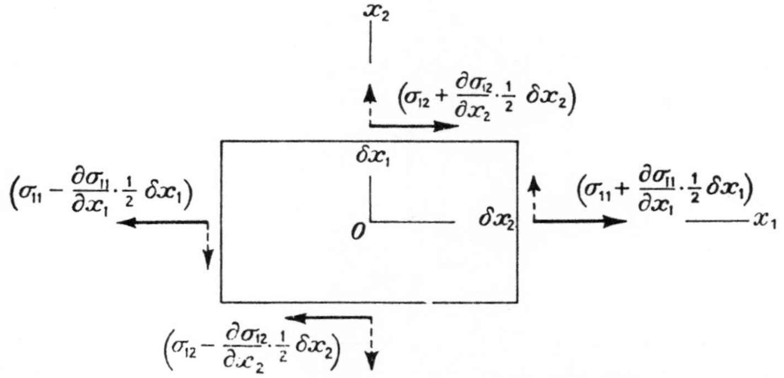

Fig. 5.3. Inhomogeneous stress: illustrating the forces on an element which have components parallel to Ox_1. The forces on the face perpendicular to Ox_3 are not shown.

Similarly, for the two faces perpendicular to Ox_3 we find

$$\frac{\partial \sigma_{13}}{\partial x_3}\delta x_1\, \delta x_2\, \delta x_3.$$

If there is a body-force with a component in the Ox_1 direction of g_1 per unit mass (gravity, for example), we have as the equation of motion

$$\frac{\partial \sigma_{11}}{\partial x_1} + \frac{\partial \sigma_{12}}{\partial x_2} + \frac{\partial \sigma_{13}}{\partial x_3} + \rho g_1 = \rho \ddot{u}_1,$$

where ρ is the density and $\ddot{u}_1$ is the acceleration in the Ox_1 direction.

Resolution of forces in the Ox_2 and Ox_3 directions yields two similar equations. Thus, finally, we may write

$$\frac{\partial \sigma_{ij}}{\partial x_j} + \rho g_i = \rho \ddot{u}_i, \tag{2}$$

a fundamental equation which connects the spatial variation of stress in a body with the accelerations of its elements; it forms the starting-point for the study of elastic waves in solid bodies. (Note that we have introduced a new idea here: the differentiation of a second-rank tensor.) If all parts of the body are in statical equilibrium, equations (2) take the simpler form

$$\frac{\partial \sigma_{ij}}{\partial x_j} + \rho g_i = 0, \tag{2a}$$

known as the *equations of equilibrium* and much used in the theory of elasticity.

Let us now write down the equation of motion for rotation of the element about the Ox_1 axis. The couple (anticlockwise in Fig. 5.4) due to the shear components of stress on the two faces perpendicular to Ox_2 is

$$\left(\sigma_{32} - \frac{\partial\sigma_{32}}{\partial x_2}\cdot\tfrac{1}{2}\delta x_2\right)\delta x_1\,\delta x_3\cdot\tfrac{1}{2}\delta x_2 + \left(\sigma_{32} + \frac{\partial\sigma_{32}}{\partial x_2}\cdot\tfrac{1}{2}\delta x_2\right)\delta x_1\,\delta x_3\cdot\tfrac{1}{2}\delta x_2$$
$$= \sigma_{32}\,\delta x_1\,\delta x_2\,\delta x_3.$$

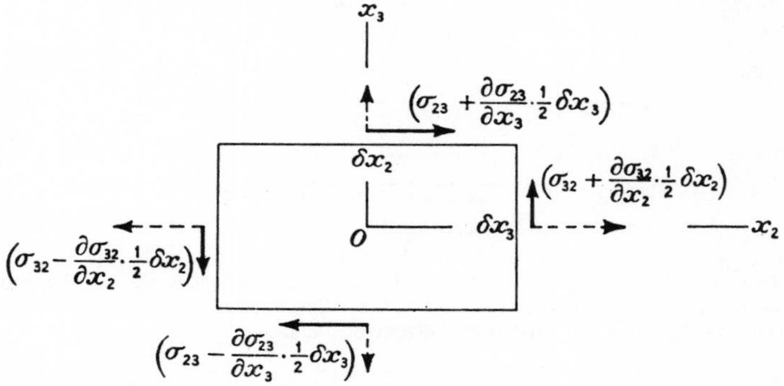

FIG. 5.4. Inhomogeneous stress: illustrating the forces on an element which have moments about the axis Ox_1.

The couple due to the shear components σ_{23} is similarly

$$-\sigma_{23}\,\delta x_1\,\delta x_2\,\delta x_3.\dagger$$

The equation of motion for rotation about Ox_1 is therefore

$$(\sigma_{32}-\sigma_{23})\delta x_1\,\delta x_2\,\delta x_3 + G_1\,\delta x_1\,\delta x_2\,\delta x_3 = I_1\,\ddot{\theta}_1,$$

where I_1 is the moment of inertia, and $\ddot{\theta}_1$ the anticlockwise angular acceleration, about Ox_1. G_1 in this equation represents a body-torque. I_1 is of order of magnitude $\rho\,\delta x^5$. Hence, as the element becomes infinitesimally small, unless

$$\sigma_{32}-\sigma_{23}+G_1 = 0, \tag{3}$$

$\ddot{\theta}_1$ must increase without limit, as $1/\delta x^2$. In a continuous material such behaviour is impossible, and so we conclude that (3) holds.

† There will also be contributions from the normal components σ_{22} and σ_{33}, due to the fact that the resultant normal forces on the faces do not pass exactly through the mid-points of the faces. However, the lever arm of each of these forces will be an order of magnitude smaller than the lever arm for the shear forces and so the terms may be neglected. (These forces actually cancel in pairs to the first approximation and so their total contribution to the couple is an order of magnitude smaller still.) A similar consideration shows that we may also neglect the couple due to the slightly off-centre disposition of the shear components σ_{21} and σ_{31}.

The equations of motion for rotations about Ox_2 and Ox_3 respectively give in a similar way

$$\sigma_{13} - \sigma_{31} + G_2 = 0, \tag{4}$$

and

$$\sigma_{21} - \sigma_{12} + G_3 = 0. \tag{5}$$

A distributed body-torque, that is, a torque proportional to volume exerted by long-range forces, such as is represented by G_1, G_2, G_3, occurs when an anisotropic crystal becomes polarized or magnetized in a field, as we have seen in Chapter III, § 3 and Chapter IV, § 5†. Equations (3) to (5) show that the stress tensor σ_{ij} is then not symmetrical, but when body-torques are absent it is and we have simply

$$\sigma_{ij} = \sigma_{ji}. \tag{6}$$

Our treatment of elasticity (Ch. VIII) assumes (6), but in fact it is still perfectly valid even in the presence of body-torques (Tiffen and Stevenson 1956) provided that, in the statement of Hooke's law, σ_{ij} is interpreted as being not the stress tensor itself but its symmetrical part $\frac{1}{2}(\sigma_{ij} + \sigma_{ji})$. Similar considerations apply to piezoelectricity (Ch. VII) and photoelasticity (Ch. XIII). (See also pp. 315–317.)

2. Proof that the σ_{ij} form a tensor

We now prove that the components of stress σ_{ij} defined in §§ 1.1 and 1.2 form a second-rank tensor. We know (Ch. I, § 3) that if a set of quantities T_{ij} relate the components of two vectors p_i, q_i by an equation of the form

$$p_i = T_{ij} q_j,$$

the T_{ij} obey the tensor transformation law, and hence form a tensor. We accordingly prove that the σ_{ij} relate two vectors by an equation of this type.

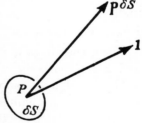

Select any small surface element of area δS containing a point P within the stressed body. Draw a unit vector $\mathbf{l}$ perpendicular to it. Let the force transmitted across the area be denoted by $\mathbf{p}\,\delta S$ (Fig. 5.5). The force is taken to be that which is exerted by the material on the positive side of the area (defined by the direction of $\mathbf{l}$) upon the material on the negative side. We have to ask: as $\mathbf{l}$ is altered in direction so that the surface element takes up different orientations, but always passing through P, how does $\mathbf{p}\,\delta S$ change? To answer this question we assume at first that the stress is homogeneous, that there are no body-forces, and that the body is in equilibrium. We consider the equilibrium of

Fig. 5.5. The force transmitted across a small surface element δS in a stressed body.

† A small body-torque occurs during the phenomenon of optical activity (Ch. XIV).

the tetrahedron-shaped element of the body $OABC$ shown in Fig. 5.6. ABC represents our variable surface element perpendicular to $\mathbf{l}$ and the force transmitted across it is $\mathbf{p} \times$ (area ABC). The forces on the three

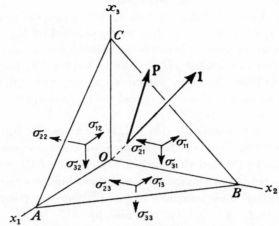

Fig. 5.6. The surface tractions on the faces of a tetrahedron bounded by the face ABC and the three coordinate planes.

faces at right angles may be specified by the stress components σ_{ij}, as shown. Resolving forces parallel to Ox_1 we have

$$p_1 . ABC = \sigma_{11} . BOC + \sigma_{12} . AOC + \sigma_{13} . AOB,$$

or $$p_1 = \sigma_{11} l_1 + \sigma_{12} l_2 + \sigma_{13} l_3.$$

Similarly, $$p_2 = \sigma_{21} l_1 + \sigma_{22} l_2 + \sigma_{23} l_3,$$

and $$p_3 = \sigma_{31} l_1 + \sigma_{32} l_2 + \sigma_{33} l_3.$$

Hence, we may write $$p_i = \sigma_{ij} l_j. \tag{7}$$

When the stress is not homogeneous, when body-forces are acting, and when the body is not in statical equilibrium, equations (7) still hold for any given point, for it is easy to see that the extra terms that enter become negligible as the tetrahedron is made vanishingly small.

Since σ_{ij} relates the two vectors p_i and l_j in a linear way it is a tensor. Equation (6) shows that it is a symmetrical tensor, and consequently it may be referred to its principal axes (Ch. I, § 4.1). Thus, on transformation to the principal axes, we have

$$\begin{bmatrix} \sigma_{11} & \sigma_{12} & \sigma_{31} \\ \sigma_{12} & \sigma_{22} & \sigma_{23} \\ \sigma_{31} & \sigma_{23} & \sigma_{33} \end{bmatrix} \rightarrow \begin{bmatrix} \sigma_1 & 0 & 0 \\ 0 & \sigma_2 & 0 \\ 0 & 0 & \sigma_3 \end{bmatrix};$$

σ_1, σ_2, σ_3 are the *principal stresses*.

When the principal stress directions are chosen as axes the shear stress components vanish, and the forces on the faces of a unit cube cut from the body with edges parallel to the axes are as shown in Fig. 5.7.

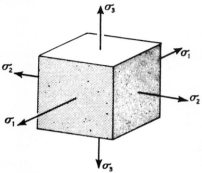

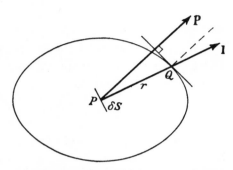

FIG. 5.7. The forces on the faces of a unit cube cut with its edges parallel to the three principal stress directions.

FIG. 5.8. Illustrating how the direction of the resultant force $\mathbf{p} \, \delta S$ transmitted across a small area δS may be found by using the radius-normal property of the stress quadric. $\mathbf{l}$ is the unit vector normal to δS. $\mathbf{p}$ is perpendicular to the tangent plane to the quadric at Q. The plane of the figure is that central cross-section of the quadric which contains both $\mathbf{p}$ and $\mathbf{l}$. The element δS and the tangent plane are seen edge on.

3. The stress quadric

The representation quadric for σ_{ij} (Ch. I, § 4) is called simply the *stress quadric*. Its equation is

$$\sigma_{ij} x_i x_j = 1,$$

or, referred to principal axes,

$$\sigma_1 x_1^2 + \sigma_2 x_2^2 + \sigma_3 x_3^2 = 1.$$

The lengths of the semi-axes are therefore $1/\sqrt{\sigma_1}$, $1/\sqrt{\sigma_2}$, $1/\sqrt{\sigma_3}$. Since σ_1, σ_2, σ_3 may each be positive or negative, the quadric may be a real or imaginary ellipsoid or a hyperboloid.

The direction of the resultant force $\mathbf{p} \, \delta S$ transmitted across a small area δS may be found from the stress quadric by the radius-normal property (p. 28). Draw (Fig. 5.8) a radius vector of length r parallel to $\mathbf{l}$, the unit vector normal to δS. Let it cut the surface of the quadric in Q. Then $\mathbf{p}$ is parallel to the normal to the quadric at Q. Whenever Q lies on one of the three principal axes, $\mathbf{p}$ is parallel to $\mathbf{l}$; that is, there are no shear components.

By the property described on pp. 26, 27, the length of the radius vector r gives the normal stress σ transmitted across the element in Fig. 5.8:

thus $\sigma = 1/r^2$. Alternatively, σ is given analytically (Ch. I, § 6.2) by

$$\sigma = \sigma_{ij} l_i l_j;$$

or, if all components are referred to the principal stress axes,

$$\sigma = \sigma_1 l_1^2 + \sigma_2 l_2^2 + \sigma_3 l_3^2.$$

4. Special forms of the stress tensor

We give now some of the forms taken by the stress tensor, referred to its principal axes, in special cases.

(i) *Uniaxial stress*, σ.
$$\begin{bmatrix} \sigma & 0 & 0 \\ 0 & 0 & 0 \\ 0 & 0 & 0 \end{bmatrix}.$$

An example is the stress in a long, vertical rod loaded by hanging a weight on the end. A non-uniform distribution of uniaxial stress occurs in the pure bending of a long bar.

(ii) *Biaxial stress*.
$$\begin{bmatrix} \sigma_1 & 0 & 0 \\ 0 & \sigma_2 & 0 \\ 0 & 0 & 0 \end{bmatrix}.$$

An example of a non-uniform distribution of biaxial stress is the stress

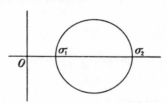

FIG. 5.9. The Mohr circle for a
state of biaxial stress.

in a thin plate loaded by forces and couples applied to its edges. The Mohr circle for a state of biaxial stress is shown in Fig. 5.9.

(iii) *Triaxial stress*. This is an alternative name for the most general stress system with three non-zero principal stresses.

(iv) *Hydrostatic pressure*, p.
$$\begin{bmatrix} -p & 0 & 0 \\ 0 & -p & 0 \\ 0 & 0 & -p \end{bmatrix} \text{ or } -p\delta_{ij}.$$

(v) *Pure shear stress*.
$$\begin{bmatrix} -\sigma & 0 & 0 \\ 0 & \sigma & 0 \\ 0 & 0 & 0 \end{bmatrix}.$$

This is a special case of biaxial stress. A non-uniform distribution of pure shear stress occurs in a long rod subjected to pure torsion. The Mohr circle construction, illustrated for this case in Fig. 5.10 a, shows at once that if the axes are turned through 45° about Ox_3 the normal stresses vanish (hence the name 'pure shear stress') and the tensor takes the form

$$\begin{bmatrix} 0 & \sigma & 0 \\ \sigma & 0 & 0 \\ 0 & 0 & 0 \end{bmatrix}.$$

Ox_3 is the *axis of shear*. Figs. 5.10 b, c show the forces acting on the faces of two elements in the two different orientations.

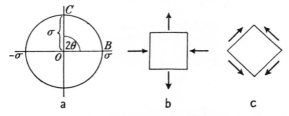

Fig. 5.10. Pure shear stress: (a) the Mohr circle, (b) and (c) the forces on an element in the orientations corresponding to points B and C respectively.

5. Difference between the stress tensor and tensors representing crystal properties

We conclude this chapter by pointing out an important distinction between the stress tensor and all the other second-rank tensors so far introduced. Tensors which measure crystal properties (such as the permittivity and the magnetic susceptibility, represented by quadrics) have definite orientations within a crystal, and, as we have seen, they must conform to the crystal symmetry. They are called *matter tensors*. The stress tensor, on the other hand, in common with the strain tensor of the next chapter, can have any orientation within a crystal, and it can exist just as well in isotropic bodies like glass as in anisotropic crystals. The stress tensor does not represent a crystal property but is akin to a 'force' impressed on the crystal; in this respect it is like an electric field, which can, of course, have an arbitrary direction in a crystal. Such tensors are called *field tensors*.

SUMMARY

THE stress at a point P in a material may be defined in the following way (§ 2). Let δS be the area of an element of surface passing through P. Draw a unit vector $\mathbf{l}$ perpendicular to it. Let the force transmitted across the area be $\mathbf{p}\,\delta S$ in the following sense: $\mathbf{p}\,\delta S$ is the force exerted by the material on the positive side of the area (defined by the direction of $\mathbf{l}$) upon the material on the negative side. Then statical considerations (or dynamical considerations if the body is not in statical equilibrium) show that, as $\delta S \to 0$, $\mathbf{p}$ is connected with $\mathbf{l}$ by the relation

$$p_i = \sigma_{ij} l_j,$$

where the σ_{ij} are coefficients. $[\sigma_{ij}]$ is a second-rank tensor and is called the *stress* at the point.

The meanings of the nine components σ_{ij} may be appreciated by considering the forces on the faces of a cube within the stressed body with edges parallel to the axes x_i (Fig. 5.1). The components with $i = j$ are the *normal components* of stress (positive for tensile stresses) and the components with $i \neq j$ are the *shear components* (§ 1.1).

The *equations of motion for translation* of a small element are (§ 1.2)

$$\frac{\partial \sigma_{ij}}{\partial x_j} + \rho g_i = \rho \ddot{u}_i,$$

or if all parts of the body are in statical equilibrium,

$$\frac{\partial \sigma_{ij}}{\partial x_j} + \rho g_i = 0,$$

where ρ is the density and g_i is the body-force per unit mass.

The *equations of motion for rotation* are (§ 1.2)

$$\left.\begin{array}{l} \sigma_{32} - \sigma_{23} + G_1 = 0 \\ \sigma_{13} - \sigma_{31} + G_2 = 0 \\ \sigma_{21} - \sigma_{12} + G_3 = 0 \end{array}\right\},$$

where G_i is the body-torque per unit volume. Hence, in the absence of body-torques,

$$\sigma_{ij} = \sigma_{ji}.$$

Since $[\sigma_{ij}]$ is a symmetrical tensor (in the absence of body-torques), it may be referred to its principal axes (§ 2) and may be represented by a quadric, $\sigma_{ij} x_i x_j = 1$. Stress is not a crystal property like the other second-rank tensors (magnetic susceptibility, permittivity, etc.) introduced so far (§ 5); it is akin to a 'force' impressed on the crystal. Accordingly, the stress quadric does not have to conform to the crystal symmetry.

EXERCISE 5.1. Show that a general stress may, by a suitable choice of axes, be expressed as the sum of (1) a hydrostatic stress (i.e. of the form $\sigma \delta_{ij}$) and (2) a shear stress (i.e. a stress whose normal components are all zero).

VI

THE STRAIN TENSOR AND
THERMAL EXPANSION

THE problem of specifying the state of deformation of a solid body, which we take up in this chapter, may be approached by considering first the simpler one-dimensional and two-dimensional cases.

1. One-dimensional strain

Fig. 6.1 a shows an extendible string. We mark an origin O, fixed in space, and then stretch the string. After stretching (Fig. 6.1 b), an arbitrary point P on the string moves to P'. Let

$$OP = x \quad \text{and} \quad OP' = x + u.$$

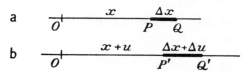

FIG. 6.1. The deformation of an extendible
string: (a) unstretched, (b) stretched.

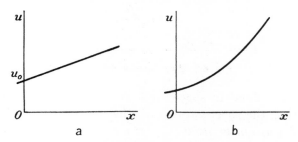

FIG. 6.2. The displacement u as a function of x in an extended
string; (a) homogeneous stretching, (b) inhomogeneous stretching.

The variation of the displacement u with x is shown in Figs. 6.2 a and b. In Fig. 6.2 a, where u is a linear function of x, the string is stretched homogeneously; Fig. 6.2 b illustrates the more general case of inhomogeneous stretching. Let a point Q, close to P, move to Q' during the stretching and let $PQ = \Delta x$. Then $P'Q' = \Delta x + \Delta u$. In studying strain we are not concerned with the actual displacement of points but with their displacements relative to one another. The strain of the section PQ is defined as:

$$\frac{\text{increase in length}}{\text{original length}} = \frac{P'Q' - PQ}{PQ} = \frac{\Delta u}{\Delta x}.$$

The *strain at the point P* is defined as:

$$e = \lim_{\Delta x \to 0} \frac{\Delta u}{\Delta x} = \frac{du}{dx}. \tag{1}$$

The strain at any point is thus defined simply as the slope of the curves in Figs. 6.2 a and b; it is the rate of change of displacement with distance and is a dimensionless quantity. Clearly, with this definition, the position of the origin is immaterial.

For a *homogeneous* strain, e is constant and equation (1) integrates to

$$u = u_0 + ex, \tag{2}$$

where u_0 is the displacement of the point at the origin.

2. Two-dimensional strain

We now consider how to specify the deformation of an extendible plane sheet. Following the same procedure as in § 1, we chose an

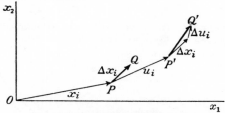

FIG. 6.3. Two-dimensional strain.

origin (Fig. 6.3) fixed in space, and we study how the displacement of the points of the sheet varies with their coordinates. *Henceforth we restrict the discussion to small displacements.* Let the point P, whose coordinates referred to axes fixed in space are (x_1, x_2) before deformation, move to P', with coordinates $(x_1 + u_1, x_2 + u_2)$. The vector u_i is therefore the displacement of P. To specify the strain at this point of the sheet we first define four quantities

$$e_{11} = \frac{\partial u_1}{\partial x_1}, \quad e_{12} = \frac{\partial u_1}{\partial x_2}, \quad e_{21} = \frac{\partial u_2}{\partial x_1}, \quad e_{22} = \frac{\partial u_2}{\partial x_2},$$

or, collectively, $\qquad e_{ij} = \frac{\partial u_i}{\partial x_j} \quad (i, j = 1, 2).$ $\qquad\qquad$ (3)

The e_{ij} are all dimensionless quantities small compared with 1. To find their geometrical meanings we may consider a point Q lying near P, such that $\mathbf{PQ} = [\Delta x_i]$. After deformation Q moves to Q', and clearly $\mathbf{P'Q'}$ is the sum of two vectors, $[\Delta x_i] + [\Delta u_i]$. $[\Delta u_i]$ is the difference in displacement between the two points P and Q originally separated by $[\Delta x_i]$. Then, since the components of $[u_i]$ are functions of position, we may write

$$\left. \begin{aligned} \Delta u_1 &= \frac{\partial u_1}{\partial x_1}\Delta x_1 + \frac{\partial u_1}{\partial x_2}\Delta x_2 \\ \Delta u_2 &= \frac{\partial u_2}{\partial x_1}\Delta x_1 + \frac{\partial u_2}{\partial x_2}\Delta x_2 \end{aligned} \right\}, \tag{4}$$

or, briefly,

$$\Delta u_i = \frac{\partial u_i}{\partial x_j}\Delta x_j = e_{ij}\,\Delta x_j. \tag{5}$$

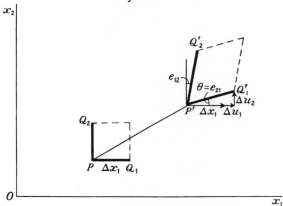

FIG. 6.4. The meanings of the strain components in two-dimensional strain.

As $[\Delta u_i]$ and $[\Delta x_j]$ are both vectors it follows (p. 15) that $[e_{ij}]$ is a tensor.

Let us now (Fig. 6.4) take two special positions of the vector $[\Delta x_i]$, first parallel to Ox_1 (PQ_1) and then parallel to Ox_2 (PQ_2), and in this manner find out how a rectangular element at P is distorted. For PQ_1 we put $\Delta x_2 = 0$, and equations (4) become

$$\left. \begin{aligned} \Delta u_1 &= \frac{\partial u_1}{\partial x_1}\Delta x_1 = e_{11}\,\Delta x_1 \\ \Delta u_2 &= \frac{\partial u_2}{\partial x_1}\Delta x_1 = e_{21}\,\Delta x_1 \end{aligned} \right\}.$$

The meanings of Δu_1 and Δu_2 are indicated in Fig. 6.4. It will be seen that e_{11} measures the extension per unit length of PQ_1 resolved along Ox_1, for

$$\frac{\Delta u_1}{\Delta x_1} = \frac{\partial u_1}{\partial x_1} = e_{11}.$$

e_{21} measures the anticlockwise rotation of PQ_1; for the angle through which it turns is given by

$$\tan \theta = \frac{\Delta u_2}{\Delta x_1 + \Delta u_1}.$$

Since we are only considering small displacements, u_1 and u_2 are small

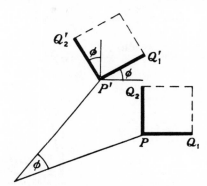

Fig. 6.5. Displacements in a small rigid-body rotation of a sheet in its own plane.

compared with x_1, and hence Δu_1 and Δu_2 are small compared with Δx_1. Thus

$$\theta = \frac{\Delta u_2}{\Delta x_1} = e_{21}.$$

In a similar way, e_{22} equals the extension per unit length of PQ_2 in the Ox_2 direction, and e_{12} measures the (small) clockwise rotation of PQ_2 to $P'Q_2'$.

It may now be asked: is the tensor $[e_{ij}]$ a satisfactory measure of the strain at the point P? For an affirmative answer it is clearly necessary that, when there is no distortion, all four components of $[e_{ij}]$ should vanish. This does not in fact happen. For consider a simple rigid-body rotation of the sheet in its own plane, anticlockwise, through a *small* angle ϕ (Fig. 6.5). The rotations of PQ_1 and PQ_2 are both ϕ anticlockwise, and hence, from the geometrical meanings of the e_{ij} established above,

$$[e_{ij}] = \begin{bmatrix} 0 & -\phi \\ \phi & 0 \end{bmatrix}.$$

There is no distortion of the sheet but $[e_{ij}]$ does not vanish. To avoid this difficulty we have to find a way of subtracting the part of $[e_{ij}]$ corresponding to a rigid-body rotation.

Now any second-rank tensor can be expressed as the sum of a symmetrical and an antisymmetrical tensor. We do this for $[e_{ij}]$ and write

$$e_{ij} = \epsilon_{ij} + \varpi_{ij},$$

where $\qquad \epsilon_{ij} = \tfrac{1}{2}(e_{ij} + e_{ji})$ and $\varpi_{ij} = \tfrac{1}{2}(e_{ij} - e_{ji}).$

$[\epsilon_{ij}]$ so defined is a symmetrical tensor, for

$$\epsilon_{ij} = \tfrac{1}{2}(e_{ji} + e_{ij}) = \epsilon_{ji};$$

and $[\varpi_{ij}]$ so defined is an antisymmetrical tensor, for

$$\varpi_{ij} = -\tfrac{1}{2}(e_{ji} - e_{ij}) = -\varpi_{ji}.$$

We see above that the tensor $[e_{ij}]$ giving a pure rotation is antisym-

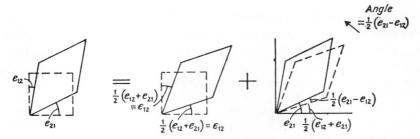

Fig. 6.6. A two-dimensional illustration of the proposition: a general deformation (left-hand diagram) equals a strain (centre diagram) plus a rotation (right-hand diagram).

metrical. We therefore define the *symmetrical* part of $[e_{ij}]$, that is $[\epsilon_{ij}]$, as the *strain*. Thus, in full,

$$\begin{bmatrix} \epsilon_{11} & \epsilon_{12} \\ \epsilon_{12} & \epsilon_{22} \end{bmatrix} = \begin{bmatrix} e_{11} & \tfrac{1}{2}(e_{12} + e_{21}) \\ \tfrac{1}{2}(e_{12} + e_{21}) & e_{22} \end{bmatrix}.$$

This division of $[e_{ij}]$ into two parts is illustrated in Fig. 6.6. The diagonal components of $[\epsilon_{ij}]$ are the extensions per unit length parallel to Ox_1 and Ox_2. ϵ_{12} measures the *tensor shear strain*; if two line elements are drawn parallel to Ox_1 and Ox_2 in the undeformed body the angle between them after deformation is $\tfrac{1}{2}\pi - 2\epsilon_{12}$ (centre diagram of Fig. 6.6). Note particularly that the tensor shear strain ϵ_{12} is *one-half* of the change in angle between the two elements.

2.1. Homogeneous two-dimensional strain. When the distortion is homogeneous the e_{ij} components are all constants and equations (3) integrate to

$$u_i = (u_0)_i + e_{ij} x_j \quad (i,j = 1,2), \tag{6}$$

where $(u_0)_i$ is the displacement of the point at the origin.†

If a curve $f(x_1, x_2) = 0$ is drawn on the sheet before deformation, it becomes $f(x_1', x_2') = 0$ after deformation, where

$$x_i' = x_i + u_i = (u_0)_i + x_i + e_{ij} x_j.$$

This is a linear substitution and it follows that, during the deformation,
  (1) a straight line remains a straight line,
  (2) parallel lines remain parallel,
  (3) all straight lines drawn in the same direction are extended or contracted in the same ratio,
  (4) an ellipse becomes a different ellipse and, in particular, a circle becomes an ellipse.

3. Three-dimensional strain

In specifying the strain of a three-dimensional body, such as a crystal, the method is essentially similar to that used in the two preceding sections. The variation of the displacement u_i with position x_i in the body is used to define nine tensor components

$$e_{ij} = \frac{\partial u_i}{\partial x_j} \quad (i,j = 1,2,3).$$

These carry the following meanings:
  e_{11}, e_{22}, e_{33} are the extensions per unit length parallel to Ox_1, Ox_2, Ox_3 respectively,
  e_{12} is the rotation about Ox_3 towards Ox_1 of a line element parallel to Ox_2,
  e_{21} is the rotation about Ox_3 towards Ox_2 of a line element parallel to Ox_1,
and similarly for the other e_{ij}.

If the body merely undergoes a rotation with no strain, the corresponding $[e_{ij}]$ is antisymmetrical, as we now prove. Choose the origin on the axis of rotation. Then

$$u_i = e_{ij} x_j.$$

† Equation (6) need not be subject to the restriction to small displacements imposed on p. 94. (6) *defines* a homogeneous deformation for displacements of any magnitude. Accordingly, the results (1) to (4) which follow are true for displacements of any magnitude.

In a pure rotation the displacement of any point is perpendicular to its radius vector. Therefore, $u_i x_i = 0$ (scalar product), or

$$e_{ij} x_i x_j = 0. \tag{7}$$

Since equation (7) is true for all x_i, the coefficients on the left-hand side must all be zero. Hence

$$e_{ij} = 0, \quad \text{if} \quad i = j,$$

and $$e_{ij} = -e_{ji}, \quad \text{if} \quad i \neq j,$$

which are the conditions that $[e_{ij}]$ should be antisymmetrical. This result might have been anticipated by considering that a general rotation is represented by an axial vector (Ch. II, § 2), and an axial vector is equivalent to an antisymmetrical second-rank tensor. The three independent components of $\varpi_{ij} = \frac{1}{2}(e_{ij} - e_{ji})$ are in fact the components of the axial vector that defines the rotation.

The *strain tensor* $[\epsilon_{ij}]$ is defined as the symmetrical part of $[e_{ij}]$:

$$\epsilon_{ij} = \frac{1}{2}(e_{ij} + e_{ji}). \tag{8}$$

In full,

$$\begin{bmatrix} \epsilon_{11} & \epsilon_{12} & \epsilon_{31} \\ \epsilon_{12} & \epsilon_{22} & \epsilon_{23} \\ \epsilon_{31} & \epsilon_{23} & \epsilon_{33} \end{bmatrix} = \begin{bmatrix} e_{11} & \frac{1}{2}(e_{12}+e_{21}) & \frac{1}{2}(e_{13}+e_{31}) \\ \frac{1}{2}(e_{12}+e_{21}) & e_{22} & \frac{1}{2}(e_{23}+e_{32}) \\ \frac{1}{2}(e_{13}+e_{31}) & \frac{1}{2}(e_{23}+e_{32}) & e_{33} \end{bmatrix}.$$

The diagonal components of ϵ_{ij} are the *stretches* or *tensile strains*. The other components measure the shear strains. Just as in two dimensions, if two line elements are drawn parallel to Ox_1 and Ox_2 in the undeformed body, the angle between them after deformation is $\frac{1}{2}\pi - 2\epsilon_{12}$. ϵ_{23} and ϵ_{31} may be interpreted in a similar way.

3.1. Homogeneous three-dimensional strain. As in two dimensions, when the distortion is homogeneous all the e_{ij} components are constants, and we may write

$$u_i = (u_0)_i + e_{ij} x_j \quad (i, j = 1, 2, 3), \tag{9}$$

where $(u_0)_i$ is the displacement of the point at the origin. Splitting e_{ij} into two parts, we then have

$$u_i = (u_0)_i + \varpi_{ij} x_j + \epsilon_{ij} x_j. \tag{10}$$

It is convenient when visualizing strain to separate the part of the displacement that is due to a rigid body translation and rotation, represented by the first two terms on the right-hand side of (10), from the part that is due to strain, represented by the last term. Accordingly, we write

$$\bar{u}_i = \epsilon_{ij} x_j,$$

thereby expressing the fact that ϵ_{ij} connects the displacement $\bar{u}_i$ of a point, due to strain, with the position vector of the point.

Since strain is a symmetrical tensor it may be referred to its principal axes. The shear components then vanish and we have

$$\begin{bmatrix} \epsilon_{11} & \epsilon_{12} & \epsilon_{31} \\ \epsilon_{12} & \epsilon_{22} & \epsilon_{23} \\ \epsilon_{31} & \epsilon_{23} & \epsilon_{33} \end{bmatrix} \rightarrow \begin{bmatrix} \epsilon_1 & 0 & 0 \\ 0 & \epsilon_2 & 0 \\ 0 & 0 & \epsilon_3 \end{bmatrix}.$$

The geometrical meanings of the *principal strains*, ϵ_1, ϵ_2, ϵ_3, may be seen by taking a unit cube (Fig. 6.7) with edges parallel to the principal axes; on straining, the edges remain at right angles and the lengths of the edges become $(1+\epsilon_1)$, $(1+\epsilon_2)$, $(1+\epsilon_3)$. Notice that it is *not* true to say that the principal axes of strain are in directions which remain unchanged by deformation. This would only be true if the rotation ϖ_{ij} was zero. *The defining property of the principal axes is that they are three mutually perpendicular directions in the body which remain mutually perpendicular during the deformation.*

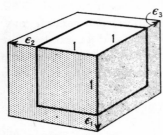

FIG. 6.7. The strain of a unit cube cut with its edges parallel to the three principal axes of strain. Only the displacements ($\bar{u}_i$) due to strain are illustrated in the figure; those due to rigid-body translation and rotation are not shown.

The change in volume of the unit cube in Fig. 6.7, called the *dilation*, is

$$\Delta = (1+\epsilon_1)(1+\epsilon_2)(1+\epsilon_3) - 1 = \epsilon_1 + \epsilon_2 + \epsilon_3,$$

since the ϵ's are small. Referred to general axes the dilation is given by $\Delta = \epsilon_{ii}$ and is an invariant.

The strain quadric, the strain ellipsoid, and the stretch in an arbitrary direction. The strain quadric has the equation

$$\epsilon_{ij} x_i x_j = 1.$$

This is identical with the quadric $e_{ij} x_i x_j = 1$, as may be seen by writing out the terms. Thus, the representation quadric for $[e_{ij}]$ represents only its symmetrical part.

The strain in an arbitrary direction, in the sense defined in Chapter I, § 6.1, is the extension per unit length, or the *stretch*, of a line drawn originally in that direction. Fig. 6.8a makes this clear. The vector **l** represents a line of unit length in the unstrained body. On straining, it changes both in length and direction. The displacement of its end due to strain is

$$\bar{u}_i = \epsilon_{ij} l_j.$$

The stretch ϵ is equal to the component of $\bar{u}$ resolved in the 1 direction and is therefore
$$\epsilon = \dot{u}_i l_i = \epsilon_{ij} l_i l_j,$$
or, in terms of the principal strains,
$$\epsilon = \epsilon_1 l_1^2 + \epsilon_2 l_2^2 + \epsilon_3 l_3^2.$$
The same situation is represented by the strain quadric in Fig. 6.8 b. The radius vector OP in the direction of 1 is of length $1/\sqrt{\epsilon}$ and the

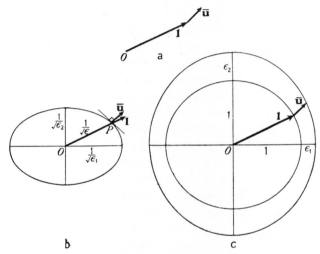

Fig. 6.8. Homogeneous three-dimensional strain. (a) The extremity of the unit vector 1 in the unstrained body is displaced by strain by the vector $\bar{u}$. (b) The direction of $\bar{u}$ is obtained from that of 1 by the radius-normal property of the strain quadric. The direction of $\bar{u}$ is normal to the quadric at P. The stretch ϵ in the direction 1 is given by $OP = 1/\sqrt{\epsilon}$. (c) Illustrating the corresponding deformation of a unit sphere.

direction of $\bar{u}$ is that of the normal at P (Ch. I, § 7). In Fig. 6.8 b, P lies in the plane of Ox_1 and Ox_2, but the construction holds, of course, for all positions.

For comparison with the shape of the strain quadric the deformation of a unit sphere with the same values of ϵ_1 and ϵ_2 is shown in Fig. 6.8 c. To find the equation of the deformed sphere we substitute into its equation, namely
$$x_1^2 + x_2^2 + x_3^2 = 1,$$
the values $x_1' = x_1(1+\epsilon_1)$, $x_2' = x_2(1+\epsilon_2)$, $x_3' = x_3(1+\epsilon_3)$, and obtain the ellipsoid
$$\frac{x_1'^2}{(1+\epsilon_1)^2} + \frac{x_2'^2}{(1+\epsilon_2)^2} + \frac{x_3'^2}{(1+\epsilon_3)^2} = 1. \tag{11}$$
This is usually known as the *strain ellipsoid*; it is to be carefully

distinguished from the strain quadric. Since the principal strains $\epsilon_1, \epsilon_2, \epsilon_3$ can be either positive or negative, the strain quadric,

$$\epsilon_1 x_1^2 + \epsilon_2 x_2^2 + \epsilon_3 x_3^2 = 1, \qquad (12)$$

can be a real or an imaginary ellipsoid or a hyperboloid; the surface given by (11), on the other hand, is always an ellipsoid.

The 'engineering strains'. The strain tensor $[\epsilon_{ij}]$ is frequently written as

$$\begin{bmatrix} \epsilon_x & \tfrac{1}{2}\gamma_{xy} & \tfrac{1}{2}\gamma_{zx} \\ \tfrac{1}{2}\gamma_{xy} & \epsilon_y & \tfrac{1}{2}\gamma_{yz} \\ \tfrac{1}{2}\gamma_{zx} & \tfrac{1}{2}\gamma_{yz} & \epsilon_z \end{bmatrix},$$

so that, for example, $\gamma_{xy} = 2\epsilon_{12}$. With this definition, γ_{xy} equals the decrease in angle between two lines originally parallel to the Ox and Oy axes (Fig. 6.9). The γ's are often called the 'shear components of strain', the 'shear strains', or simply the 'shears', but it is to be clearly understood that, on account of the additional factors of 2, they are not equal to the off-diagonal components of the strain tensor ϵ_{ij}. To avoid ambiguity we may bow to engineering practice and call the γ's the *shear strains* or the *(engineering) shear strains* and distinguish ϵ_{23}, ϵ_{31} and ϵ_{12} as the *tensor shear strains*. It should be noticed that the array of engineering strains

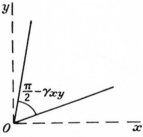

FIG. 6.9. Illustrating the meaning of the engineering shear strain component γ_{xy}.

$$\begin{matrix} \epsilon_x & \gamma_{xy} & \gamma_{zx} \\ \gamma_{xy} & \epsilon_y & \gamma_{yz} \\ \gamma_{zx} & \gamma_{yz} & \epsilon_z \end{matrix}$$

does *not* form a tensor. If one wishes to transform strain components to other axes it is simplest to work with the tensor components ϵ_{ij}, for which the transformation rule is the standard law for second-rank tensors (p. 11). The transformation rule in terms of the ϵ_x, γ_{xy},... notation is naturally clumsier in form by reason of the factors of 2 which occur in it.

The Mohr circle construction (Ch. II, § 4) is particularly useful in transforming strain components and, again, it is best to use it with the tensor strain components, to which it applies directly without any modification. By introducing factors of 2 the construction can of course be made to work for the engineering strains instead, but such a modification is really unnecessary; to avoid confusion the reader is recommended to use the construction only for true tensor components.

Plane strain and pure shear. When one principal strain is zero a body is said to be in a state of *plane strain.* *Pure shear* is a special case of plane strain. The strain tensor for a pure shear about Ox_3 has the form

$$\begin{bmatrix} 0 & \epsilon & 0 \\ \epsilon & 0 & 0 \\ 0 & 0 & 0 \end{bmatrix}.$$

Referred to the principal axes, by rotation through 45° about Ox_3, the tensor is

$$\begin{bmatrix} -\epsilon & 0 & 0 \\ 0 & \epsilon & 0 \\ 0 & 0 & 0 \end{bmatrix}.$$

Pure shear strain is analogous to the pure shear stress discussed on pp. 90, 91; the Mohr circle construction for the 45° rotation of axes

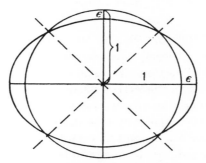

FIG. 6.10. The deformation of a unit
sphere by pure shear.

is identical in the two cases (Fig. 5.10 a). An important characteristic of pure shear strain is that the dilation is zero. The deformation of a sphere under these conditions is illustrated in Fig. 6.10. Since ϵ is small compared with 1, it is clear by symmetry that the extension in the intermediate 45° directions is zero.

Simple shear, as contrasted with pure shear, is often used to mean the distortion shown on the left-hand side of Fig. 6.11. The strain involved is identical with that of pure shear; the only difference is that simple shear includes a rotation. Fig. 6.11 shows this diagrammatically.

3.2. Extension to inhomogeneous strain. The results of § 3.1 have been derived for homogeneous strain, but their extension to cover inhomogeneous deformation is quite straightforward. Each point of an inhomogeneously strained body may be regarded as surrounded by a small region in which the strain is sensibly homogeneous. Three principal strain directions may then be defined for each point; they will, in general, be different for every point of the body. In the same way the strain at each point of the body may be characterized by a strain quadric, which will vary, in size, shape and orientation, with the position of the point.

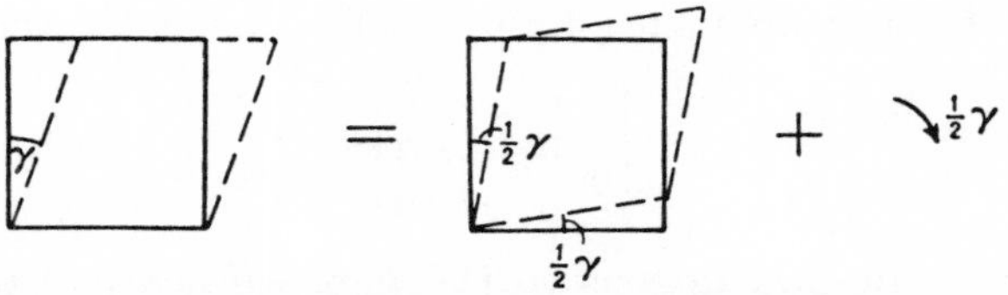

FIG. 6.11. Illustrating that a simple shear (left-hand diagram) equals a pure shear (centre diagram) plus a rotation (right-hand diagram).

4. Strain and crystal symmetry

We conclude this general study of strain by making a distinction that is obvious but important. The strain of a crystal is not a property in the same sense that the dielectric constant, for example, is a property. The strain is a response of the crystal to an influence. The influence may be a stress (elasticity) or it may be an electric field (piezoelectricity). In both these cases the magnitudes and directions of the principal strains are determined by the magnitude and orientation of the influence, as well as by the physical properties and the symmetry of the crystal. The strain tensor, therefore, like the stress tensor, does not have to conform to the crystal symmetry (unless the influence itself conforms). However, a strain may also be caused by a temperature change (thermal expansion, § 5). In this case the influence has no orientation (it is represented by a scalar), and so the resulting strain must conform to the symmetry of the crystal. A very similar situation arises with electric polarization (Ch. IV). The polarization does not have to conform to the crystal symmetry when it is produced by a field in a general direction; but when it is produced by a change of temperature (pyroelectricity) it does have to conform.

Before passing on to thermal expansion we shall summarize the main facts about strain obtained so far.

SUMMARY OF §§ 1–4

Displacements and strains assumed to be small throughout.

1. General analysis of strain. If u_i represents the displacement of a point x_j in a deformed body, we define the tensor $[e_{ij}]$ by

$$e_{ij} = \frac{\partial u_i}{\partial x_j}.$$

The symmetrical part of this tensor, with components

$$\epsilon_{ij} = \tfrac{1}{2}(e_{ij} + e_{ji}),$$

represents the *strain* at the point; the antisymmetrical part, with components

$$\varpi_{ij} = \tfrac{1}{2}(e_{ij} - e_{ji}),$$

represents the *rotation*.

The components ϵ_{11}, ϵ_{22}, ϵ_{33} are the *stretches* or *tensile strains*; they are the extensions that elements of unit length drawn originally parallel to Ox_1, Ox_2, Ox_3, respectively, undergo during the strain. ϵ_{23}, ϵ_{31}, ϵ_{12} are the *tensor shear strains*; $2\epsilon_{23}$ equals the change in angle between two elements drawn parallel to Ox_2 and Ox_3 before deformation (positive if the angle decreases), and similarly for ϵ_{31} and ϵ_{12}. The principal strains ϵ_1, ϵ_2, ϵ_3, at the point are the principal components of $[\epsilon_{ij}]$. The principal directions of strain define three mutually perpendicular directions at the point which remain mutually perpendicular during the strain.

The strain at a point is characterized by the *strain quadric*, whose equation is

$$\epsilon_{ij} x_i x_j = 1,$$

or, referred to the principal strain directions as axes,

$$\epsilon_1 x_1^2 + \epsilon_2 x_2^2 + \epsilon_3 x_3^2 = 1.$$

$\gamma_{yz} = 2\epsilon_{23}$, $\gamma_{zx} = 2\epsilon_{31}$, $\gamma_{xy} = 2\epsilon_{12}$ are the *(engineering) shear strains*. The array formed from the tensile strains and the engineering shear strains is not a tensor.

2. Homogeneous strain. For a homogeneous strain

$$u_i = (u_0)_i + e_{ij} x_j = (u_0)_i + \varpi_{ij} x_j + \epsilon_{ij} x_j,$$

and the displacement due to strain may be defined as

$$\bar{u}_i = \epsilon_{ij} x_j.$$

A unit cube in the unstrained body with its edges parallel to the principal strain directions becomes a rectangular parallelepiped with edges of lengths $(1+\epsilon_1)$, $(1+\epsilon_2)$, $(1+\epsilon_3)$. The *dilation* (increase in volume) is $(\epsilon_1 + \epsilon_2 + \epsilon_3)$ or ϵ_{ii}.

The stretch in an arbitrary direction is

$$\epsilon = \epsilon_{ij} l_i l_j \quad \text{or} \quad \epsilon = \epsilon_1 l_1^2 + \epsilon_2 l_2^2 + \epsilon_3 l_3^2.$$

The direction of $\bar{\mathbf{u}}$ and the value of ϵ for a given direction in the body may be obtained from the radius-normal and the radius vector properties of the strain quadric.

A unit sphere is deformed into the ellipsoid

$$\frac{x_1^2}{(1+\epsilon_1)^2} + \frac{x_2^2}{(1+\epsilon_2)^2} + \frac{x_3^2}{(1+\epsilon_3)^2} = 1,$$

known as the *strain ellipsoid*.

This analysis for homogeneous strain applies also to any sufficiently small volume of an inhomogeneously strained body.

3. Strain and crystal symmetry. The strain of a crystal is not a crystal property but, rather, the response of the crystal to an influence. The strain produced by a stress (elasticity) or an electric field (piezoelectricity) need not conform to the symmetry of the crystal unless the **stress** or the field itself conforms. The strain caused by a temperature change (thermal expansion), on the other hand, must conform to the crystal symmetry.

EXERCISE 6.1. A small deformation of a certain crystal is defined by the tensor

$$[e_{ij}] = \begin{bmatrix} 8 & -1 & -1 \\ 1 & 6 & 0 \\ -5 & 0 & 2 \end{bmatrix} \times 10^{-6}.$$

Determine the strain $[\epsilon_{ij}]$, and the rotation $[\varpi_{ij}]$. Hence find the magnitudes and directions of the principal strains ϵ_1, ϵ_2, ϵ_3, the axis of rotation and the angle of rotation.

---

5. Thermal expansion

If the temperature of a crystal is changed, the resulting deformation may be specified by the strain tensor $[\epsilon_{ij}]$. When a small temperature change ΔT takes place uniformly throughout the crystal the deformation is homogeneous, and it is found that all the components of $[\epsilon_{ij}]$ are proportional to ΔT; thus

$$\epsilon_{ij} = \alpha_{ij}\Delta T, \tag{13}$$

where the α_{ij} are constants, the *coefficients of thermal expansion*. Since $[\epsilon_{ij}]$ is a tensor, so also is $[\alpha_{ij}]$, and, moreover, since $[\epsilon_{ij}]$ is symmetrical so also is $[\alpha_{ij}]$. The thermal expansion tensor $[\alpha_{ij}]$ may therefore be referred to its principal axes; and equations (13) then simplify to

$$\epsilon_1 = \alpha_1\Delta T, \quad \epsilon_2 = \alpha_2\Delta T, \quad \epsilon_3 = \alpha_3\Delta T, \tag{14}$$

where α_1, α_2, α_3 are the *principal expansion coefficients*. It follows that if a sphere is drawn in a crystal it becomes, on change of temperature, an ellipsoid (the strain ellipsoid) with axes proportional to $(1+\alpha_1\Delta T)$, $(1+\alpha_2\Delta T)$, $(1+\alpha_3\Delta T)$.

The representation quadric (Ch. I, § 4) for thermal expansion has the equation
$$\alpha_{ij}x_i x_j = 1,$$
or, referred to the principal axes,

$$\alpha_1 x_1^2 + \alpha_2 x_2^2 + \alpha_3 x_3^2 = 1.$$

The shape and orientation of the quadric is subject to restrictions imposed by the symmetry of the crystal, according to Neumann's Principle (p. 20). The forms it may take in the different crystal systems have been listed in Table 3, Chapter I. It is worth noting that, since the thermal expansion of a crystal must possess the symmetry of the crystal, it cannot destroy any symmetry elements (we exclude phase transformations from this argument). This is why the class of a crystal does not depend on the temperature.

The coefficient of volume (bulk) expansion is $(\alpha_1+\alpha_2+\alpha_3)$ or, in general, α_{ii}, which is an invariant.

The principal thermal expansion coefficients for most substances are all positive, and the thermal expansion quadric is accordingly an ellipsoid, but in a few crystals some coefficients are negative (e.g. calcite, beryl, silver iodide). Some measured values of expansion coefficients

TABLE 6

Principal thermal expansion coefficients of crystals

Unit $= 10^{-6}(^\circ\text{C})^{-1}$

Crystal	System	Temperature	α_1	α_2	α_3
Gypsum . .	monoclinic	40° C	1·6	42	29
Aragonite . .	orthorhombic	40° C	35	17	10
Zinc . . .	hexagonal	60° K	−2		55
		150° K	8		65
		300° K	13		64
Quartz . .	trigonal	Room temp.	13		8
Calcite . .	,,	40° C	−5·6		25
Rutile . .	tetragonal	40° C	7·1		9·2
Copper . .	cubic	Room temp.	16		
Diamond . .	,,	,,	0·89		
Sodium chloride	,,	,,	40		

Values from *International Critical Tables* (1929).

are shown in Table 6. The coefficients often depend markedly on the temperature (theoretically they tend to zero at 0° K).

Thermal expansion is sometimes measured by finding the increase in thickness of a parallel-sided plate when it is heated.† It is fairly obvious that this method measures the coefficient of expansion normal to the plate, but the deformations involved when the plate has an arbitrary crystallographic orientation are not simple. In Fig. 6.12 the plate is supposed to be resting on a plane bed, and the point O in the lower surface remains fixed. Take axes as shown, with Ox_3 normal to the

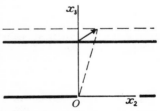

FIG. 6.12. The thermal expansion of a parallel-sided plate (greatly exaggerated).

† The experimental method is described by Tutton (1922). The X-ray method, in which the change in spacing of a set of parallel crystal planes is measured, is formally equivalent.

plate. Since lines drawn parallel to Ox_1 and Ox_2 remain parallel to the bed the deformation is of the form

$$[e_{ij}] = \begin{bmatrix} e_{11} & e_{12} & e_{13} \\ e_{21} & e_{22} & e_{23} \\ 0 & 0 & e_{33} \end{bmatrix}.$$

The measured expansion is $e_{33} = \epsilon_{33} = \alpha_{33}\Delta T$, by (8) and (13). α_{33} is the coefficient of expansion along Ox_3 in the sense defined in Chapter I, § 6.1. It should be noticed that a line drawn parallel to Ox_3 rotates during the expansion, as is shown by the presence of the components e_{13} and e_{23}.

EXERCISE 6.2. Derive an expression in terms of the α_{ij} for the angle of rotation of a line drawn parallel to Ox_3 in Fig. 6.12.

Cone of zero expansion in calcite. In calcite (trigonal) the expansion quadric is a surface of revolution about the triad axis. It will be seen from the table (p. 107) that the expansion coefficient parallel to the triad axis is positive, but that the other two (equal) coefficients are negative. Hence, the length of one axis of the expansion quadric is real and that of the other two is imaginary. The quadric is therefore a hyperboloid of revolution of two sheets (Fig. 1.5 c, p. 18). The asymptotic cone gives a family of directions for which the radius vector $1/\sqrt{\alpha}$ is infinite, that is, for which the coefficient of expansion is zero. The angle θ between these directions and the Ox_3 axis is easily found. If (l_1, l_2, l_3) are the direction cosines of one of the lines of zero expansion we have

$$(l_1^2 + l_2^2)\alpha_1 + l_3^2\alpha_3 = (1 - l_3^2)\alpha_1 + l_3^2\alpha_3 = 0.$$

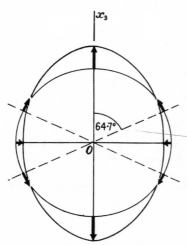

FIG. 6.13. The thermal deformation (greatly exaggerated) of a unit sphere drawn in a calcite crystal, showing the unstretched cone.

Since $l_3 = \cos\theta$, we obtain

$$\tan^2\theta = -\alpha_3/\alpha_1,$$

and, inserting the values given in Table 6, we find

$$\theta = 64{\cdot}7°.$$

The thermal deformation of a unit sphere drawn in a calcite crystal is illustrated, greatly exaggerated, in Fig. 6.13. It may be noticed that, although lines at $64{\cdot}7°$ to Ox_3 are unchanged in length during the heating, they do nevertheless alter their directions relative to Ox_3, for the normal to the expansion quadric at the corresponding points is perpendicular to the radius vector.

EXERCISE 6.3. Show that the magnitude of the rotation just mentioned is $\sqrt{(-\alpha_1\alpha_3)} = 2{\cdot}4$ seconds/° C.

SUMMARY OF § 5

Thermal expansion. If the temperature of a crystal is raised uniformly by an amount ΔT, the crystal undergoes a homogeneous strain given by

$$\epsilon_{ij} = \alpha_{ij}\Delta T,$$

where the α_{ij} are the coefficients of thermal expansion. $[\alpha_{ij}]$ is a symmetrical second-rank tensor. There are, accordingly, three principal expansion coefficients and directions. The strains along these directions are $\epsilon_1 = \alpha_1 \Delta T$, $\epsilon_2 = \alpha_2 \Delta T$, $\epsilon_3 = \alpha_3 \Delta T$. The thermal expansion quadric

$$\alpha_1 x_1^2 + \alpha_2 x_2^2 + \alpha_3 x_3^2 = 1,$$

and the strain produced by thermal expansion, must conform to the crystal symmetry, in accordance with Neumann's Principle.

EXERCISES ON THERMAL EXPANSION

EXERCISE 6.4 (graphical). A monoclinic crystal (afwillite) has the following cell dimensions:

$$a = 16\cdot21, \qquad b = 5\cdot63, \qquad c = 13\cdot23 \text{ Å}, \qquad \beta = 134° 48'.$$

X-ray measurements of the coefficients of thermal expansion normal to certain planes $h0l$ gave the following results:

h	0	l	α		h	0	l	α
$\bar{4}$	0	14	$13\cdot7 \times 10^{-6}$ per ° C		20	0	$\bar{8}$	$24\cdot3 \times 10^{-6}$ per ° C
0	0	12	$17\cdot1$		16	0	$\overline{12}$	$11\cdot8$
6	0	8	$21\cdot4$		8	0	$\overline{12}$	$8\cdot2$
18	0	$\bar{4}$	$28\cdot2$					

Draw to scale the axes Ox and Oz with the lengths a and c, and insert graphically the normals to all the planes $h0l$ listed above. Along these normals, on a suitable scale, plot points at distances $\pm 1/\sqrt{\alpha}$ from the origin. Hence draw the principal section of the thermal expansion quadric parallel to (010). Determine from it the directions of the principal axes relative to Ox and Oz and the magnitudes of the principal thermal expansion coefficients α_1 and α_3.

EXERCISE 6.5. The following thermal expansion measurements were made on a monoclinic crystal. Values of $10^6\alpha$ in (° C)$^{-1}$:

parallel to Oy (diad axis), 41·0;

     „ Oz 32·0;

perpendicular to Oz and Oy, 15·0;

     „   „ Oy and at 45° to Oz (measured clockwise about Oy), 16·0.

Calculate the three principal expansion coefficients and the angles between the principal expansion directions and Oz. Illustrate the answer by a Mohr circle diagram (compare the example on p. 161).

EXERCISE 6.6. An orthorhombic crystal (aragonite) has

$$a:b:c = 0\cdot6224:1:0\cdot7206.$$

On heating from 0° C to 100° C the angle (100):(110) decreases by 1·14′ and the angle (001):(011) increases by 2·84′. The coefficient of bulk expansion is $62\cdot0 \times 10^{-6}$ per ° C. Calculate the three principal thermal expansion coefficients.

VII

PIEZOELECTRICITY
THIRD-RANK TENSORS

1. The direct piezoelectric effect

IF a stress is applied to certain crystals they develop an electric moment whose magnitude is proportional to the applied stress. This is known as the *direct piezoelectric effect*. For example, if a uniaxial tensile stress σ is applied along one of the diad axes of a quartz crystal (class *32*), the magnitude of the electric moment per unit volume, or the polarization charge per unit area, is given by

$$P = d\sigma, \tag{1}$$

where d is a constant, called a *piezoelectric modulus*. As is implied by this equation, a change from a tensile stress to an equal compressive stress reverses the direction of the polarization.

In general, as we have seen in Chapter V, a state of stress is specified by a second-rank tensor with nine components, while the polarization of a crystal, being a vector, is specified by three components. It is found that when a general stress σ_{ij} acts on a piezoelectric crystal each component of the polarization P_i is linearly related to all the components of σ_{ij}, just as in a dielectric, for example, each component of P_i is linearly related to all three components of E_i.† We accordingly write for P_1,

$$\begin{aligned}
P_1 = \ & d_{111}\,\sigma_{11} + d_{112}\,\sigma_{12} + d_{113}\,\sigma_{13} + \\
& + d_{121}\,\sigma_{21} + d_{122}\,\sigma_{22} + d_{123}\,\sigma_{23} + \\
& + d_{131}\,\sigma_{31} + d_{132}\,\sigma_{32} + d_{133}\,\sigma_{33},
\end{aligned} \tag{2}$$

where the d's are constant coefficients, and two similar equations for P_2 and P_3. By using the summation convention (Ch. I, § 1.1) and taking the terms on the right of equation (2) one line at a time, we may abbreviate the equation to

$$P_1 = d_{11k}\,\sigma_{1k} + d_{12k}\,\sigma_{2k} + d_{13k}\,\sigma_{3k},$$

which may be written still more shortly as

$$P_1 = d_{1jk}\,\sigma_{jk}.$$

---

† If the crystal has a spontaneous polarization, P_i must be interpreted as the change in the polarization caused by the stress.

Similarly, $P_2 = d_{2jk}\,\sigma_{jk}$, and $P_3 = d_{3jk}\,\sigma_{jk}$. Thus, the general statement of the relationship between P_i and σ_{ij} is

$$P_i = d_{ijk}\,\sigma_{jk}. \tag{3}$$

The d_{ijk} are the *piezoelectric moduli*.

Let us consider the meaning of the various moduli. If a uniaxial tensile stress given by σ_{11} is applied to the crystal, the resulting polarization has components

$$P_1 = d_{111}\,\sigma_{11}, \quad P_2 = d_{211}\,\sigma_{11}, \quad P_3 = d_{311}\,\sigma_{11}.$$

So, in principle, by measuring P_1, P_2 and P_3, the values of d_{111}, d_{211} and d_{311} might be found. In the same way, by imagining uniaxial tensile stresses applied in turn along Ox_2 and Ox_3, physical meanings may be attached to the other d_{ijk} coefficients in which $j = k$.

Suppose now that a shear stress about Ox_3 is applied. We recall from (6) (Ch. V) and the remark that follows it that, even with body-torques, equal components σ_{12} and σ_{21} will both be present. Then we have

$$P_1 = d_{112}\,\sigma_{12} + d_{121}\,\sigma_{21} = (d_{112} + d_{121})\sigma_{12},$$

and similar equations for P_2 and P_3. $(d_{112} + d_{121})$ thus has a definite physical meaning, but it is impossible to devise an experiment by which d_{112} may be separated from d_{121}. This element of arbitrariness in the physical meanings of the d_{ijk} is removed by putting $d_{112} = d_{121}$, and, in general,

$$d_{ijk} = d_{ikj}. \tag{4}$$

We are permitted, but at present we are not compelled, to do this. We shall see in § 3 that (4) is consistent with later developments.

The 27 coefficients d_{ijk} form an example of a *third-rank tensor*. The definition of a third-rank tensor is obtained by an extension of the definition for tensors of the first and second ranks which we gave in Chapter I, § 3. Tensors were there defined by means of their transformation laws. A physical quantity represented by a set of numbers p_i was defined as a first-rank tensor (vector) if the p_i transformed according to the law

$$p_i' = a_{ij} p_j; \tag{13, Ch. I}$$

the transformation law for a second-rank tensor $[T_{ij}]$ was

$$T_{ij}' = a_{ik} a_{jl} T_{kl}. \tag{22, Ch. I}$$

In a similar way the 27 numbers T_{ijk} representing a physical quantity are said to form a third-rank tensor if they transform on change of axes to T_{ijk}', where

$$T_{ijk}' = a_{il} a_{jm} a_{kn} T_{lmn}. \tag{5}$$

As with tensors of lower rank (Ch. I, § 3), we emphasize that this definition implies that there is some test, independent of the law (5), for deciding how the components transform on change of axes. If it were not so, the definition would be merely circular.

The transformation laws for first- and second-rank tensors are the same, respectively, as those for the coordinates x_i themselves, and for products of the coordinates $x_i x_j$ (p. 13). In the same way, (5) is the transformation law for third-order products of the type $x_i x_j x_k$. This is obvious at once from the law (13), Chapter I, applied to the members of the product in turn; for

$$x_i' x_j' x_k' = a_{il} x_l . a_{jm} x_m . a_{kn} x_n,$$

or, rearranging, $\qquad (x_i x_j x_k)' = a_{il} a_{jm} a_{kn} (x_l x_m x_n).$ \hfill (6)

Thus, any given tensor component transforms like the corresponding product. For example, T_{112} transforms like $x_1^2 x_2$. However, in making this statement it is to be understood that, when (6) is written out in full for a particular component, the order of the x's in the various terms on the right is preserved. Thus, the coefficient of the term in $x_1 x_2 x_3$ is to be identified with the coefficient of T_{123}, and is not to be grouped with the coefficients of $x_2 x_1 x_3$, $x_2 x_3 x_1$, etc.

We now prove that the d_{ijk} of equation (3) obey the transformation law (5). Suppose that the crystal is subjected to a certain fixed stress, which is given by components σ_{ij} when referred to axes Ox_i, and by components σ_{ij}' when referred to Ox_i'. The polarization is given by P_i when referred to Ox_i, and by P_i' when referred to Ox_i'. The general form of the relationship (3) is the same whatever reference axes are chosen for the polarization and the stress. Hence, for the components referred to Ox_i', we may write

$$P_i' = d_{ijk}' \sigma_{jk}',$$ \hfill (7)

where the d_{ijk}' are a different set of 27 coefficients. We have to ask how the d_{ijk}' are related to the d_{ijk}.

The P_i are related to the P_i' by the equation

$$P_i' = a_{il} P_l,$$ \hfill (8)

and the σ_{ij}' to the σ_{ij} by

$$\sigma_{mn} = a_{jm} a_{kn} \sigma_{jk}'.$$ \hfill (9)

Equations (8), (3) and (9) form the scheme

$$P' \xrightarrow{\;(8)\;} P \xrightarrow{\;(3)\;} \sigma \xrightarrow{\;(9)\;} \sigma',$$

where $\longrightarrow$ means 'in terms of'. Hence P' may be expressed in terms of σ' as

$$P'_i = a_{il}P_l = a_{il}d_{lmn}\sigma_{mn} = a_{il}d_{lmn}a_{jm}a_{kn}\sigma'_{jk}. \tag{10}$$

Comparing (10) with (7) we find

$$d'_{ijk} = a_{il}a_{jm}a_{kn}d_{lmn}. \tag{11}$$

It follows that the d_{ijk} transform according to equation (5), and therefore constitute a third-rank tensor.

The above proof of the tensor character of the d_{ijk} is a general one. It may be shown in the same way that a set of quantities B_{ijk} which relate two tensors A_i and C_{jk} by the equation

$$A_i = B_{ijk}C_{jk} \tag{12}$$

form a third-rank tensor.

2. Reduction in the number of independent moduli. Matrix notation

A general third-rank tensor, as we have seen, has $3^3 = 27$ independent components. When the components are written out in full they form, not a square array as with a second-rank tensor, but an array in the shape of a cube. If the first suffix in d_{ijk} refers to the layer in which this particular component lies, the second to the row and the third to the column, the three layers are:

1st layer			*2nd layer*			*3rd layer*			
$i = 1$			$i = 2$			$i = 3$			
d_{111}	d_{112}	d_{113}	d_{211}	d_{212}	d_{213}	d_{311}	d_{312}	d_{313}	(13)
(d_{121})	d_{122}	d_{123}	(d_{221})	d_{222}	d_{223}	(d_{321})	d_{322}	d_{323}	
(d_{131})	(d_{132})	d_{133}	(d_{231})	(d_{232})	d_{233}	(d_{331})	(d_{332})	d_{333}	

The fact that d_{ijk} is symmetrical in j and k eliminates as independent components the coefficients shown in brackets, leaving 18 independent d_{ijk}; it also facilitates the use of a more concise notation known as the *matrix notation*.

Up to this point all the equations have been developed in full tensor notation, because only in this way can their true character, and particularly their transformation properties, be displayed. But when calculating in particular problems it is advantageous to reduce the number of suffixes as much as possible. This is done by defining new symbols d_{11}, d_{12} etc. In terms of these new symbols the array (13) appearing above is rewritten as:

d_{11}	$\frac{1}{2}d_{16}$	$\frac{1}{2}d_{15}$	d_{21}	$\frac{1}{2}d_{26}$	$\frac{1}{2}d_{25}$	d_{31}	$\frac{1}{2}d_{36}$	$\frac{1}{2}d_{35}$	
	d_{12}	$\frac{1}{2}d_{14}$		d_{22}	$\frac{1}{2}d_{24}$		d_{32}	$\frac{1}{2}d_{34}$	(14)
		d_{13}			d_{23}			d_{33}	

Thus, for instance, we define $d_{21} = d_{211}$ and $d_{14} = 2d_{123}$. It will be seen that the first suffix is the same in the two notations, but the second and third suffixes in the full tensor notation are replaced in the new notation by a single suffix running from 1 to 6, as follows:

tensor notation	11	22	33	23, 32	31, 13	12, 21
matrix notation	1	2	3	4	5	6

The last two suffixes in the tensor notation correspond to those of the stress components [see equation (3)]; so, for consistency, we make the following change in the notation for the stress components

$$\begin{bmatrix} \sigma_{11} & \sigma_{12} & \sigma_{31} \\ \sigma_{12} & \sigma_{22} & \sigma_{23} \\ \sigma_{31} & \sigma_{23} & \sigma_{33} \end{bmatrix} \rightarrow \begin{bmatrix} \sigma_1 & \sigma_6 & \sigma_5 \\ \sigma_6 & \sigma_2 & \sigma_4 \\ \sigma_5 & \sigma_4 & \sigma_3 \end{bmatrix}. \tag{15}$$

(The symbols σ_1, σ_2, σ_3 were also used to denote principal stresses, but it will always be clear which meaning is intended.)

In the new notation we may write equation (2) in the form

$$P_1 = \quad d_{11}\sigma_1 + \tfrac{1}{2}d_{16}\sigma_6 + \tfrac{1}{2}d_{15}\sigma_5 +$$
$$+ \tfrac{1}{2}d_{16}\sigma_6 + \quad d_{12}\sigma_2 + \tfrac{1}{2}d_{14}\sigma_4 +$$
$$+ \tfrac{1}{2}d_{15}\sigma_5 + \tfrac{1}{2}d_{14}\sigma_4 + \quad d_{13}\sigma_3,$$

or $\quad P_1 = d_{11}\sigma_1 + d_{12}\sigma_2 + d_{13}\sigma_3 + d_{14}\sigma_4 + d_{15}\sigma_5 + d_{16}\sigma_6,$ \hfill (16)

and there are two similar equations for P_2 and P_3; with the dummy suffix convention they may be written

$$P_i = d_{ij}\sigma_j \quad (i = 1, 2, 3; j = 1, 2, ..., 6). \tag{17}$$

There is sometimes confusion about the $\frac{1}{2}$'s that enter the definitions of the d_{ij}. The reason for introducing the $\frac{1}{2}$'s in the array (14) is to avoid having factors of 2 in the last three terms of equations such as (16). It is impossible to avoid having them in one place or the other. We prefer to have the $\frac{1}{2}$'s in (14) in order that (16) and the two companion equations for P_2 and P_3 may be written in the compact form (17).†

---

† This procedure agrees with what is recommended in *Standards on piezoelectric crystals*, 1949 (the report of a committee published in the United States by the Institute of Radio Engineers). It also agrees with the convention established by Voigt (1910). Contrary to our convention Voigt counts compressive stresses as positive, but he introduces a minus sign into equation (17); as a result Voigt's d_{ij} are identical with our d_{ij}.

Wooster (1938) uses a different notation: our d_{ij} are equal to Wooster's q_{ji} for $j = 1, 2, 3$, but are equal to $2q_{ji}$ for $j = 4, 5, 6$. This conclusion results from assuming that Wooster counts tensile stresses as positive; it is true that he defines compressive stress components as positive (p. 232), but this seems to be an oversight, for throughout the rest of his book he treats them as positive when tensile.

The array of d_{ij} written out as

$$\begin{pmatrix} d_{11} & d_{12} & d_{13} & d_{14} & d_{15} & d_{16} \\ d_{21} & d_{22} & d_{23} & d_{24} & d_{25} & d_{26} \\ d_{31} & d_{32} & d_{33} & d_{34} & d_{35} & d_{36} \end{pmatrix} \tag{18}$$

is a matrix. It will be observed that the rows of this matrix correspond to the layers in the arrangement (13).

The matrix notation has the advantage of greater compactness than the tensor notation, and it makes it easy to display the coefficients on a plane diagram; that is why it is used; but it must always be remembered that in spite of their appearance, with two suffixes, *the d_{ij} do not transform like the components of a second-rank tensor.*

3. The converse piezoelectric effect

When an electric field is applied in a piezoelectric crystal the shape of the crystal changes slightly. This is known as the *converse piezoelectric effect*; its existence is a thermodynamic consequence of the direct effect. It is found that there is a linear relation between the components of the vector E_i giving the electric field intensity within the crystal and the components of the strain tensor ϵ_{ij} which describe the change of shape. Moreover, *the coefficients connecting the field and the strain in the converse effect are the same as those connecting the stress and the polarization in the direct effect.* Explicitly, when the direct effect is written

$$P_i = d_{ijk}\,\sigma_{jk}, \tag{3}$$

the converse effect is written

$$\epsilon_{jk} = d_{ijk}\,E_i. \tag{19}$$

The proof of this equality of the coefficients is based on thermodynamical reasoning and is given in Chapter X. Since $\epsilon_{jk} = \epsilon_{kj}$ it follows that d_{ijk} is symmetrical in j and k, as was anticipated in § 1.

To write (19) in the matrix notation we must use one suffix instead of two for the strain components. We substitute as follows:

$$\begin{bmatrix} \epsilon_{11} & \epsilon_{12} & \epsilon_{31} \\ \epsilon_{12} & \epsilon_{22} & \epsilon_{23} \\ \epsilon_{31} & \epsilon_{23} & \epsilon_{33} \end{bmatrix} \rightarrow \begin{bmatrix} \epsilon_1 & \tfrac{1}{2}\epsilon_6 & \tfrac{1}{2}\epsilon_5 \\ \tfrac{1}{2}\epsilon_6 & \epsilon_2 & \tfrac{1}{2}\epsilon_4 \\ \tfrac{1}{2}\epsilon_5 & \tfrac{1}{2}\epsilon_4 & \epsilon_3 \end{bmatrix}. \tag{20}$$

When (19) is written out and put into the matrix notation, the equation for ϵ_{11}, which is

$$\epsilon_{11} = d_{111}E_1 + d_{211}E_2 + d_{311}E_3,$$

becomes $\qquad \epsilon_1 = d_{11}E_1 + d_{21}E_2 + d_{31}E_3.$

The equation for ϵ_{23}, which is

$$\epsilon_{23} = d_{123}E_1 + d_{223}E_2 + d_{323}E_3,$$

becomes $$\tfrac{1}{2}\epsilon_4 = \tfrac{1}{2}d_{14}E_1 + \tfrac{1}{2}d_{24}E_2 + \tfrac{1}{2}d_{34}E_3.$$

The $\tfrac{1}{2}$'s cancel on both sides and, in general, we have

$$\epsilon_j = d_{ij}E_i \quad (i = 1, 2, 3; j = 1, 2, ..., 6). \tag{21}$$

The reason for the $\tfrac{1}{2}$'s in the substitution (20) is now clear. Since we are already committed to having factors of $\tfrac{1}{2}$ in the definitions of the d_{ij} (p. 113) it is necessary to have them also in the definitions of the ϵ_i; otherwise equation (21) would have $\tfrac{1}{2}$'s in front of some of the terms, and we could not write it as a summation of similar terms. It will be noticed that, by a happy chance, the factors of $\tfrac{1}{2}$ in the definitions of the ϵ_i make ϵ_4, ϵ_5, ϵ_6 the (engineering) shear strain components introduced on p. 102.

The following scheme summarizes the piezoelectric equations in the matrix notation. Read horizontally by rows it gives the direct effect, and read vertically by columns it gives the converse effect.

$$
\begin{array}{cc|cccccc}
 & & \epsilon_1 & \epsilon_2 & \epsilon_3 & \epsilon_4 & \epsilon_5 & \epsilon_6 \\
 & & \sigma_1 & \sigma_2 & \sigma_3 & \sigma_4 & \sigma_5 & \sigma_6 \\
\hline
E_1 & P_1 & d_{11} & d_{12} & d_{13} & d_{14} & d_{15} & d_{16} \\
E_2 & P_2 & d_{21} & d_{22} & d_{23} & d_{24} & d_{25} & d_{26} \\
E_3 & P_3 & d_{31} & d_{32} & d_{33} & d_{34} & d_{35} & d_{36}
\end{array}
$$

Table 7 compares the equations defining the direct and converse piezoelectric effects in the tensor and the matrix notation.

TABLE 7

Defining equations for the direct and converse piezoelectric effects

	Tensor notation $(i, j, k = 1, 2, 3)$	Matrix notation $(i = 1, 2, 3; j = 1, 2, ..., 6)$
Direct effect . . .	$P_i = d_{ijk}\,\sigma_{jk}$	$P_i = d_{ij}\,\sigma_j$
Converse effect . .	$\epsilon_{jk} = d_{ijk}\,E_i$	$\epsilon_j = d_{ij}\,E_i$

4. Reduction in the number of independent moduli by crystal symmetry

4.1. Pure symmetry arguments. Crystal symmetry, when it is present, reduces the number of independent piezoelectric moduli still further. As an extreme example let us consider the effect of a centre of symmetry. Suppose a crystal with a centre of symmetry were subjected to a general stress and became polarized. Now imagine that the whole system, crystal plus stress, is inverted through the centre of symmetry. The stress, being centrosymmetrical, will be unchanged; so also will the crystal; but the polarization will be reversed in direction. We are left with the same crystal as before under the same stress but with the reverse polarization. This situation is only possible if the polarization is zero. Hence, *a crystal with a centre of symmetry cannot be piezoelectric.*

The vanishing and the equality of many of the moduli in the various classes can be deduced by similar arguments from symmetry, without

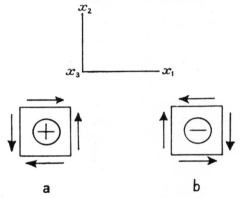

FIG. 7.1. Illustrating that in class *4* the modulus d_{36} vanishes.

recourse to any analytical work. As an example, consider the modulus d_{36} in class *4*. This measures the component P_3 for a shear stress about Ox_3. In Fig. 7.1 *a* the Ox_3 axis points out of the diagram. Let us suppose that a shear stress in the sense indicated produces a dipole parallel to Ox_3 whose positive end is uppermost. Changing the sign of the stress must change the sign of the dipole, giving the situation shown in Fig. 7.1 *b*. But since the crystal has a 4-fold axis the crystal in Fig. 7.1 *b* is stressed in exactly the same way as in Fig. 7.1 *a*; it is merely turned through 90°. The polarizations must therefore be the same in the two cases, and this is only possible if they are zero. Hence $d_{36} = 0$.

EXERCISE 7.1. Prove by symmetry arguments that in class *m*, with *m* perpendicular to Ox_2, $d_{14} = d_{16} = d_{21} = d_{22} = d_{23} = d_{25} = d_{34} = d_{36} = 0$.

EXERCISE 7.2. Show by a symmetry argument that, in class *4*, $d_{14} = -d_{25}$.

4.2. Analytical methods. Pure symmetry arguments like those used in § 4.1 give an elegant way of finding some of the relations between the moduli, but it is difficult to be sure that they are exhaustive. We therefore describe now a more systematic method of approach. The principle of the method is to transform the axes of reference of the tensor by one of the symmetry elements possessed by the crystal. Then the coefficients describing the effect must be the same after transformation as before.

(i) *Centre of symmetry.* To illustrate the method let us first consider again a crystal possessing a centre of symmetry. The transformation matrix is
$$a_{ij} = -\delta_{ij}.$$
The transformed piezoelectric moduli are thus, by equation (11),
$$d'_{ijk} = a_{il} a_{jm} a_{kn} d_{lmn} = -\delta_{il} \delta_{jm} \delta_{kn} d_{lmn} = -d_{ijk},$$
by the substitution property of δ_{ij}. But since the crystal has a centre of symmetry
$$d'_{ijk} = d_{ijk}.$$
Therefore,
$$d_{ijk} = 0.$$

With other symmetry operations the working is not quite as short as this. It is a considerable help to use a slightly different symbolism for the transformations and to use what is known as the *direct inspection method*, due to Fumi (1952 *a*, *b*, *c*) (Fieschi and Fumi 1953). The method is best explained by taking the same example of a centre of symmetry and working out its consequences for the moduli in the new notation.

The operation of the centre of symmetry on the axes of reference is as follows:
$$x_1 \rightarrow -x_1, \quad x_2 \rightarrow -x_2, \quad x_3 \rightarrow -x_3, \tag{22}$$
where $\rightarrow$ here means 'transforms to'. (22) may be alternatively interpreted as expressing that the coordinates (x_1, x_2, x_3) of a general point transform to $(-x_1, -x_2, -x_3)$. Now we have seen (p. 112) that tensor components transform like the corresponding products of coordinates. As an example, d_{122} transforms like the product $x_1 x_2^2$. By making the substitution indicated in (22), it is seen that $x_1 x_2^2$ transforms as follows:
$$x_1 x_2^2 \rightarrow -x_1 x_2^2.$$
Hence
$$d_{122} \rightarrow -d_{122}.$$
But since the crystal possesses a centre of symmetry the transformation does not change the components: d_{122} must transform into itself. Hence
$$d_{122} \rightarrow d_{122},$$
and so
$$d_{122} = 0.$$

It is easy to see that the same reasoning applies to all the other components, and we have the general result, as before, that $d_{ijk} = 0$.

(ii) *Diad axis.* To illustrate the direct inspection method further let us consider what effect a diad axis has on the moduli. Suppose the diad axis is parallel to x_3. Then the transformation is

$$x_1 \to -x_1, \quad x_2 \to -x_2, \quad x_3 \to x_3,$$

or, more compactly,

$$1 \to -1, \quad 2 \to -2, \quad 3 \to 3. \tag{23}$$

We must now take the moduli one by one, and transform them according to (23). If the sign changes the modulus must be zero; if the sign remains the same the modulus survives. Thus, for example, $d_{133} \to -d_{133}$, and so $d_{133} = 0$; but $d_{123} \to d_{123}$, and so d_{123} survives. Clearly, only those d_{ijk} which have a single 3 or three 3's will remain. The non-vanishing moduli are therefore those in bold type in the array:

$$
\begin{array}{ccc}
\boldsymbol{d_{111}} \;\; d_{112} \;\; \boldsymbol{d_{113}} & d_{211} \;\; d_{212} \;\; \boldsymbol{d_{213}} & \boldsymbol{d_{311}} \;\; \boldsymbol{d_{312}} \;\; d_{313} \\
\quad d_{122} \;\; \boldsymbol{d_{123}} & \quad d_{222} \;\; \boldsymbol{d_{223}} & \quad \boldsymbol{d_{322}} \;\; d_{323} \\
\quad\quad d_{133} & \quad\quad d_{233} & \quad\quad \boldsymbol{d_{333}} \;.
\end{array} \tag{24}
$$

In the matrix (two-suffix) notation the corresponding moduli are

$$
\begin{pmatrix}
0 & 0 & 0 & d_{14} & d_{15} & 0 \\
0 & 0 & 0 & d_{24} & d_{25} & 0 \\
d_{31} & d_{32} & d_{33} & 0 & 0 & d_{36}
\end{pmatrix}. \tag{25}
$$

The correspondence between schemes (24) and (25) is most readily seen by reading each *layer* of scheme (24) in the order

and writing it as a *row* of the matrix scheme.

(iii) *Example of a crystal class.* The method may now be applied to a crystal class, and we select $\bar{4}2m$, the class of ammonium dihydrogen phosphate (ADP), as an example.

The symmetry elements are shown on a stereogram in Fig. 7.2 a. The $\bar{4}$ axis and the diad axis parallel to x_1 give all the necessary symmetry information. The $\bar{4}$ axis parallel to x_3 includes a diad axis parallel to x_3, and we have already found that only the moduli in bold type in (24) survive the operation of a diad parallel to x_3. Therefore only these need be considered. The $\bar{4}$ axis† transforms the axes as follows:

$$1 \to -2, \quad 2 \to 1, \quad 3 \to -3.$$

† Operation of 90° rotation followed by inversion.

Hence,

$$d_{113} \to -d_{223}, \qquad d_{213} \to \ d_{123}, \qquad d_{311} \to -d_{322}, \qquad d_{312} \to \ d_{321},$$
$$d_{123} \to \ d_{213}, \qquad d_{223} \to -d_{113}, \qquad\qquad\qquad\qquad d_{322} \to -d_{311},$$
$$d_{333} \to -d_{333}.$$

$$(26)$$

Thus, for example, d_{113} transforms into $-d_{223}$. But we know that, since the crystal possesses this $\bar{4}$ axis, d_{113} must transform into itself. Hence

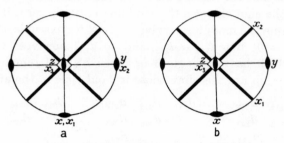

Fig. 7.2. The point-group symmetry elements of class $\bar{4}2m$ showing (a) the conventional setting of the axes Ox_1, Ox_2, Ox_3, and (b) the setting obtained by a rotation of 45° about Ox_3.

$d_{113} = -d_{223}$. In a similar way all the $\to$ signs in (26) may be read as equality signs.

Now consider the diad parallel to x_1. We saw in (ii) above that a diad parallel to x_3 removed all moduli having no 3's or two 3's. Hence a diad parallel to x_1 removes all moduli having no 1's or two 1's. Therefore, only the following moduli remain,

$$d_{123} = d_{213}, \qquad d_{312} \ (= d_{321}).$$

In the two suffix notation these are

$$d_{14} = d_{25}, \qquad d_{36};$$

and the matrix is

$$\begin{pmatrix} 0 & 0 & 0 & d_{14} & 0 & 0 \\ 0 & 0 & 0 & 0 & d_{14} & 0 \\ 0 & 0 & 0 & 0 & 0 & d_{36} \end{pmatrix}.$$

Numerical example. The matrix elements of (d_{ij}) for ammonium dihydrogen phosphate (ADP) at 0° C in rationalized m.k.s. units (coulombs/newton) have the values (Mason 1946, Van Dyke and Gordon 1950):

$$\begin{pmatrix} 0 & 0 & 0 & 0\cdot17 & 0 & 0 \\ 0 & 0 & 0 & 0 & 0\cdot17 & 0 \\ 0 & 0 & 0 & 0 & 0 & 5\cdot17 \end{pmatrix} \times 10^{-11}.$$

Notice that this arrangement of moduli only holds for the particular setting of crystal axes relative to x_1, x_2, x_3 that is shown in Fig. 7.2 a. If the axes are rotated through 45° about x_3, so that x_1 is now perpendicular to a plane of symmetry (Fig. 7.2 b), the matrix takes the new form

$$\begin{pmatrix} 0 & 0 & 0 & 0 & 0{\cdot}17 & 0 \\ 0 & 0 & 0 & -0{\cdot}17 & 0 & 0 \\ 2{\cdot}6 & -2{\cdot}6 & 0 & 0 & 0 & 0 \end{pmatrix} \times 10^{-11}.$$

(iv) *Trigonal and hexagonal systems.* The direct inspection method used above is the most convenient one to use in deducing the conditions imposed on the moduli in all the classes except those belonging to the trigonal and hexagonal systems. This is because, in all classes except the trigonal and hexagonal ones, the symmetry elements simply change the *order* of the axes: that is to say, each axis transforms either into itself or into one of the other axes (possibly with its direction reversed) and not into some intermediate position. Now in class *6*, for example, it is impossible to choose Cartesian axes with this property; x_1 would transform into a linear *combination* of x_1, x_2, x_3. In classes *3* and *3m*, it is true, Cartesian axes *can* be chosen with the necessary property, but they are not the axes conventionally used. Therefore, for the trigonal and hexagonal systems some recourse has to be made to a more comprehensive analytical method. We now give this in outline. The principle is exactly the same as before: transform the axes by the operation of a symmetry element of the crystal, and equate the old moduli with the new ones.

We consider the operation of a 6-fold and a 3-fold axis parallel to x_3. The transformation matrices for rotations through 60° and 120°, respectively, about Ox_3 are

$$(a_{ij}) = \begin{pmatrix} \tfrac{1}{2} & \tfrac{1}{2}\sqrt{3} & 0 \\ -\tfrac{1}{2}\sqrt{3} & \tfrac{1}{2} & 0 \\ 0 & 0 & 1 \end{pmatrix} \quad \text{and} \quad (a_{ij}) = \begin{pmatrix} -\tfrac{1}{2} & \tfrac{1}{2}\sqrt{3} & 0 \\ -\tfrac{1}{2}\sqrt{3} & -\tfrac{1}{2} & 0 \\ 0 & 0 & 1 \end{pmatrix}.$$

When these transformations are applied to the d_{ijk} the result is a set of simultaneous equations. We find, for example, that after using the symmetry property of d_{ijk}, d'_{111} is a function of the six d_{ijk}: d_{111}, d_{112}, d_{122}, d_{211}, d_{212}, d_{222}. By putting $d'_{111} = d_{111}$ an equation is obtained connecting these six d_{ijk}. By proceeding similarly for d'_{112}, d'_{122}, d'_{211}, d'_{212} and d'_{222} five more simultaneous equations may be written down for the same six d_{ijk}. Their solution gives, for a 3-fold axis,

$$d_{111} = -d_{122} = -d_{212}; \quad d_{222} = -d_{112} = -d_{211},$$

and, for a 6-fold axis,

$$d_{111} = d_{112} = -d_{122} = d_{211} = d_{212} = d_{222} = 0. \qquad (27)$$

The conditions on the other moduli are derived in a straightforward way and are found to be, for a 3-fold or a 6-fold axis,

$$d_{113} = d_{223}, \quad d_{123} = -d_{213}, \quad d_{311} = d_{322},$$
$$d_{133} = d_{233} = d_{312} = d_{313} = d_{323} = 0.$$

We have thus found the conditions on the moduli for classes *6* and *3*. The other non-centrosymmetrical hexagonal and trigonal classes are obtained by adding planes of symmetry and diad axes to these two classes. The further restrictions on the moduli which these planes and diad axes impose are readily found by the direct inspection method.

(v) *Application of group theory.* The total number of independent tensor components necessary for specifying a particular tensor property in any crystal class may also be found by applying group theory; see Higman (1955) (see also Bhagavantam and Venkatarayudu 1951 and Jahn 1949). This provides a useful check on the results given by other methods. Group theory as applied in the above publications does not reveal which of the moduli are independent, but only the *total number* of independent ones. The *listing* of the independent components in each class by group theoretical methods has been studied by Fumi (1952*d*) and by Fieschi and Fumi (1953), who give further references (see also Jahn 1937). Group theory is a useful method when a number of related classes have to be studied at once.

5. Results for all the crystal classes

We give in Table 8 the results obtained by applying the methods of § 4.2 to all the crystal classes in turn. Each of the arrays in brackets in the table represents the (d_{ij}) matrix written out in the form (18). By referring to the key at the beginning of the table it will be seen that a light dot denotes a modulus that is zero. A heavy dot denotes a non-zero modulus. A line joining two heavy dots denotes that the two moduli are numerically equal. When a heavy dot is joined by a line to an open circle it means that the two moduli concerned are numerically equal but opposite in sign. Thus in class *4*, $d_{31} = d_{32}$, $d_{15} = d_{24}$ and $d_{14} = -d_{25}$. The symbol of a double circle is always joined to a heavy dot. If x is the numerical value of the modulus denoted by the heavy dot, the value of the modulus denoted by the double circle is $-2x$. Thus, in class *3*, $d_{16} = -2d_{22}$ and $d_{26} = -2d_{11}$ (in this last case the joining line passes through an open circle). The number of independent moduli in each class is shown in brackets after the matrix. The relation of the reference axes to the symmetry elements conforms, except where otherwise stated, to the conventions recommended in *Standards on piezoelectric crystals* (1949). These conventions are summarized in Appendix B (p. 282).

TABLE 8

Form of the (d_{ij}) matrix

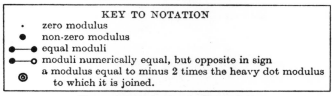

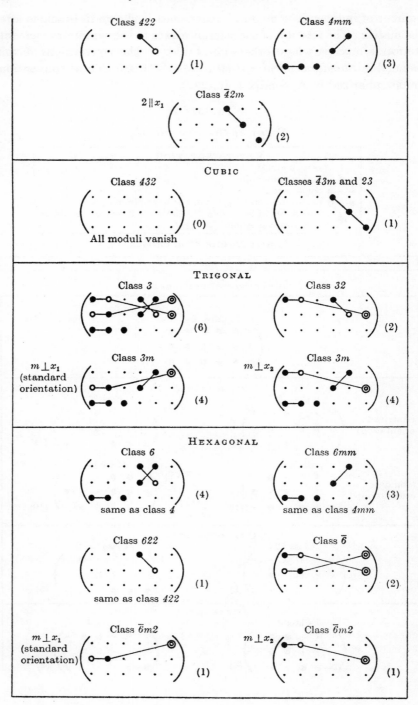

The physical significance of the matrices in Table 8 is best appreciated by working out the polarization produced in a chosen crystal class by various simple stress systems. The reader is recommended to do this. We select the piezoelectric behaviour of quartz, which is of considerable practical importance, for more detailed study.

Quartz. The room temperature form of quartz belongs to class *32*. The (d_{ij}) matrix is accordingly

$$\begin{array}{cccccc} \sigma_1 & \sigma_2 & \sigma_3 & \sigma_4 & \sigma_5 & \sigma_6 \end{array}$$

$$\begin{array}{c} P_1 \\ P_2 \\ P_3 \end{array} \begin{pmatrix} d_{11} & -d_{11} & 0 & d_{14} & 0 & 0 \\ 0 & 0 & 0 & 0 & -d_{14} & -2d_{11} \\ 0 & 0 & 0 & 0 & 0 & 0 \end{pmatrix},$$

the P's and σ's being inserted for guidance.

If a tensile stress σ_1 is applied parallel to x_1, which is a diad axis of the crystal (Fig. 7.3), equation (17) shows that the components of polarization are given by the moduli in the first column of the matrix; thus

$$P_1 = d_{11}\sigma_1, \quad P_2 = 0, \quad P_3 = 0.$$

The polarization is therefore directed along x_1.†

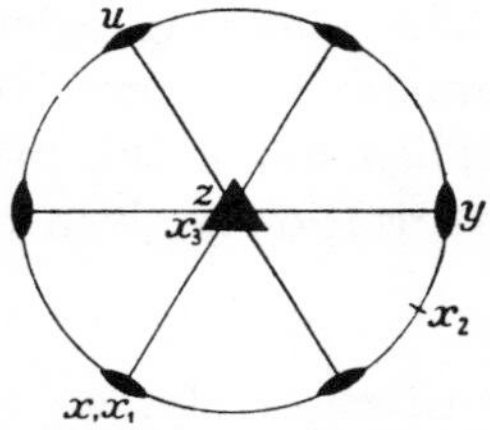

On the other hand, a tensile stress σ_2 along x_2 produces no polarization parallel to itself, but it does produce a polarization along x_1:

FIG. 7.3. The point-group symmetry elements of α-quartz (class *32*) and the conventional setting of the axes Ox_1, Ox_2, Ox_3.

$$P_1 = -d_{11}\sigma_2, \quad P_2 = 0, \quad P_3 = 0$$

(second column of the matrix). A given polarization along the diad axis x_1 may therefore be produced either by a tensile stress along x_1 or by an equal compressive stress along x_2. The diad axis of quartz is known as the *electric axis*.

It is seen by looking along the top row of the matrix of moduli that a polarization along x_1 can also be produced by a shear stress σ_4 about x_1; thus, for this stress,

$$P_1 = d_{14}\sigma_4, \quad P_2 = 0, \quad P_3 = 0.$$

A polarization along x_2 (second row of the matrix) may evidently be produced in two ways: (i) by a shear σ_5 about x_2, giving

$$P_1 = 0, \quad P_2 = -d_{14}\sigma_5, \quad P_3 = 0,$$

† The *sense* of the x_1 axis is conventionally fixed (*Standards on piezoelectric crystals,* 1949) by ruling that in right-handed quartz the positive end of the x_1 axis develops a negative charge on extension (and therefore a positive charge on compression). In left-handed quartz the positive end of the x_1 axis develops a positive charge on extension. This convention evidently makes d_{11} negative for right-handed quartz and positive for left-handed quartz.

or (ii) by a shear σ_6 about x_3, giving

$$P_1 = 0, \quad P_2 = -2d_{11}\sigma_6, \quad P_3 = 0.$$

The fact that all the moduli on the bottom row of the matrix are zero shows that no possible condition of stressing can produce a polarization along the triad axis x_3.

The measured values of the d_{ij} for right-handed quartz (calculated from Cady 1946, p. 219) are

$$\begin{pmatrix} -2\cdot3 & 2\cdot3 & 0 & -0\cdot67 & 0 & 0 \\ 0 & 0 & 0 & 0 & 0\cdot67 & 4\cdot6 \\ 0 & 0 & 0 & 0 & 0 & 0 \end{pmatrix} \times 10^{-12}$$

in rationalized m.k.s. units (coulombs/newton).

The following calculation illustrates the size of the direct effect. Suppose a compressive stress of 1 kg/cm² ($= 9\cdot81 \times 10^4$ newtons/m²) is applied along the diad axis of a right-handed quartz crystal. The polarization is given by

$$P_1 = d_{11}\sigma_1 = (-2\cdot3 \times 10^{-12}) \times (-9\cdot81 \times 10^4) = 2\cdot3 \times 10^{-7} \text{ coulombs/m}^2.$$

Alternatively, we may calculate the corresponding converse effect (§ 3). Suppose an electric field of 100 volt/cm ($= 10^4$ volts/m) acts along the diad axis x_1. Then the strain along this axis is a contraction and is given by equation (21) as

$$\epsilon_1 = d_{11}E_1 = -2\cdot3 \times 10^{-12} \times 10^4 = -2\cdot3 \times 10^{-8}.$$

This field evidently produces an equal extension along x_2 and a shear about x_1 (see the first row of the matrix) for we have

$$\epsilon_2 = d_{12}E_1 = -d_{11}E_1 = 2\cdot3 \times 10^{-8},$$

and

$$\epsilon_4 = d_{14}E_1 = -0\cdot67 \times 10^{-8}.$$

6. Representation surfaces

As we have seen in Chapter I, § 4, a second-rank tensor, provided it is symmetrical, may be represented by a quadric surface (ellipsoid or hyperboloid). Representation surfaces may also be given for third-rank tensors, but they are not as simple as those for second-rank symmetrical tensors; and, in fact, it is not possible to represent a third-rank tensor completely by one surface only.

A surface which is useful in practice is defined as follows. Suppose that a plate (Fig. 7.4 a) is cut perpendicular to an arbitrary direction denoted by Ox'_1 and is subjected to a normal tensile stress σ'_{11} on its surface. A polarization will be set up with components in all three directions Ox_1', Ox_2', Ox_3'. The polarization component in the Ox_i' direction will be given by

$$P_i' = d_{i11}' \sigma_{11}',$$

and, in particular, $P_1' = d_{111}' \sigma_{11}'.$

d_{111}' therefore measures the 'longitudinal' piezoelectric effect; it gives the charge per unit area produced on the surface of the plate divided by the normal force per unit area applied to the surface of the plate.

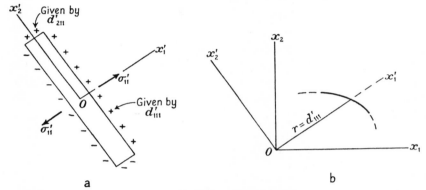

FIG. 7.4. Illustrating (a) the longitudinal piezoelectric effect, and (b) the definition of the longitudinal piezoelectric surface.

Considerations of symmetry show that a longitudinal effect can only occur along a direction which is polar in the sense defined on p. 79. A surface for which the radius vector in the Ox_1' direction is equal to d_{111}' is called the *longitudinal piezoelectric surface* (Fig. 7.4 b). (This surface also gives the tensile strain produced in the direction of an applied field, since, by the converse effect, $\epsilon_{11}' = d_{111}' E_1'$.) It is evident that many other surfaces could be defined to represent particular components of polarization resulting from particular directions of stressing (or particular components of strain resulting from a field in a particular direction).

Form of the longitudinal piezoelectric surface for quartz. To find the shape of the longitudinal piezoelectric surface for quartz we proceed as follows. The radius vector in the arbitrary direction Ox_1' is

$$r = d_{111}' = a_{1i} a_{1j} a_{1k} d_{ijk},$$

where, as before, Ox_i are the axes related to the crystal symmetry for which the scheme of moduli given in § 5 applies. The only non-vanishing moduli in class 32 are d_{111}, d_{122}, d_{123}, d_{212} and d_{213}; so we have

$$r = a_{11}a_{11}a_{11}d_{111} + a_{11}a_{12}a_{12}d_{122} + 2a_{11}a_{12}a_{13}d_{123} +$$
$$+ 2a_{12}a_{11}a_{12}d_{212} + 2a_{12}a_{11}a_{13}d_{213}.$$

We replace (a_{11}, a_{12}, a_{13}), which are the direction cosines of Ox_1' relative

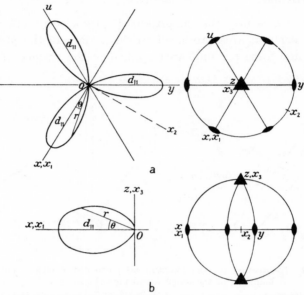

Fig. 7.5. The longitudinal piezoelectric surface for quartz: (a) the central section perpendicular to the 3-fold axis Ox_3, and (b) the section containing Ox_3 and one of the 2-fold axes, Ox_1. Stereograms of the symmetry elements in these two orientations are also shown.

to Ox_1, Ox_2, Ox_3, by (l, m, n) for shortness, and make use of the relations between the moduli shown in § 5. Thus we obtain

$$r = l(l^2 - 3m^2)d_{11} \qquad (28)$$

as the equation of the surface. Notice that the equation only involves one of the two independent moduli.

The section of the surface by the plane containing Ox_1 and Ox_2 is found, in polar coordinates, by putting $l = \cos\theta$, $m = \sin\theta$, where θ is the angle between the radius vector and Ox_1. Then

$$r = d_{11}\cos\theta(\cos^2\theta - 3\sin^2\theta) = d_{11}\cos 3\theta.$$

The section is drawn in Fig. 7.5a for a positive value of d_{11}. As we have seen (footnote, p. 125), this is appropriate for left-handed quartz. Thus, in such a crystal, a tensile stress along Ox_1 produces a positive charge towards the positive end of the Ox_1 axis. As the direction of the applied tensile stress moves round anticlockwise, the longitudinal component of the polarization evidently decreases, until it reaches zero when θ in Fig. 7.5 a equals 30°. The longitudinal component of polarization then changes sign and when $\theta = 60°$, so that the tensile stress is again along a diad axis, it reaches a maximum in the $+u$ direction. The figure may also be readily interpreted in terms of the converse effect.

The section of the surface by the plane containing Ox_1 and Ox_3 is obtained by putting $m = 0$ and $l = \cos\theta$ in equation (28), where θ is again the angle between the radius vector and Ox_1. We have

$$r = d_{11}\cos^3\theta,$$

which gives the section shown in Fig. 7.5 b. The shape of the whole representation surface has been likened to that of three almonds whose pointed ends meet in the trigonal axis. We can see immediately from the shape of the surface that, for example, there is no longitudinal piezoelectric effect when a quartz crystal is compressed along the triad axis. This is also obvious from symmetry. The longitudinal effect is evidently greatest when the crystal is compressed along a diad axis (electric axis).

Longitudinal piezoelectric surfaces for other crystals.† The form of the longitudinal piezoelectric surface for crystals of other classes may be obtained in a similar way. The surface must conform to the point-group symmetry. For example, the surface for zinc blende, which belongs to the cubic class $\bar{4}3m$, has four lobes projecting in the directions of the four triad axes. Symmetry and the matrix in Table 8 both show that in this class pressure along a diad axis can produce no longitudinal effect. (It may be noted, however, that shearing about a diad axis would produce an effect.)

Exercise 7.3. Show that the longitudinal piezoelectric surface for classes 422 and 622 vanishes in all directions.

Exercise 7.4. By using Table 8, find which classes may show a piezoelectric effect under hydrostatic pressure, and hence verify the components of the pyro-electric vector **p** given in Chapter IV, § 7.

Exercise 7.5. A constant uniaxial pressure is applied to a quartz crystal perpendicular to the triad axis. If now the direction of the pressure rotates about the triad axis, show that the piezoelectric polarization vector remains of constant magnitude and rotates twice as fast about the triad axis but in the opposite sense.

---

† For further details see the books by Voigt (1910, pp. 864 et seq.) and Wooster (1938, pp. 215–17).

SUMMARY

Definition of a third-rank tensor. A set of 27 numbers T_{ijk} which represent a physical quantity are said to form a third-rank tensor if they transform according to the equation

$$T'_{ijk} = a_{il} a_{jm} a_{kn} T_{lmn}.$$

This is the same as the transformation equation for products of coordinates $x_i x_j x_k$.

The direct piezoelectric effect is given by

$$P_i = d_{ijk} \sigma_{jk}, \tag{3}$$

where the d_{ijk} are the piezoelectric moduli; they form a third-rank tensor.

Since σ_{jk} is the symmetric part of the stress tensor we may put $d_{ijk} = d_{ikj}$. This reduces the number of independent d_{ijk} to 18.

Matrix notation. The second and third suffixes in d_{ijk}, and both suffixes of σ_{jk}, are abbreviated into a single suffix running from 1 to 6, thus:

tensor notation	11	22	33	23, 32	31, 13	12, 21
matrix „	1	2	3	4	5	6

We also introduce factors of 2 as follows:

$$d_{ijk} = d_{in} \quad \text{when } n = 1, 2 \text{ or } 3,$$

and

$$2d_{ijk} = d_{in} \quad \text{when } n = 4, 5 \text{ or } 6.$$

We may then write (3) as

$$P_i = d_{ij}\sigma_j \quad (i = 1, 2, 3; j = 1, 2, ..., 6). \tag{17}$$

The coefficients d_{ij} form a 3-row, 6-column, matrix (d_{ij}).

The converse piezoelectric effect is given by

$$\epsilon_{jk} = d_{ijk} E_i, \tag{19}$$

with the same d_{ijk} as in the direct effect. The two suffixes in the ϵ_{jk} are abbreviated into a single one according to the above scheme by writing $\epsilon_{jk} = \epsilon_n$ when $n = 1$, 2 or 3 and $2\epsilon_{jk} = \epsilon_n$ when $n = 4, 5$ or 6. (19) is then written

$$\epsilon_j = d_{ij} E_i \quad (i = 1, 2, 3; j = 1, 2, ..., 6). \tag{21}$$

Symmetry requirements. A crystal with a centre of symmetry cannot be piezoelectric. The restrictions imposed by crystal symmetry on the moduli d_{ij} in the non-centrosymmetrical crystal classes may often be obtained by pure symmetry arguments. The *direct inspection method* provides a systematic approach, except for the trigonal and hexagonal systems, where it is necessary to use a more comprehensive analytical method. Group theory may also be used for this purpose.

A third-rank tensor such as d_{ijk} cannot be completely represented by a single surface. The *longitudinal piezoelectric surface* represents the component of polarization developed parallel to an applied tensile stress. The radius vector in any direction is directly proportional to the longitudinal effect in that direction. A longitudinal piezoelectric effect can only occur along a polar direction.

GENERAL REFERENCES

The two most comprehensive books on piezoelectricity are (see also pp. 313–19):

CADY, W. G. (1946) *Piezoelectricity*, New York: McGraw-Hill.

MASON, W. P. (1950) *Piezoelectric crystals and their applications to ultrasonics*, New York: van Nostrand.

VIII

ELASTICITY. FOURTH-RANK TENSORS

1. Hooke's Law

A SOLID body changes its shape when subjected to a stress. Provided the stress is below a certain limiting value, the *elastic limit*, the strain is recoverable, that is to say, the body returns to its original shape when the stress is removed. It is further observed (Hooke's Law) that for sufficiently small stresses the amount of strain is proportional to the magnitude of the applied stress. For example, suppose a bar of an isotropic solid is loaded in pure tension so that the tensile stress is σ. The longitudinal strain ϵ equals $\Delta l/l$, where Δl is the increase in length and l is the original length. Hooke's Law states that

$$\epsilon = s\sigma,$$

where s is a constant. s is called the *elastic compliance constant* or, shortly, the *compliance*, for this particular arrangement of stress and strain directions. As an alternative we could write

$$\sigma = c\epsilon, \qquad c = 1/s,$$

where c is the *elastic stiffness constant*, or the *stiffness*. c is also Young's Modulus.†

These statements and definitions must now be generalized. We have seen (Chs. V and VI) that a homogeneous stress and a homogeneous strain are each specified, in general, by second-rank tensors. It is found that, if a general homogeneous stress σ_{ij} is applied to a crystal, the resulting homogeneous strain ϵ_{ij} is such that each component is linearly

---

† The use of the terms *compliance* and *stiffness* accords with current American usage. The corresponding terms used by many English authors are, respectively, *elastic modulus* and *elastic constant*. If this latter system is adopted we have the confusing fact that Young's Modulus is not an elastic modulus but an elastic constant. Another reason for preferring the words 'compliance' and 'stiffness' is that they are descriptive: a stiff crystal has a high value of its *stiffness*, a compliant crystal has a high value of its *compliance*. It is unfortunate, but a fact easily memorized, that the initial letters of 'stiffness' and 'compliance' are the reverse of the corresponding symbols.

Symbol	American authors and this book	English authors	Dimensions
s	compliance	modulus	stress^{-1}
c	stiffness	constant	stress

related to all the components of the stress. Thus, for example,

$$\epsilon_{11} = \quad s_{1111}\sigma_{11} + s_{1112}\sigma_{12} + s_{1113}\sigma_{13} +$$
$$+ s_{1121}\sigma_{21} + s_{1122}\sigma_{22} + s_{1123}\sigma_{23} +$$
$$+ s_{1131}\sigma_{31} + s_{1132}\sigma_{32} + s_{1133}\sigma_{33},$$

and eight similar equations for the other eight components of ϵ_{ij}, where the s's are constants. The generalized form of Hooke's Law may therefore be written

$$\epsilon_{ij} = s_{ijkl}\sigma_{kl}; \tag{1}$$

the s_{ijkl} are the *compliances* of the crystal. Equation (1) stands for nine equations, each with nine terms on the right-hand side. There are 81 s_{ijkl} coefficients.

If we apply only one component of stress, say σ_{11}, equations (1) imply that all the strain components, not just ϵ_{11}, may be different from zero. It follows that, if a rectangular block of crystal is loaded by a uniaxial tension applied parallel to one set of edges, it will not only stretch in the direction of the tension but it may also shear so that all the angles between the edges become different from right angles. As a corollary, if we try to bend a bar of crystal by applying pure bending couples to its ends, it will, in general, twist as well as bend. Correspondingly, if we try to twist a rod of crystal by applying pure twisting couples to its ends it will, in general, bend as well as twist.

As an alternative to equations (1) the stresses may be expressed in terms of the strains by the equations,

$$\sigma_{ij} = c_{ijkl}\epsilon_{kl}, \tag{2}$$

where the c_{ijkl} are the 81 *stiffness constants* of the crystal. If the relations (1) were solved as a set of simultaneous equations for the σ_{ij}, a set of solutions of the form (2) would be obtained, the coefficients c_{ijkl} being functions of the s_{ijkl}. The form of the relation between the c_{ijkl} and the s_{ijkl} is treated in § 7 and in Chapter IX, § 4.2.

The physical meaning of the s_{ijkl} may be appreciated by imagining the crystal to be subjected to various simple stress conditions. We recall from p. 87 that σ_{ij} may always be taken as symmetrical, even when body-torques are present. Hence, if a shear stress about Ox_3 were applied, both σ_{12} and σ_{21} would be present and we should have

$$\epsilon_{11} = s_{1112}\sigma_{12} + s_{1121}\sigma_{21} = (s_{1112} + s_{1121})\sigma_{12}.$$

s_{1112} and s_{1121} always occur together; it follows that it is in principle impossible to devise an experiment by which s_{1112} can be separated from s_{1121} and, in general, by which s_{ijkl} can be separated from s_{ijkl}. A similar situation arose with the piezoelectric moduli in

§ 1. Therefore, to avoid an arbitrary constant we set the two components equal:

$$s_{ijkl} = s_{ijlk}. \tag{3}$$

If, on the other hand, a uniaxial tension were applied parallel to Ox_3 the components of strain would be given by

$$\epsilon_{11} = s_{1133}\,\sigma_{33}, \quad \epsilon_{22} = s_{2233}\,\sigma_{33}, \quad \text{etc.}$$

In particular,

$$\epsilon_{12} = s_{1233}\,\sigma_{33} \quad \text{and} \quad \epsilon_{21} = s_{2133}\,\sigma_{33}.$$

But, from the definition of the components of the strain tensor (Ch. VI), $\epsilon_{12} = \epsilon_{21}$. Hence, $s_{1233} = s_{2133}$, and, in general, by considering other special cases, we see that

$$s_{ijkl} = s_{jikl}. \tag{4}$$

On account of the relations (3) and (4), only 36 of the 81 components s_{ijkl} are independent.

To attach physical meanings to the c_{ijkl} in equation (2) we have to imagine a set of stress components applied to the crystal and chosen in such a way that all the components of strain, except for one normal component or a pair of shear components, vanish. Thus the requisite stresses to produce tensor shear strain components ϵ_{12}, ϵ_{21} are

$$\sigma_{ij} = c_{ij12}\,\epsilon_{12} + c_{ij21}\,\epsilon_{21} = (c_{ij12} + c_{ij21})\epsilon_{12}.$$

Again we put the pairs of coefficients that always occur together equal to one another. Then, in general,

$$c_{ijkl} = c_{ijlk}. \tag{5}$$

By considering special cases, as with the s_{ijkl}, we also find that

$$c_{ijkl} = c_{jikl}. \tag{6}$$

Again, the equations (5) and (6) reduce the number of independent c_{ijkl} from 81 to 36.

We shall now show that the 81 compliances s_{ijkl} form a *fourth-rank tensor*. A fourth-rank tensor is defined, like tensors of lower rank, by its transformation law. The 81 numbers T_{ijkl} representing a physical quantity are said to form a fourth-rank tensor if they transform on change of axes to T'_{ijkl}, where

$$T'_{ijkl} = a_{im}\,a_{jn}\,a_{ko}\,a_{lp}\,T_{mnop}. \tag{7}$$

To prove that the s_{ijkl} form such a tensor we proceed as follows. We have

$$\epsilon'_{ij} = a_{ik}\,a_{jl}\,\epsilon_{kl}, \tag{8}$$

$$\epsilon_{kl} = s_{klmn}\,\sigma_{mn}, \tag{9}$$

$$\sigma_{mn} = a_{om}\,a_{pn}\,\sigma'_{op}. \tag{10}$$

Hence, combining these three equations, which form the scheme

$$\epsilon' \xrightarrow{(8)} \epsilon \xrightarrow{(9)} \sigma \xrightarrow{(10)} \sigma',$$

where $\rightarrow$ means 'in terms of', we obtain

$$\epsilon'_{ij} = a_{ik}\, a_{jl}\, s_{klmn}\, a_{om}\, a_{pn}\, \sigma'_{op}.$$

But we have
$$\epsilon'_{ij} = s'_{ijop}\, \sigma'_{op},$$
and so, by comparing coefficients,

$$s'_{ijop} = a_{ik}\, a_{jl}\, a_{om}\, a_{pn}\, s_{klmn}.$$

On interchanging the dummy suffixes this becomes

$$s'_{ijkl} = a_{im}\, a_{jn}\, a_{ko}\, a_{lp}\, s_{mn\acute{o}p}, \tag{11}$$

which is the necessary transformation law. It is worth noting, as a reminder of the economy of the dummy suffix notation, that equation (11) typifies 3^4 equations each with 3^4 terms on the right-hand side, making a total of $3^8 = 6{,}561$ terms in all.

The above proof is a general one. If two second-rank tensors A_{ij} and B_{kl} are related by the equation

$$A_{ij} = C_{ijkl}\, B_{kl},$$

the quantities C_{ijkl} form a fourth-rank tensor. It follows that the elastic stiffness constants c_{ijkl} also form a fourth-rank tensor.

2. The matrix notation

The symmetry of s_{ijkl} and c_{ijkl} in the first two and the last two suffixes makes it possible to use the matrix notation introduced in the preceding chapter. Both the stress components and the strain components are written, as before, with a single suffix running from 1 to 6:

$$\begin{bmatrix} \sigma_{11} & \sigma_{12} & \sigma_{31} \\ \sigma_{12} & \sigma_{22} & \sigma_{23} \\ \sigma_{31} & \sigma_{23} & \sigma_{33} \end{bmatrix} \rightarrow \begin{bmatrix} \sigma_1 & \sigma_6 & \sigma_5 \\ \sigma_6 & \sigma_2 & \sigma_4 \\ \sigma_5 & \sigma_4 & \sigma_3 \end{bmatrix}, \qquad \begin{bmatrix} \epsilon_{11} & \epsilon_{12} & \epsilon_{31} \\ \epsilon_{12} & \epsilon_{22} & \epsilon_{23} \\ \epsilon_{31} & \epsilon_{23} & \epsilon_{33} \end{bmatrix} \rightarrow \begin{bmatrix} \epsilon_1 & \tfrac{1}{2}\epsilon_6 & \tfrac{1}{2}\epsilon_5 \\ \tfrac{1}{2}\epsilon_6 & \epsilon_2 & \tfrac{1}{2}\epsilon_4 \\ \tfrac{1}{2}\epsilon_5 & \tfrac{1}{2}\epsilon_4 & \epsilon_3 \end{bmatrix}. \tag{12}$$

In the s_{ijkl} and the c_{ijkl} the first two suffixes are abbreviated into a single one running from 1 to 6, and the last two are abbreviated in the same way, according to the scheme,

tensor notation	11	22	33	23, 32	31, 13	12, 21
matrix notation	1	2	3	4	5	6

At the same time factors of 2 and 4 are introduced as follows:

$$s_{ijkl} = s_{mn} \text{ when } m \text{ and } n \text{ are 1, 2 or 3,}$$
$$2s_{ijkl} = s_{mn} \text{ when either } m \text{ or } n \text{ are 4, 5 or 6,}$$
$$4s_{ijkl} = s_{mn} \text{ when both } m \text{ and } n \text{ are 4, 5 or 6.}$$

Now consider equation (1) written out for ϵ_{11} and ϵ_{23}:

$$\epsilon_{11} = s_{1111}\,\sigma_{11} + s_{1112}\,\sigma_{12} + s_{1113}\,\sigma_{13} +$$
$$+ s_{1121}\,\sigma_{21} + s_{1122}\,\sigma_{22} + s_{1123}\,\sigma_{23} +$$
$$+ s_{1131}\,\sigma_{31} + s_{1132}\,\sigma_{32} + s_{1133}\,\sigma_{33};$$

$$\epsilon_{23} = s_{2311}\,\sigma_{11} + s_{2312}\,\sigma_{12} + s_{2313}\,\sigma_{13} +$$
$$+ s_{2321}\,\sigma_{21} + s_{2322}\,\sigma_{22} + s_{2323}\,\sigma_{23} +$$
$$+ s_{2331}\,\sigma_{31} + s_{2332}\,\sigma_{32} + s_{2333}\,\sigma_{33}.$$

In the matrix notation these two equations become

$$\epsilon_1 = s_{11}\,\sigma_1 + \tfrac{1}{2}s_{16}\,\sigma_6 + \tfrac{1}{2}s_{15}\,\sigma_5 +$$
$$+ \tfrac{1}{2}s_{16}\,\sigma_6 + s_{12}\,\sigma_2 + \tfrac{1}{2}s_{14}\,\sigma_4 +$$
$$+ \tfrac{1}{2}s_{15}\,\sigma_5 + \tfrac{1}{2}s_{14}\,\sigma_4 + s_{13}\,\sigma_3;$$

$$\tfrac{1}{2}\epsilon_4 = \tfrac{1}{2}s_{41}\,\sigma_1 + \tfrac{1}{4}s_{46}\,\sigma_6 + \tfrac{1}{4}s_{45}\,\sigma_5 +$$
$$+ \tfrac{1}{4}s_{46}\,\sigma_6 + \tfrac{1}{2}s_{42}\,\sigma_2 + \tfrac{1}{4}s_{44}\,\sigma_4 +$$
$$+ \tfrac{1}{4}s_{45}\,\sigma_5 + \tfrac{1}{4}s_{44}\,\sigma_4 + \tfrac{1}{2}s_{43}\,\sigma_3;$$

or
$$\epsilon_1 = s_{1j}\,\sigma_j \quad \text{and} \quad \epsilon_4 = s_{4j}\,\sigma_j.$$

In general, therefore, equation (1) takes the shorter form

$$\epsilon_i = s_{ij}\,\sigma_j \quad (i,j = 1,2,...,6). \tag{13}$$

The reason for introducing the 2's and 4's into the definitions of the s_{ij} is to avoid the appearance of 2's and 4's in equation (13) and to make it possible to write this equation in a compact form.†

For the c_{ijkl} no factors of 2 or 4 are necessary. For if we write simply

$$c_{ijkl} = c_{mn} \quad (i,j,k,l = 1,2,3;\ m,n = 1,...,6),$$

it may be shown by writing out some typical members that equations (2) take the form

$$\sigma_i = c_{ij}\,\epsilon_j \quad (i,j = 1,2,...,6). \tag{14}$$

The arrays of s_{ij} and c_{ij} written out in squares, thus:

$$
\begin{pmatrix}
s_{11} & s_{12} & s_{13} & s_{14} & s_{15} & s_{16} \\
s_{21} & s_{22} & s_{23} & s_{24} & s_{25} & s_{26} \\
s_{31} & s_{32} & s_{33} & s_{34} & s_{35} & s_{36} \\
s_{41} & s_{42} & s_{43} & s_{44} & s_{45} & s_{46} \\
s_{51} & s_{52} & s_{53} & s_{54} & s_{55} & s_{56} \\
s_{61} & s_{62} & s_{63} & s_{64} & s_{65} & s_{66}
\end{pmatrix}
\quad \text{and} \quad
\begin{pmatrix}
c_{11} & c_{12} & c_{13} & c_{14} & c_{15} & c_{16} \\
c_{21} & c_{22} & c_{23} & c_{24} & c_{25} & c_{26} \\
c_{31} & c_{32} & c_{33} & c_{34} & c_{35} & c_{36} \\
c_{41} & c_{42} & c_{43} & c_{44} & c_{45} & c_{46} \\
c_{51} & c_{52} & c_{53} & c_{54} & c_{55} & c_{56} \\
c_{61} & c_{62} & c_{63} & c_{64} & c_{65} & c_{66}
\end{pmatrix}
\tag{15}
$$

are matrices, (s_{ij}) and (c_{ij}). As with the piezoelectric moduli we add the reminder that, in spite of their appearance with two suffixes, the s_{ij} and c_{ij} are not the components, and so do not transform like the components, of a second-rank tensor. To transform them to other axes it is necessary to go back to the tensor notation.

† The placing of the 2's and 4's in the definitions of the s_{ij} rather than in equation (13) conforms to established practice, to the recommendations in *Standards on piezo-electric crystals* (1949), and to the usage introduced by Voigt (1910). Wooster (1938) adopts the reverse procedure. Because of the two possible definitions care is needed when looking up numerical values.

3. The energy of a strained crystal

Consider a crystal which in the unstrained state has the form of a unit cube, and suppose it is subjected to a small homogeneous strain with components ϵ_i. Now let the strain components all be changed to $\epsilon_i + d\epsilon_i$. We prove that the work done by the stress components σ_i acting on the cube faces is

$$dW = \sigma_i \, d\epsilon_i \quad (i = 1, 2, ..., 6). \tag{16}$$

First suppose that the strain component ϵ_1 is increased to $\epsilon_1 + d\epsilon_1$, while the other strain components, and the position of the centre of the cube, remain unaltered. The two faces perpendicular to Ox_1 will move outwards by amounts $\frac{1}{2}d\epsilon_1$; the other four faces will simply increase in area, but the positions of their centres will be unchanged. The work done by the forces on these last four faces is therefore zero. The work done on the faces perpendicular to Ox_1 equals their displacement multiplied by the normal component of the force on them; it is therefore $2\sigma_1 . \frac{1}{2}d\epsilon_1 = \sigma_1 \, d\epsilon_1$. This is the term with $i = 1$ in (16); the terms with $i = 2$ and $i = 3$ are obtained in a similar way.

Now let the cube be sheared by making the two faces perpendicular to Ox_2 move in opposite directions parallel to Ox_3, so as to increase the strain component ϵ_4 to $\epsilon_4 + d\epsilon_4$. In this deformation (simple shear) the mid-points of the faces perpendicular to Ox_2 each move a distance $\frac{1}{2}d\epsilon_4$. The component of force on the faces in this direction is σ_4. The work done by the forces is therefore $2\sigma_4 . \frac{1}{2}d\epsilon_4 = \sigma_4 \, d\epsilon_4$. The terms with $i = 5$ and $i = 6$ in (16) are obtained in a similar way.

It is readily shown that the corresponding equation to (16) in tensor notation is

$$dW = \sigma_{ij} \, d\epsilon_{ij} \quad (i, j = 1, 2, 3).$$

Expression (16) is analogous to the expressions for the work of magnetization, Chapter III, (11), and of polarization, Chapter IV, (9). Each has the form, for unit volume, of a 'force' (H_i, E_i or σ_i) multiplied by a small 'displacement' (dB_i, dD_i or $d\epsilon_i$). Just as in the magnetic and electrical cases we go on to prove that the matrix connecting the two quantities involved, in this case (c_{ij}), is symmetrical.

If the deformation process is isothermal and reversible the work done is equal to the increase in the free energy $d\Psi$ and we may write, per unit volume,

$$d\Psi = dW = \sigma_i \, d\epsilon_i. \tag{17}$$

If Hooke's Law (14) is obeyed this becomes

$$d\Psi = c_{ij} \epsilon_j \, d\epsilon_i. \tag{18}$$

Hence,
$$\frac{\partial \Psi}{\partial \epsilon_i} = c_{ij}\,\epsilon_j$$

(the argument is similar to that in Ch. III, § 2). Differentiating both sides of this equation with respect to ϵ_j we have

$$\frac{\partial}{\partial \epsilon_j}\left(\frac{\partial \Psi}{\partial \epsilon_i}\right) = c_{ij}.$$

But since Ψ is a function only of the state of the body, specified by the strain components, the order of differentiation is immaterial, and the left-hand side of this equation is symmetrical with respect to i and j. Hence

$$c_{ij} = c_{ji}. \tag{19}$$

It follows from the form of the relationship between the c_{ij} and the s_{ij} (Ch. IX, § 4.2) that

$$s_{ij} = s_{ji}. \tag{20}$$

The symmetry of the (c_{ij}) and (s_{ij}) matrices implied by relations (19) and (20) further reduces the number of independent stiffness constants and compliances from 36 to 21.

Integrating equation (18) and using (19) we find that the work necessary to produce a strain ϵ_i, called the *strain energy*, is

$$\tfrac{1}{2}c_{ij}\,\epsilon_i\,\epsilon_j \tag{21}$$

per unit volume of the crystal [cf. Ch. III, (17)].

4. The effect of crystal symmetry

The presence of symmetry in the crystal reduces still further the number of independent s_{ij} and c_{ij}. It should first be noticed that elasticity is a centrosymmetrical property. By this it is meant that, if the reference axes are transformed by the operation of a centre of symmetry, the components of s_{ijkl} and c_{ijkl} remain unaltered. The proof is simple. The elements a_{ij} of the transformation matrix are equal to $-\delta_{ij}$. We have, therefore, from (11),

$$s'_{ijkl} = \delta_{im}\delta_{jn}\delta_{ko}\delta_{lp}\,s_{mnop} = s_{ijkl},$$

by the substitution property of δ_{ij} (p. 35); and similarly for c_{ijkl}. However, symmetry elements other than a centre do, in general, impose conditions on the constants, which we must now consider. The methods used for finding the conditions are exactly the same as those used for the piezoelectric moduli.

(i) *Pure symmetry arguments.* The conditions on the s_{ij} and c_{ij} can often be deduced purely from symmetry arguments without analysis, just as with the piezoelectric moduli. Consider the compliance s_{34} in

the orthorhombic class *222* for instance. It measures the extension in
the Ox_3 direction when the crystal is sheared about the Ox_1 direction,
as in Fig. 8.1 *a*. Now operate on the whole system, crystal plus shearing
forces, with a diad axis parallel to Ox_2. The crystal remains unaltered
since its symmetry includes this diad axis; so does the extension parallel
to Ox_3. The forces on the faces, however, are changed to those shown
in Fig. 8.1 *b*. We therefore have the same crystal, still extended in the
Ox_3 direction, but now under the reverse forces. This situation is only

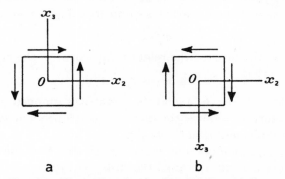

FIG. 8.1. Illustrating that in class *222* the compliance
s_{34} vanishes.

possible if the extension is zero. Hence $s_{34} = 0$. Similar symmetry
arguments may be framed for most of the compliances in the various
classes.

EXERCISE 8.1. Deduce purely from symmetry arguments the form of the
compliance matrix for class *4*.

(ii) *Direct inspection method*. The direct inspection method, described
on pp. 118–20, gives the quickest way of finding the independent coefficients in all classes except those of the trigonal and hexagonal systems.
One example will suffice to illustrate it; we choose class $\bar{4}$.

With the $\bar{4}$ axis parallel to x_3, the axes transform as follows:

$$1 \to 2, \qquad 2 \to -1, \qquad 3 \to -3.$$

Hence in the *four-suffix notation* the pairs of suffixes transform as
follows:

$$11 \to 22, \quad 22 \to 11, \quad 33 \to 33, \quad 23 \to 13, \quad 31 \to -32, \quad 12 \to -21.$$

In the *two-suffix notation* these transformations are:

$$1 \to 2, \quad 2 \to 1, \quad 3 \to 3, \quad 4 \to 5, \quad 5 \to -4, \quad 6 \to -6.$$

The array of suffixes in the matrix written out in the usual order (15) thus transforms to:

$$
\begin{array}{cccccc}
22 & 21 & 23 & 25 & -24 & -26 \\
   & 11 & 13 & 15 & -14 & -16 \\
   &   & 33 & 35 & -34 & -36 \\
   &   &   & 55 & -54 & -56 \\
   &   &   &   & 44 & 46 \\
   &   &   &   &   & 66
\end{array}
$$

We have omitted the lower left-hand half of the array, since the matrix is symmetrical. Equating this array, component by component, with the original one, the relations between the components are at once seen to be

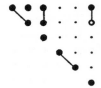

The notation used here has already been described (see the key to Table 9): a light dot denotes a component that is zero; a heavy dot denotes a non-zero component; a line joining two heavy dots means that the two components are numerically equal; a heavy dot and an open circle linked together denote components that are numerically equal, but opposite in sign.

(iii) *Results for all the crystal classes.* For the trigonal and hexagonal systems an analytical method has to be used, just as for the piezoelectric moduli. Direct inspection may be used in all other cases, and the number of independent components may be checked by group theory. The results, for both the s and the c matrices, are given in Table 9. A full key to the notation appears at the head of the table. To illustrate the meaning of the notation we may refer to classes 3 and $\bar{3}$, where

$$s_{15} = -s_{25}, \; s_{46} = 2s_{25}, \text{ and } s_{66} = 2(s_{11}-s_{12});$$

$$c_{15} = -c_{25}, \; c_{46} = c_{25}, \text{ and } c_{66} = \tfrac{1}{2}(c_{11}-c_{12}).$$

The number of independent components is given in brackets after each matrix. The orientation of the axes conforms to the conventions in *Standards on piezoelectric crystals* (1949) (Appendix B, p. 282) except where otherwise stated.

TABLE 9

Form of the (s_{ij}) and (c_{ij}) matrices

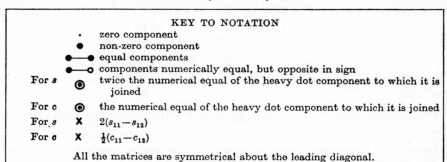

KEY TO NOTATION

·	zero component
●	non-zero component
●——●	equal components
●——○	components numerically equal, but opposite in sign
For s ◉	twice the numerical equal of the heavy dot component to which it is joined
For c ◉	the numerical equal of the heavy dot component to which it is joined
For s ✗	$2(s_{11}-s_{12})$
For c ✗	$\frac{1}{2}(c_{11}-c_{12})$

All the matrices are symmetrical about the leading diagonal.

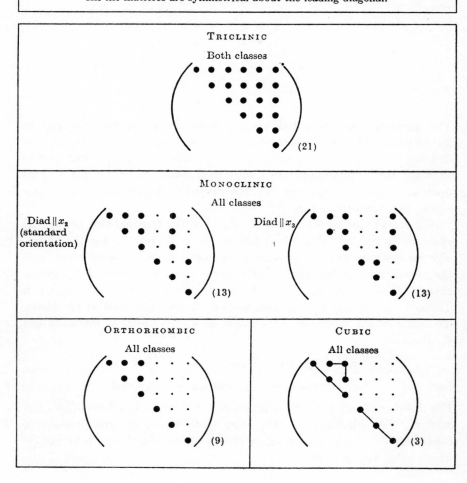

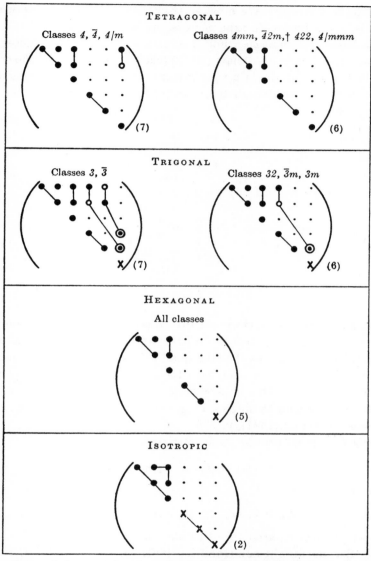

The form of the matrix given in the table for a completely isotropic material is obtained from the cubic matrix by requiring that the components should be unaltered by rotations of 45° about the reference axes. It may be verified that the form so given is unaltered by any rotation of axes. For second-rank tensor properties cubic crystals were

† The same matrix holds for both possible orientations of class $\bar{4}2m$ ($2 \parallel x_1$ and $m \perp x_1$) since the addition of a centre of symmetry makes the two orientations indistinguishable.

found to be isotropic. We now see that the elastic properties of cubic crystals, given by fourth-rank tensors, are not isotropic.

(iv) *Numerical example.* For a numerical example we choose ammonium dihydrogen phosphate (ADP) (class $\bar{4}2m$). This was the crystal for which piezoelectric data were given on p. 120. The components of the (s_{ij}) and (c_{ij}) matrices in m.k.s. units† for ADP at $0°$ C are measured (Mason 1946) as:

$$(s_{ij}) = \begin{pmatrix} 1\cdot8 & 0\cdot7 & -1\cdot1 & 0 & 0 & 0 \\ & 1\cdot8 & -1\cdot1 & 0 & 0 & 0 \\ & & 4\cdot3 & 0 & 0 & 0 \\ & & & 11\cdot3 & 0 & 0 \\ & & & & 11\cdot3 & 0 \\ & & & & & 16\cdot2 \end{pmatrix} \times 10^{-11},$$

$$(c_{ij}) = \begin{pmatrix} 0\cdot71 & -0\cdot20 & 0\cdot13 & 0 & 0 & 0 \\ & 0\cdot71 & 0\cdot13 & 0 & 0 & 0 \\ & & 0\cdot30 & 0 & 0 & 0 \\ & & & 0\cdot088 & 0 & 0 \\ & & & & 0\cdot088 & 0 \\ & & & & & 0\cdot070 \end{pmatrix} \times 10^{11}.$$

4.1. Further restrictions on the constants. The strain energy of a crystal given by (21) must be positive, for otherwise the crystal would be unstable. This means that the quadratic form (21) must be positive definite, that is, greater than zero for all real values of the ϵ_{ij} unless all the ϵ_{ij} are zero. This implies further restrictions on the s_{ij} and c_{ij}, which may be found by standard algebraical methods; see, for example, the book by Ferrar (1941), p. 138.

For a hexagonal crystal these restrictions on the c_{ij} are:

$$c_{44} > 0, \quad c_{11} > |c_{12}|, \quad (c_{11}+c_{12})c_{33} > 2c_{13}^2.$$

For a cubic crystal:

$$c_{44} > 0, \quad c_{11} > |c_{12}|, \quad c_{11}+2c_{12} > 0.$$

The compliances s_{ij} are subject to identical restrictions.

4.2. Stress-strain relations for isotropic materials. Using the (s_{ij}) matrix given in Table 9 for an isotropic material, we may express the s_{ij} in terms of more familiar quantities, such as Young's Modulus and the Rigidity Modulus. First we write out the equations for the strain components in terms of the stress components, and, by their

---

† The m.k.s. units of c_{ij} and s_{ij} are newtons/m² and m²/newton respectively. 1 newton $= 10^5$ dynes; 1 newton/m² $= 10$ dynes/cm².

side, for comparison, we write the same equations in a form frequently used in elasticity textbooks:

$$\left.\begin{aligned}
\epsilon_1 &= s_{11}\sigma_1 + s_{12}\sigma_2 + s_{12}\sigma_3 \\[6pt]
\epsilon_2 &= s_{12}\sigma_1 + s_{11}\sigma_2 + s_{12}\sigma_3 \\[6pt]
\epsilon_3 &= s_{12}\sigma_1 + s_{12}\sigma_2 + s_{11}\sigma_3 \\[6pt]
\epsilon_4 &= 2(s_{11}-s_{12})\sigma_4 \\[6pt]
\epsilon_5 &= 2(s_{11}-s_{12})\sigma_5 \\[6pt]
\epsilon_6 &= 2(s_{11}-s_{12})\sigma_6
\end{aligned}\right\} ; \quad
\left.\begin{aligned}
\epsilon_1 &= \frac{1}{E}\{\sigma_1 - \nu(\sigma_2+\sigma_3)\} \\[6pt]
\epsilon_2 &= \frac{1}{E}\{\sigma_2 - \nu(\sigma_3+\sigma_1)\} \\[6pt]
\epsilon_3 &= \frac{1}{E}\{\sigma_3 - \nu(\sigma_1+\sigma_2)\} \\[6pt]
\epsilon_4 &= \frac{1}{G}\sigma_4 \\[6pt]
\epsilon_5 &= \frac{1}{G}\sigma_5 \\[6pt]
\epsilon_6 &= \frac{1}{G}\sigma_6
\end{aligned}\right\}.$$

E is Young's Modulus, G is the Rigidity Modulus and ν is Poisson's Ratio. Comparing coefficients we have

$$s_{11} = 1/E, \quad s_{12} = -\nu/E \quad \text{and} \quad 2(s_{11}-s_{12}) = 1/G, \tag{22}$$

from which follows the relation

$$G = E/\{2(1+\nu)\}.$$

The equations for the stresses in terms of the strains, using the stiffness constants, may be compared with the same equations as they are usually written in books on elasticity in the λ, μ notation. Thus:

$$\left.\begin{aligned}
\sigma_1 &= c_{11}\epsilon_1 + c_{12}\epsilon_2 + c_{12}\epsilon_3 \\
\sigma_2 &= c_{12}\epsilon_1 + c_{11}\epsilon_2 + c_{12}\epsilon_3 \\
\sigma_3 &= c_{12}\epsilon_1 + c_{12}\epsilon_2 + c_{11}\epsilon_3 \\
\sigma_4 &= \tfrac{1}{2}(c_{11}-c_{12})\epsilon_4 \\
\sigma_5 &= \tfrac{1}{2}(c_{11}-c_{12})\epsilon_5 \\
\sigma_6 &= \tfrac{1}{2}(c_{11}-c_{12})\epsilon_6
\end{aligned}\right\} ; \quad
\left.\begin{aligned}
\sigma_1 &= (2\mu+\lambda)\epsilon_1 + \qquad \lambda\epsilon_2 + \qquad \lambda\epsilon_3 \\
\sigma_2 &= \qquad \lambda\epsilon_1 + (2\mu+\lambda)\epsilon_2 + \qquad \lambda\epsilon_3 \\
\sigma_3 &= \qquad \lambda\epsilon_1 + \qquad \lambda\epsilon_2 + (2\mu+\lambda)\epsilon_3 \\
\sigma_4 &= \mu\epsilon_4 \\
\sigma_5 &= \mu\epsilon_5 \\
\sigma_6 &= \mu\epsilon_6
\end{aligned}\right\} ;$$

whence $\qquad\qquad c_{11} = 2\mu+\lambda \quad \text{and} \quad c_{12} = \lambda.$

5. Representation surfaces and Young's Modulus

No single surface can represent the elastic behaviour of a crystal completely. A surface that is useful in practice is one that shows the variation of Young's Modulus with direction. A bar of the crystal is supposed to be cut with its length parallel to some arbitrary direction, Ox_1', and loaded in simple tension. As we have seen, the tension produces, in general, not only longitudinal and lateral strains but shear strains as well. The Young's Modulus for the direction of the tension is defined as the ratio of the longitudinal stress to the longitudinal strain, that is, $1/s'_{11}$. Surfaces for which the radius vector in the direction Ox'_1 is proportional either to s'_{11} or to $1/s'_{11}$ are commonly used.

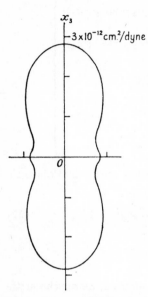

As an example we may take a crystal of zinc (hexagonal). Writing

$$s'_{1111} = a_{1m}a_{1n}a_{1p}a_{1q}s_{mnpq},$$

and using the matrix of compliances given in Table 9, we find, after some reduction and changing to the contracted notation,

$$s'_{11} = s_{11}(1-a_{13}^2)^2 + s_{33}a_{13}^4 + (s_{44}+2s_{13})(1-a_{13}^2)a_{13}^2,$$

or

$$s'_{11} = s_{11}\sin^4\theta + s_{33}\cos^4\theta + (s_{44}+2s_{13})\sin^2\theta\cos^2\theta, \tag{23}$$

where θ is the angle between the arbitrary direction Ox'_1 and the crystallographic z-axis, Ox_3. The s'_{11} or $1/s'_{11}$ surface is then one of revolution about Ox_3. The compliances are (Goens 1933):

$$s_{11} = 8\cdot4, \qquad s_{12} = 1\cdot1, \qquad s_{13} = -7\cdot8,$$

FIG. 8.2. A central section of a representation surface for Young's Modulus in zinc. The length of the radius vector is proportional to s'_{11}, that is, to the reciprocal of Young's Modulus. (After Goens 1933.)

$$s_{33} = 28\cdot7, \qquad s_{44} = 26\cdot4\times10^{-12} \text{ m}^2/\text{newton;†}$$

a section of the s'_{11} surface is shown in Fig. 8.2.

We collect here the expressions for the reciprocal of Young's Modulus in the direction of the unit vector l_i in the various crystal systems.

Triclinic system

$$\begin{aligned}
l_1^4 s_{11} &+ 2l_1^2 l_2^2 s_{12} + 2l_1^2 l_3^2 s_{13} + (2l_1^2 l_2 l_3 s_{14}) + \quad 2l_1^3 l_3 s_{15} + \quad (2l_1^3 l_2 s_{16}) + \\
&+ \quad l_2^4 s_{22} + 2l_2^2 l_3^2 s_{23} + \quad (2l_2^3 l_3 s_{24}) + 2l_1 l_2^2 l_3 s_{25} + \quad (2l_1 l_2^3 s_{26}) + \\
&+ \quad\quad l_3^4 s_{33} + \quad (2l_2 l_3^3 s_{34}) + \quad 2l_1 l_3^3 s_{35} + (2l_1 l_2 l_3^2 s_{36}) + \\
&+ \quad\quad\quad l_2^2 l_3^2 s_{44} + (2l_1 l_2 l_3^2 s_{45}) + 2l_1 l_2^2 l_3 s_{46} + \\
&+ \quad\quad\quad\quad l_1^2 l_3^2 s_{55} + (2l_1^2 l_2 l_3 s_{56}) + \\
&+ \quad\quad\quad\quad\quad l_1^2 l_2^2 s_{66}.
\end{aligned}$$

Monoclinic system (standard orientation, p. 284).

As above but omitting terms in brackets

† $10 \text{ m}^2/\text{newton} = 1 \text{ cm}^2/\text{dyne}.$

Orthorhombic system

$$l_1^4 s_{11} + 2l_1^2 l_2^2 s_{12} + 2l_1^2 l_3^2 s_{13} +$$
$$+ \; l_2^4 s_{22} + 2l_2^2 l_3^2 s_{23} +$$
$$+ \; l_3^4 s_{33} + l_2^2 l_3^2 s_{44} + l_1^2 l_3^2 s_{55} + l_1^2 l_2^2 s_{66}.$$

Tetragonal system. *Classes* 4, $\bar{4}$, 4/m

$$(l_1^4 + l_2^4)s_{11} + l_3^4 s_{33} + l_1^2 l_2^2 (2s_{12} + s_{66}) + l_3^2 (1 - l_3^2)(2s_{13} + s_{44}) + [2l_1 l_2 (l_1^2 - l_2^2)s_{16}].$$

Classes 4mm, $\bar{4}$2m, 422, 4/mmm

As above but omitting the term in square brackets.

Cubic system

$$s_{11} - 2(s_{11} - s_{12} - \tfrac{1}{2}s_{44})(l_1^2 l_2^2 + l_2^2 l_3^2 + l_3^2 l_1^2).$$

Trigonal system. *Classes* 3, $\bar{3}$

$$(1 - l_3^2)^2 s_{11} + l_3^4 s_{33} + l_3^2 (1 - l_3^2)(2s_{13} + s_{44}) + 2l_2 l_3 (3l_1^2 - l_2^2)s_{14} + [2l_1 l_3 (3l_2^2 - l_1^2)s_{25}].$$

Classes 3m, 32, $\bar{3}$m

As above but omitting the term in square brackets.

Hexagonal system

$$(1 - l_3^2)^2 s_{11} + l_3^4 s_{33} + l_3^2 (1 - l_3^2)(2s_{13} + s_{44}).$$

Notice that in the cubic system Young's Modulus is not isotropic. The variation with direction depends on $(l_1^2 l_2^2 + l_2^2 l_3^2 + l_3^2 l_1^2)$. This quantity is zero for the directions of the cube axes $\langle 100 \rangle$ and has its maximum value of $\tfrac{1}{3}$ in the $\langle 111 \rangle$ directions. Hence, if $(s_{11} - s_{12} - \tfrac{1}{2}s_{44})$ is positive (as it is for all cubic metals except molybdenum), Young's Modulus is a maximum in the $\langle 111 \rangle$ directions and a minimum in the $\langle 100 \rangle$ directions. A surface for which the radius vector is directly proportional to Young's Modulus would then have the form of a cube with rounded corners and depressions at the centres of the faces. Its central sections normal to $\langle 111 \rangle$ are readily seen to be circles. $(s_{11} - s_{12} - \tfrac{1}{2}s_{44}) = 0$ is the condition for elastic isotropy. If $(s_{11} - s_{12} - \tfrac{1}{2}s_{44})$ is negative, Young's Modulus is a minimum for $\langle 111 \rangle$ and a maximum for $\langle 100 \rangle$. The Young's Modulus surface then has protuberances along the cube axes.†

6. Volume and linear compressibility of a crystal

(i) *Volume compressibility.* We calculate the proportional decrease in volume of a crystal when subjected to unit hydrostatic pressure, that is, its *volume compressibility.* In equation (1) put $\sigma_{kl} = -p\delta_{kl}$. Then

$$\epsilon_{ij} = -ps_{ijkl}\delta_{kl} = -ps_{ijkk}. \tag{24}$$

† Representation surfaces both for Young's Modulus and for the twisting modulus of a cylinder are illustrated in the books by Wooster (1938) and by Schmid and Boas (1950).

For the dilation Δ (p. 100) we have

$$\Delta = \epsilon_{ii} = -ps_{iikk};$$

and so the volume compressibility, $-\Delta/p$, is s_{iikk}. This is another example of an invariant formed from a tensor. In the matrix notation the volume compressibility is

$$s_{11}+s_{22}+s_{33}+2(s_{12}+s_{23}+s_{31}), \tag{25}$$

and is thus the sum of the nine coefficients in the upper left-hand corner of the compliance matrix. For a cubic crystal it is evidently $3(s_{11}+2s_{12})$. This last expression also holds for an isotropic material; but for this case it is customary to define the reciprocal of the volume compressibility as the *Bulk Modulus*

$$K = 1/\{3(s_{11}+2s_{12})\} = E/\{3(1-2\nu)\},$$

from equations (22).

(ii) *Linear compressibility.* The linear compressibility of a crystal is the relative decrease in length of a line when the crystal is subjected to unit hydrostatic pressure. In general it varies with direction. Under pressure p the stretch of a line in the direction of the unit vector l_i is (p. 101)

$$\epsilon_{ij} l_i l_j = -ps_{ijkk} l_i l_j$$

from (24), and so the linear compressibility is

$$\beta = s_{ijkk} l_i l_j. \tag{26}$$

Written out in the matrix notation for the seven crystal systems the expressions for β are:

Triclinic system

$$\begin{aligned}\beta = (s_{11}+s_{12}+s_{13})l_1^2 + (s_{16}+s_{26}+s_{36})l_1 l_2 + (s_{15}+s_{25}+s_{35})l_3 l_1 + \\ + (s_{12}+s_{22}+s_{23})l_2^2 + (s_{14}+s_{24}+s_{34})l_2 l_3 + \\ + (s_{13}+s_{23}+s_{33})l_3^2.\end{aligned}$$

Monoclinic system (standard orientation, p. 284)

$$\begin{aligned}\beta = (s_{11}+s_{12}+s_{13})l_1^2 + (s_{12}+s_{22}+s_{23})l_2^2 + (s_{13}+s_{23}+s_{33})l_3^2 + \\ + (s_{15}+s_{25}+s_{35})l_3 l_1.\end{aligned}$$

Orthorhombic system

$$\beta = (s_{11}+s_{12}+s_{13})l_1^2 + (s_{12}+s_{22}+s_{23})l_2^2 + (s_{13}+s_{23}+s_{33})l_3^2.$$

Tetragonal, trigonal and hexagonal systems. (*All classes*)

$$\beta = (s_{11}+s_{12}+s_{13}) - (s_{11}+s_{12}-s_{13}-s_{33})l_3^2.$$

Cubic system

$$\beta = s_{11}+2s_{12}.$$

Thus, the linear compressibility in the optically uniaxial systems is

rotationally symmetrical about the unique axis. In the cubic system
the linear compressibility is isotropic: a sphere of a cubic crystal under
hydrostatic pressure remains a sphere.

EXERCISE 8.2. Prove that the volume change of a cubic crystal under uniaxial tension T is independent of the direction of the tension and is given by
$(s_{11}+2s_{12})T$.

7. Relations between the compliances and the stiffnesses

Explicit general equations for the s_{ij} in terms of the c_{ij} and vice versa
are derived in the next chapter. We give here a number of useful
relations between the s_{ij} and the c_{ij} in some of the more symmetrical
classes (Boas and Mackenzie 1950):

Trigonal system. *Classes 3m, 32, $\bar{3}m$*

$$c_{11}+c_{12} = s_{33}/s, \qquad c_{11}-c_{12} = s_{44}/s', \qquad c_{13} = -s_{13}/s,$$
$$c_{14} = -s_{14}/s', \qquad c_{33} = (s_{11}+s_{12})/s, \qquad c_{44} = (s_{11}-s_{12})/s',$$

where
$$s = s_{33}(s_{11}+s_{12})-2s_{13}^2,$$
and
$$s' = s_{44}(s_{11}-s_{12})-2s_{14}^2.$$

Tetragonal system. *Classes 4mm, $\bar{4}2m$, 422, 4/mmm*

$$c_{11}+c_{12} = s_{33}/s, \qquad c_{11}-c_{12} = 1/(s_{11}-s_{12}), \qquad c_{13} = -s_{13}/s,$$
$$c_{33} = (s_{11}+s_{12})/s, \qquad c_{44} = 1/s_{44}, \qquad c_{66} = 1/s_{66},$$

where
$$s = s_{33}(s_{11}+s_{12})-2s_{13}^2.$$

Hexagonal system. *(All classes)*

$$c_{11}+c_{12} = s_{33}/s, \qquad c_{11}-c_{12} = 1/(s_{11}-s_{12}), \qquad c_{13} = -s_{13}/s,$$
$$c_{33} = (s_{11}+s_{12})/s, \qquad c_{44} = 1/s_{44},$$

where
$$s = s_{33}(s_{11}+s_{12})-2s_{13}^2.$$

Cubic system. *(All classes)*

$$c_{11} = \frac{s_{11}+s_{12}}{(s_{11}-s_{12})(s_{11}+2s_{12})},$$
$$c_{12} = \frac{-s_{12}}{(s_{11}-s_{12})(s_{11}+2s_{12})},$$
$$c_{44} = 1/s_{44}.$$

8. Numerical values of the elastic coefficients

Some further numerical data on the elastic compliances of crystals
are given in Table 10. Many cubic crystals are markedly anisotropic
in their elastic behaviour, but tungsten and aluminium are only slightly
anisotropic. With hexagonal crystals the anisotropy is sometimes great:
for example, the linear compressibility of zinc is $1\cdot31 \times 10^{-11}$ m²/newton
parallel to the z-axis and $0\cdot175 \times 10^{-11}$ in all directions perpendicular
to it (as may be calculated from the expression for β on p. 146). For

cadmium the linear compressibilities are $1 \cdot 69$ and $0 \cdot 15 \times 10^{-13}$. For tellurium, which has a chain structure, the linear compressibility parallel to the chain axis is negative.

TABLE 10

Elasticity of Crystals

Compliances at room temperature (unit $= 10^{-11}$ m²/newton)

Crystal	Class	s_{11}	s_{12}	s_{44}	s_{33}	s_{13}	s_{14}	s_{66}
Sodium chloride	$m3m$	2·21	−0·45	7·83	..	..	..	..
Aluminium .	$m3m$	1·59	−0·58	3·52	..	..	..	..
Copper .	$m3m$	1·49	−0·63	1·33	..	..	..	..
Nickel .	$m3m$	0·799	−0·312	0·844	..	..	..	..
Tungsten . .	$m3m$	0·257	−0·073	0·660	..	..	..	..
Sodium chlorate	23	2·2	−0·6	8·6	..	..	..	..
Tin . . .	$4/mmm$	1·85	−0·99	5·70	1·18	−0·25	..	13·5
ADP	$\bar{4}2m$	1·8	0·7	11·3	4·3	−1·1	..	16·2
Zinc . .	$6/mmm$	0·84	0·11	2·64	2·87	−0·78	..	..
Cadmium . .	$6/mmm$	1·23	−0·15	5·40	3·55	−0·93	..	..
Quartz . .	32	1·27	−0·17	2·01	0·97	−0·15	−0·43	..
Tourmaline .	$3m$	0·40	−0·10	1·51	0·63	−0·016	0·058	..

Values are taken from the following sources: Boas and Mackenzie (1950), Van Dyke and Gordon (1950), and Hearmon (1946); the last of these contains a full collection of elastic data up to the end of 1944, and has been supplemented to cover the period 1945–1955 (Hearmon 1956). Other recent collections are those of Bhagavantam (1955) and Raman and Krishnamurti (1955). See also the important review article by Huntington (1958), which deals with the microscopic, atomic, theory of elastic constants as well as with the macroscopic theory.

SUMMARY

Hooke's Law for a crystal is written

$$\epsilon_{ij} = s_{ijkl}\sigma_{kl}; \qquad \sigma_{ij} = c_{ijkl}\epsilon_{kl}, \qquad\qquad (1); (2)$$

where the s_{ijkl} and c_{ijkl} are the components of fourth-rank tensors. The s_{ijkl} are the *elastic compliances* and the c_{ijkl} are the *elastic stiffnesses*. Since σ_{ij} in these equations is the symmetric part of the stress tensor it is convenient to put

$$s_{ijkl} = s_{ijlk}.$$

The symmetry of the strain tensor implies that

$$s_{ijkl} = s_{jikl}.$$

c_{ijkl} is likewise symmetrical in the first two and the last two suffixes. These relations reduce the number of independent s's and c's to 36.

   Matrix notation. We abbreviate the first two and the last two suffixes of s_{ijkl} and c_{ijkl} into single suffixes running from 1 to 6, as with the piezoelectric moduli. We introduce factors of 2 and 4 in the s_{ijkl} but not in the c_{ijkl}. We use also the single suffix notation already developed for the stress and strain components. Then (1) and (2) become

$$\epsilon_i = s_{ij}\sigma_j; \qquad \sigma_i = c_{ij}\epsilon_j \quad (i,j = 1, 2,..., 6). \qquad\qquad (13); (14)$$

(s_{ij}) and (c_{ij}) are 6-row, 6-column matrices.

Other results. The work done per unit volume when there is a small change of strain in a crystal is

$$dW = \sigma_i \, d\epsilon_i. \tag{16}$$

When the change is isothermal and reversible dW may be equated with the increase in free energy $d\Psi$. The fact that $d\Psi$ is a perfect differential, together with (14), then implies that

$$c_{ij} = c_{ji}, \qquad s_{ij} = s_{ji}; \tag{19), (20}$$

and the strain energy per unit volume is

$$\tfrac{1}{2} c_{ij} \epsilon_i \epsilon_j. \tag{21}$$

(19) and (20) reduce the number of independent compliances and stiffnesses from 36 to 21, and the number is further reduced by crystal symmetry. The condition that (21) should be positive definite imposes still further restrictions.

A crystal bar loaded in simple tension undergoes longitudinal and lateral strains and also shear strains. *Young's Modulus* for the direction of the tension is defined as the ratio of the longitudinal tension to the longitudinal strain. Young's Modulus is anisotropic for all crystal classes, including the cubic classes, and its variation with direction may be represented by a surface.

Let unit hydrostatic pressure be applied to a crystal. Then: (1) the proportional decrease in volume, the *volume compressibility*, is s_{iikk} and is the sum of the nine components in the upper left-hand corner of the (s_{ij}) matrix; (2) the proportional decrease in length of a line in the crystal in the direction l_i, the *linear compressibility*, is $\beta = s_{ijkk} l_i l_j$. The linear compressibility of cubic crystals is isotropic.

IX

THE MATRIX METHOD†

1. The matrix and tensor notations

THE matrix notation introduced in Chapters VII and VIII in the piezo-electric equations

$$P_i = d_{ij}\sigma_j, \qquad \epsilon_j = d_{ij}E_i, \tag{1}$$

and the elasticity equations

$$\epsilon_i = s_{ij}\sigma_j, \qquad \sigma_i = c_{ij}\epsilon_j, \tag{2}$$

is a convenient shorthand that for many purposes provides an alternative to the tensor notation. We may repeat that the advantage of the tensor notation is that it provides simple equations for the transformation, from one set of reference axes to another, of the coefficients describing a property; moreover, the tensor nature of the property, that is, its tensor rank and how it transforms, is indicated precisely by the number of suffixes. If, however, we always bear in mind the different transformation properties of the tensors of different rank—that κ_{ij}, d_{ij} and s_{ij}, although similar in appearance, all transform in different ways—there is no objection to reducing the number of suffixes to two. Having decided on this abbreviation we may shorten the notation still further and at the same time look at the equations from a new and fruitful viewpoint. For this purpose it is necessary to review briefly some of the rules of matrix algebra.

2. Matrix algebra‡

2.1. Linear transformations and matrix multiplication. Most of the equations describing crystal properties that we have met up to this point have formed sets of linear simultaneous equations. Matrix algebra provides a convenient way of shortening the work involved in handling such equations.

Suppose the variables $x_1, x_2,..., x_m$ are linearly related to the independent variables $y_1, y_2,..., y_n$ by the simultaneous equations

$$\left.\begin{aligned}
x_1 &= \alpha_{11}y_1 + \alpha_{12}y_2 + \cdots + \alpha_{1n}y_n \\
x_2 &= \alpha_{21}y_1 + \alpha_{22}y_2 + \cdots + \alpha_{2n}y_n \\
&\cdots\cdots\cdots\cdots\cdots\cdots \\
x_m &= \alpha_{m1}y_1 + \alpha_{m2}y_2 + \cdots + \alpha_{mn}y_n
\end{aligned}\right\},$$

† This chapter may be omitted on a first reading.
‡ Bond (1943), Ferrar (1941), Aitken (1948).

or, in the dummy suffix notation,

$$x_i = \alpha_{ij} y_j \quad (i = 1, ..., m; \, j = 1, ..., n). \tag{3}$$

Now suppose that the y_j are related in turn to a third set of variables $z_1, z_2, ..., z_p$ by the equations

$$y_j = \beta_{jk} z_k \quad (j = 1, ..., n; \, k = 1, ..., p). \tag{4}$$

Then the x_i are related to the z_k by the equations

$$x_i = \alpha_{ij} \beta_{jk} z_k. \tag{5}$$

If we now define a new set of quantities γ_{ik} by

$$\gamma_{ik} = \alpha_{ij} \beta_{jk}, \tag{6}$$

we may express the z_k directly in terms of the x_i by

$$x_i = \gamma_{ik} z_k. \tag{7}$$

Now let us consider the α's and β's written out as 'tables' thus:

$$\begin{pmatrix} \alpha_{11} & \alpha_{12} & \cdot & \cdot & \cdot & \alpha_{1n} \\ \alpha_{21} & \alpha_{22} & \cdot & \cdot & \cdot & \alpha_{2n} \\ \cdot & \cdot & \cdot & \cdot & \cdot & \cdot \\ \alpha_{m1} & \alpha_{m2} & \cdot & \cdot & \cdot & \alpha_{mn} \end{pmatrix} \quad \text{and} \quad \begin{pmatrix} \beta_{11} & \beta_{12} & \cdot & \cdot & \cdot & \beta_{1p} \\ \beta_{21} & \beta_{22} & \cdot & \cdot & \cdot & \beta_{2p} \\ \cdot & \cdot & \cdot & \cdot & \cdot & \cdot \\ \beta_{n1} & \beta_{n2} & \cdot & \cdot & \cdot & \beta_{np} \end{pmatrix}.$$

We may think of these two tables as being the quantities or *matrices* α and β. Then, if the γ's are written out as a matrix in a similar way, the typical element γ_{ik}, that is, the element in the ith row and the kth column of the matrix γ, is

$$\gamma_{ik} = \alpha_{ij} \beta_{jk} = \alpha_{i1} \beta_{1k} + \alpha_{i2} \beta_{2k} + ...$$

It is obtained by multiplying the ith row of the α matrix by the kth column of the β matrix, term by term, and summing. This is the operation known as *matrix multiplication*. We write

$$\gamma = \alpha \beta, \tag{8}$$

signifying that the matrix γ is obtained by matrix multiplication of α with β.†

It will be clear that the multiplication process will only succeed if the number of columns of the first matrix is equal to the number of rows of the second matrix. Thus, in the foregoing example α is an

† An easy way of forming the ikth element of the product of two matrices is to let the index finger of the left hand follow the ith row of the left matrix while the index finger of the right hand follows the kth column of the right matrix. The two fingers move along in step, and at each pause the two elements under the fingers are multiplied and the product is added to the accumulated sum.

$(m \times n)$ matrix, β is an $(n \times p)$ matrix and γ is an $(m \times p)$ matrix. To show this we could write conventionally

$$(m \times p) = (m \times n) \times (n \times p).$$

The n's appear in neighbouring positions on the right-hand side, and one can think of them as cancelling and leaving $(m \times p)$.

If the number of columns of a matrix **A** is equal to the number of rows of a matrix **B**, the matrices are said to be *conformable for the product* **AB**.

EXERCISE 9.1. Form the product **AB** of the two matrices,

$$\mathbf{A} = \begin{pmatrix} a_1 & c_1 \\ a_2 & c_2 \\ a_3 & c_3 \end{pmatrix}, \qquad \mathbf{B} = \begin{pmatrix} b_1 \\ b_2 \end{pmatrix}.$$

EXERCISE 9.2. Evaluate the products,

$$\begin{pmatrix} 0 & 3 & 2 \\ 1 & -1 & 3 \\ 2 & 1 & 4 \end{pmatrix} \times \begin{pmatrix} 0 & 2 \\ 3 & 1 \\ -2 & 8 \end{pmatrix}; \qquad \begin{pmatrix} 1 & 2 & -1 \\ 0 & 0 & 5 \end{pmatrix} \times \begin{pmatrix} 1 & 0 & 2 & 3 \\ 2 & 1 & 0 & 0 \\ 1 & 0 & -2 & 0 \end{pmatrix}.$$

EXERCISE 9.3. If **A** and **B** are matrices, verify that, in general, **AB** $\neq$ **BA**, by evaluating the two products

$$(2 \quad 3) \times \begin{pmatrix} 4 \\ 1 \end{pmatrix} \quad \text{and} \quad \begin{pmatrix} 4 \\ 1 \end{pmatrix} \times (2 \quad 3).$$

Returning now to equation (3), the x_i and the y_j may be considered as matrices. If the x_i are written as the elements of the m-row, single-column, matrix

$$\mathbf{x} = \begin{pmatrix} x_1 \\ x_2 \\ \cdot \\ \cdot \\ \cdot \\ x_m \end{pmatrix},$$

then, in order that α and **y** shall be conformable for the product $\alpha\mathbf{y}$, **y** must be written as the n-row, single-column matrix

$$\mathbf{y} = \begin{pmatrix} y_1 \\ y_2 \\ \cdot \\ \cdot \\ \cdot \\ y_n \end{pmatrix},$$

so that we have

$$\mathbf{x} = \alpha\mathbf{y}. \tag{9}$$

Note that the number of rows and columns form the scheme

$$(m \times 1) = (m \times n) \times (n \times 1).$$

In a similar way equation (4) may be written

$$y = \beta z, \tag{10}$$

where z is a $(p \times 1)$ matrix, and equation (5) becomes

$$x = \alpha\beta z. \tag{11}$$

2.2. Matrix addition and subtraction. If $x = Ay$ and $w = By$, where x, A, y, B and w are matrices, we see, by writing the equations as

$$x_i = A_{ij}y_j, \qquad w_i = B_{ij}y_j,$$

that

$$\overline{x_i + w_i} = (A_{ij} + B_{ij})y_j.$$

Thus it is logical to define a matrix C whose elements C_{ij} are given by

$$C_{ij} = A_{ij} + B_{ij},$$

and to call it the sum $(A+B)$. Then our operations may be written

$$x + w = Ay + By = (A+B)y = Cy.$$

Two matrices are *conformable for addition* when each has the same number of rows and each has the same number of columns.

The operation of subtraction of one matrix from another is defined in a similar way.

2.3. Summary of matrix properties. *In general, the symbols α, β, A, B etc., representing matrices, may be added, subtracted, and multiplied, but not divided, as though they represented ordinary numbers.*[†] *In multiplication, however, it must be remembered that usually $AB \neq BA$.*

3. Crystal properties in matrix notation

The equations in Chapters III to VIII describing crystal properties may all be concisely expressed in the matrix notation. Representing the electric displacement and the electric field intensity by single column matrices,

$$D = \begin{pmatrix} D_1 \\ D_2 \\ D_3 \end{pmatrix} \quad \text{and} \quad E = \begin{pmatrix} E_1 \\ E_2 \\ E_3 \end{pmatrix},$$

we may write

$$D = \kappa E, \tag{12}$$

where κ is the square matrix (3×3)

$$\begin{pmatrix} \kappa_{11} & \kappa_{12} & \kappa_{31} \\ \kappa_{12} & \kappa_{22} & \kappa_{23} \\ \kappa_{31} & \kappa_{23} & \kappa_{33} \end{pmatrix}.$$

[†] For a full justification of this statement see the book by Ferrar (1941). A further point of manipulation where matrices A, B differ from ordinary numbers a, b is that, whereas the equation $ab = 0$ implies that either a or b (or both) is zero, the equation $AB = 0$ (0 standing for the *null matrix* all of whose elements are zero) does not necessarily imply that either A or B is zero.

Similarly, for the magnetic permeability,

$$\mathbf{B} = \mu\mathbf{H}. \tag{13}$$

The pyroelectric effect given by

$$\Delta P_i = p_i\Delta T \qquad \text{[Ch. IV, (21)]}$$

may be written $$\Delta\mathbf{P} = \mathbf{p}\Delta T, \tag{14}$$

where $\Delta\mathbf{P}$ and $\mathbf{p}$ are single-column matrices. Notice here that the multiplication of a matrix by a single number (ΔT) multiplies all its elements by the number.

Similarly, thermal expansion given by

$$\epsilon_{ij} = \alpha_{ij}\Delta T \qquad \text{[Ch. VI, (13)]}$$

may be written as $$\epsilon_i = \alpha_i\Delta T \quad (i = 1,...,6),$$

or $$\boldsymbol{\epsilon} = \boldsymbol{\alpha}\Delta T, \tag{15}$$

where $\boldsymbol{\epsilon}$ is a (6×1) matrix with components as defined in Chapter VII, § 3, and $\boldsymbol{\alpha}$ is the (6×1) matrix

$$\begin{pmatrix} \alpha_1 \\ \alpha_2 \\ \alpha_3 \\ \alpha_4 \\ \alpha_5 \\ \alpha_6 \end{pmatrix} = \begin{pmatrix} \alpha_{11} \\ \alpha_{22} \\ \alpha_{33} \\ 2\alpha_{23} \\ 2\alpha_{31} \\ 2\alpha_{12} \end{pmatrix}. \tag{16}$$

In calculations on thermal expansion it is sometimes more convenient to use the matrices

$$\begin{pmatrix} \alpha_{11} & \alpha_{12} & \alpha_{31} \\ \alpha_{12} & \alpha_{22} & \alpha_{23} \\ \alpha_{31} & \alpha_{23} & \alpha_{33} \end{pmatrix} \quad \text{or} \quad \begin{pmatrix} \alpha_{11} \\ \alpha_{22} \\ \alpha_{33} \\ \alpha_{23} \\ \alpha_{31} \\ \alpha_{12} \end{pmatrix}.$$

The direct piezoelectric effect, formerly written as

$$P_i = d_{ij}\sigma_j \quad (i = 1, 2, 3; j = 1,...,6), \qquad \text{[Ch. VII, (17)]}$$

becomes $$\mathbf{P} = \mathbf{d}\boldsymbol{\sigma}, \tag{17}$$

where $\mathbf{P} = \begin{pmatrix} P_1 \\ P_2 \\ P_3 \end{pmatrix}$, $\mathbf{d} = \begin{pmatrix} d_{11} & d_{12} & d_{13} & d_{14} & d_{15} & d_{16} \\ d_{21} & d_{22} & d_{23} & d_{24} & d_{25} & d_{26} \\ d_{31} & d_{32} & d_{33} & d_{34} & d_{35} & d_{36} \end{pmatrix}$, and $\boldsymbol{\sigma} = \begin{pmatrix} \sigma_1 \\ \sigma_2 \\ \sigma_3 \\ \sigma_4 \\ \sigma_5 \\ \sigma_6 \end{pmatrix}$.

The elasticity equations are written

$$\boldsymbol{\epsilon} = \mathbf{s}\boldsymbol{\sigma}, \qquad \boldsymbol{\sigma} = \mathbf{c}\boldsymbol{\epsilon}, \qquad (18)$$

where $\boldsymbol{\epsilon}$ and $\boldsymbol{\sigma}$ are both (6×1) matrices and $\mathbf{s}$ and $\mathbf{c}$ are the (6×6) matrices of Chapter VIII, § 2.

4. Two derived matrices

4.1. The transpose of a matrix. The second of equations (1) written out in full is

$$\left.\begin{aligned}
\epsilon_1 &= d_{11} E_1 + d_{21} E_2 + d_{31} E_3 \\
\epsilon_2 &= d_{12} E_1 + d_{22} E_2 + d_{32} E_3 \\
\epsilon_3 &= d_{13} E_1 + d_{23} E_2 + d_{33} E_3 \\
\epsilon_4 &= d_{14} E_1 + d_{24} E_2 + d_{34} E_3 \\
\epsilon_5 &= d_{15} E_1 + d_{25} E_2 + d_{35} E_3 \\
\epsilon_6 &= d_{16} E_1 + d_{26} E_2 + d_{36} E_3
\end{aligned}\right\}.$$

To represent this in the short matrix notation without suffixes we first write

$$\begin{pmatrix} \epsilon_1 \\ \epsilon_2 \\ \epsilon_3 \\ \epsilon_4 \\ \epsilon_5 \\ \epsilon_6 \end{pmatrix} = \begin{pmatrix} d_{11} & d_{21} & d_{31} \\ d_{12} & d_{22} & d_{32} \\ d_{13} & d_{23} & d_{33} \\ d_{14} & d_{24} & d_{34} \\ d_{15} & d_{25} & d_{35} \\ d_{16} & d_{26} & d_{36} \end{pmatrix} \times \begin{pmatrix} E_1 \\ E_2 \\ E_3 \end{pmatrix},$$

and then contract the equation to

$$\boldsymbol{\epsilon} = \mathbf{d_t} \mathbf{E}. \qquad (19)$$

It will be seen that $\mathbf{d_t}$ stands for the same matrix as $\mathbf{d}$ but with the rows and columns interchanged. $\mathbf{d_t}$ is called the *transpose* of $\mathbf{d}$.

This definition is general: the transpose $\mathbf{A_t}$ of a matrix $\mathbf{A}$ is the matrix whose rth column is the rth row of $\mathbf{A}$.

4.2. The reciprocal matrix. Suppose that $\boldsymbol{\alpha}$ is a general square matrix $(n \times n)$ and $\mathbf{x}$ and $\mathbf{y}$ are single-column matrices $(n \times 1)$, and that

$$\mathbf{x} = \boldsymbol{\alpha}\mathbf{y}, \qquad (20)$$

or, in suffix notation,

$$x_i = \alpha_{ij} y_j \quad (i, j = 1, ..., n).$$

If the determinant of the α_{ij}, which we denote by $\Delta = |\alpha_{ij}|$, is not zero, in which case $\boldsymbol{\alpha}$ is said to be a non-singular matrix, these equations

may be solved for the y_j. We multiply the ith equation by the cofactor†
A_{ik} of the element α_{ik} and sum for i, thus:

$$A_{ik}x_i = A_{ik}\alpha_{ij}y_j \quad (k = 1,...,n)$$
$$= \Delta\delta_{kj}y_j, \text{ by a theorem on the expansion of}$$
$$\text{determinants,‡}$$
$$= \Delta y_k.$$

Therefore, changing suffixes,

$$y_i = \frac{A_{ji}}{\Delta}x_j.$$

This may be written

$$y_i = (\alpha^{-1})_{ij}x_j, \quad \text{or} \quad \mathbf{y} = \boldsymbol{\alpha}^{-1}\mathbf{x},$$

which is the natural complement to (20), if we *define* $\boldsymbol{\alpha}^{-1}$ to be the
matrix whose i, jth element $(\alpha^{-1})_{ij}$ is A_{ji}/Δ.

The i, kth element of the product $\boldsymbol{\alpha\alpha}^{-1}$ is

$$\alpha_{ij}(\alpha^{-1})_{jk} = \frac{\alpha_{ij}A_{kj}}{\Delta} = \frac{\delta_{ik}\Delta}{\Delta} = \delta_{ik}.$$

The matrix whose elements are δ_{ik} is known as a *unit matrix* or
idemfactor and is denoted by $\mathbf{I}$. Thus

$$\boldsymbol{\alpha\alpha}^{-1} = \mathbf{I}.$$

Similarly, it may be shown that $\boldsymbol{\alpha}^{-1}\boldsymbol{\alpha} = \mathbf{I}$. By virtue of these properties
$\boldsymbol{\alpha}^{-1}$ is called the *reciprocal* of $\boldsymbol{\alpha}$. It is important to notice that only
matrices which are square and non-singular have reciprocals.

It is easily seen that multiplication of any matrix by $\mathbf{I}$ leaves it
unchanged. Thus we might consider that equation (20) was solved for
$\mathbf{y}$ by multiplying through by $\boldsymbol{\alpha}^{-1}$:

$$\boldsymbol{\alpha}^{-1}\mathbf{x} = \boldsymbol{\alpha}^{-1}\boldsymbol{\alpha}\mathbf{y} = \mathbf{Iy} = \mathbf{y}.$$

As an example of a pair of reciprocal matrices we have the permit-
tivity matrix $\boldsymbol{\varkappa}$ linking $\mathbf{D}$ with $\mathbf{E}$ by the equation

$$\mathbf{D} = \boldsymbol{\varkappa}\mathbf{E},$$

and the *impermeability matrix* $\boldsymbol{\beta}$ linking $\mathbf{E}$ with $\mathbf{D}$ by

$$\mathbf{E} = \boldsymbol{\beta}\mathbf{D}, \quad \text{where} \quad \boldsymbol{\beta} = \boldsymbol{\varkappa}^{-1}. \tag{20 a}$$

---

† The cofactor of the element α_{ik} is $(-1)^{i+k}$ times the determinant which remains after
striking out the ith row and the kth column of $|\alpha_{ij}|$.

‡ The theorem is as follows. The sum of elements of a row (or column) of a determinant multiplied by their own cofactors is the value of the determinant itself; the sum
of elements of a row multiplied by the corresponding cofactors of another row is zero;
the sum of elements of a column multiplied by the corresponding cofactors of another
column is zero (Ferrar 1941, p. 30). The first of these three statements may be used
as a definition of a determinant.

Similarly, from the elastic relations

$$\epsilon = \mathsf{s}\boldsymbol{\sigma} \quad \text{and} \quad \boldsymbol{\sigma} = \mathsf{c}\boldsymbol{\epsilon},$$

we see that $\mathsf{s} = \mathsf{c}^{-1}$ and $\mathsf{c} = \mathsf{s}^{-1}$. It also follows that

$$\mathsf{sc} = \mathsf{cs} = \mathsf{I},$$

a relation which in suffix notation is

$$s_{ij}c_{jk} = c_{ij}s_{jk} = \delta_{ik},$$

or, in full tensor notation,

$$s_{ijkl}c_{klmn} = c_{ijkl}s_{klmn} = \delta_{im}\delta_{jn}. \tag{21}$$

5. The magnitude of a second-rank tensor property in an arbitrary direction

In Chapter I, § 6.2, we found that, if the tensor S_{ij} represents a certain symmetrical second-rank tensor property, the magnitude S (say) of the property in the arbitrary direction (l_1, l_2, l_3) is given by

$$S = S_{ij}l_i l_j.$$

To put this in matrix form we first rearrange the expression to bring the dummy suffixes together, thus, $l_i S_{ij} l_j$, and then we may write

$$S = \mathsf{l_t S l}, \tag{22}$$

or, in full,

$$S = (l_1 \; l_2 \; l_3) \begin{pmatrix} S_{11} & S_{12} & S_{31} \\ S_{12} & S_{22} & S_{23} \\ S_{31} & S_{23} & S_{33} \end{pmatrix} \begin{pmatrix} l_1 \\ l_2 \\ l_3 \end{pmatrix}.$$

Note that $\mathsf{l_t}$ has to be used in order to make the matrices conformable for multiplication, according to the scheme

$$(1 \times 1) = (1 \times 3) \times (3 \times 3) \times (3 \times 1).$$

6. Rotation of axes

The equations for the rotation of axes developed in Chapter I, § 2, are readily expressed in matrix form. The equations for the transformation of the coordinates of a point,

$$x'_i = a_{ij}x_j, \qquad \text{[Ch. I, (18)]}$$

become simply

$$\mathsf{x}' = \mathsf{ax}, \tag{23}$$

where x' and x are single-column matrices. The equations for the transformation of the components of a vector, namely,

$$p'_i = a_{ij}p_j,$$

become, similarly,

$$\mathsf{p}' = \mathsf{ap}. \tag{24}$$

Two successive rotations, such as $\mathbf{x}' = \mathbf{ax}$ followed by $\mathbf{x}'' = \mathbf{bx}'$, may be handled by combining them into one with the aid of matrix multiplication, thus: $\mathbf{x}'' = (\mathbf{ba})\mathbf{x}$.

The equation for the transformation of the components of a second-rank tensor such as the permittivity is, in suffix notation,

$$\kappa'_{ij} = a_{ik} a_{jl} \kappa_{kl}.$$

Before we can use the rule of matrix multiplication we have to bring the repeated suffixes together. This may be done by writing the a_{jl} term after the κ_{kl} term and transposing, thus:

$$\kappa'_{ij} = a_{ik} \kappa_{kl} (a_t)_{lj};$$

so that we may write $\qquad \mathbf{x}' = \mathbf{axa_t}.$ \hfill (25)

It is also possible to develop matrix equations for the transformation of third- and fourth-rank tensor components (Bond 1943) but the matrices involved are rather lengthy when written out.

Exercise 9.4. Refer to Exercise 2.4 (a), (b), (c) and (d) on p. 47 and the solution on p. 313. In each case rotate the axes of reference as noted in the solution and verify by matrix algebra that the given solution is correct.

7. Examples of matrix calculations

When making numerical calculations the matrix method frequently results in a great shortening of the work; we give here some practical illustrations [due to W. L. Bond (1945, unpublished)].

7.1. Principal coefficients and directions for a monoclinic crystal. Least squares method.
Suppose it is required to find the coefficients for some second-rank tensor property in a monoclinic crystal. For definiteness let us choose thermal expansion. With Ox_2 as the diad axis the coefficients may be written as the matrix

$$\alpha = \begin{pmatrix} \alpha_{11} & 0 & \alpha_{31} \\ 0 & \alpha_{22} & 0 \\ \alpha_{31} & 0 & \alpha_{33} \end{pmatrix}.$$

Notice that we do not at first know the directions of the two principal axes perpendicular to Ox_2, and so Ox_1 and Ox_3 may be chosen arbitrarily. By convention we chose Ox_3 parallel to the crystallographic z-axis, and then the direction of Ox_1 is determined.

The α_{ij} may be formed by measuring the expansions, along their lengths, of bars cut from the crystal in different orientations. The expansion per unit rise of temperature in an arbitrary direction specified by the unit vector $\mathbf{l}$, written as a single-column matrix $\mathbf{l}$, is, by equation (22),

$$\alpha_l = \mathbf{l}_t \alpha \mathbf{l}.$$

If
$$\mathbf{l} = \begin{pmatrix} 0 \\ 1 \\ 0 \end{pmatrix}, \quad \alpha_l = (0 \quad 1 \quad 0)\begin{pmatrix} \alpha_{11} & 0 & \alpha_{31} \\ 0 & \alpha_{22} & 0 \\ \alpha_{31} & 0 & \alpha_{33} \end{pmatrix}\begin{pmatrix} 0 \\ 1 \\ 0 \end{pmatrix} = \alpha_{22},$$

as is also obvious from the meaning of the α_{22} component of the matrix.

The expansion in the Ox_2 direction is a maximum or a minimum, and so it is not sensitive to a small misorientation of the crystal. α_{22} may thus be measured directly with satisfactory results. On the other hand, α_{11} and α_{33} are sensitive to small rotations about Ox_2, and so errors made in orienting specimen bars nominally parallel to Ox_1 and Ox_3 will be important. We therefore have recourse to the following method for the determination of α_{11}, α_{33} and α_{31}, which minimizes the errors.

Let bars be cut with
$$\mathbf{l} = \begin{pmatrix} \sin\theta_1 \\ 0 \\ \cos\theta_1 \end{pmatrix}. \tag{26}$$

Then, denoting the expansion in this direction by A_1, we have

$$A_1 = (\sin\theta_1 \quad 0 \quad \cos\theta_1)\begin{pmatrix} \alpha_{11} & 0 & \alpha_{31} \\ 0 & \alpha_{22} & 0 \\ \alpha_{31} & 0 & \alpha_{33} \end{pmatrix}\begin{pmatrix} \sin\theta_1 \\ 0 \\ \cos\theta_1 \end{pmatrix}$$

$$= (\overline{\sin\theta_1 . \alpha_{11} + \cos\theta_1 . \alpha_{31}} \quad 0 \quad \overline{\sin\theta_1 . \alpha_{31} + \cos\theta_1 . \alpha_{33}})\begin{pmatrix} \sin\theta_1 \\ 0 \\ \cos\theta_1 \end{pmatrix}$$

$$= \sin^2\theta_1 . \alpha_{11} + \sin 2\theta_1 . \alpha_{31} + \cos^2\theta_1 . \alpha_{33} \tag{27}$$

(the same result can, of course, be quickly obtained by working in tensor notation).

If several determinations of expansion, A_1, A_2,..., A_n, are made with different orientations, θ_1, θ_2,..., θ_n, we obtain a set of equations similar to (27) and the whole set may be written

$$\mathbf{A} = \mathbf{\Theta}\alpha, \tag{28}$$

where
$$\mathbf{A} = \begin{pmatrix} A_1 \\ A_2 \\ . \\ A_n \end{pmatrix}, \quad \mathbf{\Theta} = \begin{pmatrix} \sin^2\theta_1 & \sin 2\theta_1 & \cos^2\theta_1 \\ \sin^2\theta_2 & \sin 2\theta_2 & \cos^2\theta_2 \\ . & . & . \\ \sin^2\theta_n & \sin 2\theta_n & \cos^2\theta_n \end{pmatrix}, \quad \alpha = \begin{pmatrix} \alpha_{11} \\ \alpha_{31} \\ \alpha_{33} \end{pmatrix}.$$

Three measurements A_1, A_2, A_3 are sufficient to determine α_{11}, α_{31}, α_{33}.

In this case Θ is a square and, in general, non-singular matrix; thus, we may solve (28) to give
$$\alpha = \Theta^{-1}\mathbf{A}. \tag{29}$$

It is more accurate to measure more than three values of A_1, A_2, etc. and to combine the observations to give the best value of α. Let us therefore consider the following general problem.

General problem. z_1, z_2,..., z_q are q unknown quantities which, when multiplied by the coefficients a_1, a_2,..., a_q, give the quantity M, according to the equation
$$M = a_1 z_1 + a_2 z_2 + ... + a_q z_q.$$

With assumed values of the coefficients a_1, a_2,..., a_q we can set up an experiment and measure M. Owing to experimental errors we shall not measure the true value of M but shall find a value $M+v$, where v is the error. We could set up other experiments, p in all, with different values of the a's and write

$$\left. \begin{array}{l} v_1 = a_{11}z_1 + a_{12}z_2 + \quad . \quad . \quad . + a_{1q}z_q - M_1 \\ v_2 = a_{21}z_1 + a_{22}z_2 + \quad . \quad . \quad . + a_{2q}z_q - M_2 \\ \quad . \quad . \quad . \quad . \quad . \quad . \quad . \quad . \quad . \quad . \quad . \\ v_p = a_{p1}z_1 + a_{p2}z_2 + \quad . \quad . \quad . + a_{pq}z_q - M_p \end{array} \right\}. \tag{30}$$

If now $p > q$, there are a number of ways in which q equations can be chosen from the p equations (30) and for each set we could, by assuming the errors to be zero, solve for the q unknowns z_1, z_2,..., z_q. Thus we should obtain several sets of values of z_1, z_2,..., z_q, but the sets would not agree exactly. The principle of least squares states that the values of the z's should be such that $v_1^2 + v_2^2 + ... + v_p^2$ is a minimum [see, for example, the book by Scarborough (1950)].

Equations (30) may be written
$$v_i = a_{ij}z_j - M_i \quad (i = 1,...,p; \, j = 1,...,q), \tag{31}$$
or
$$\mathbf{v} = \mathbf{az} - \mathbf{M}, \tag{32}$$
where $\mathbf{v}$, $\mathbf{z}$ and $\mathbf{M}$ are single-column matrices, and $\mathbf{a}$ is the matrix of the a_{ij} coefficients. Since $v_i v_i$ is a minimum we have
$$v_i \frac{\partial v_i}{\partial z_j} = 0, \tag{33}$$
or, in view of (31),
$$v_i a_{ij} = 0. \tag{34}$$
Eliminating the v_i between (31) and (34), and changing the dummy suffix from j to k, we obtain
$$(a_{ik}z_k - M_i)a_{ij} = 0,$$
or
$$(a_t)_{ji}(a_{ik}z_k - M_i) = 0,$$

which may be written without suffixes as

$$\mathbf{a_t\,az - a_t\,M} = 0.$$

$(\mathbf{a_t\,a})$ is a square matrix—schematically, $(q \times p) \times (p \times q) = (q \times q)$—and has a reciprocal. Hence, multiplying before each term by $(\mathbf{a_t\,a})^{-1}$, we solve for $\mathbf{z}$ and find

$$\mathbf{z} = (\mathbf{a_t\,a})^{-1}\mathbf{a_t\,M}. \tag{35}$$

One way of looking at our procedure is this. We started with p equations with q unknowns, namely, $\mathbf{az} = \mathbf{M}$. To deal with the redundant equations we added p unknowns v_i together with q equations (34), and so obtained $p+q$ equations with $p+q$ unknowns.

The values of v_i, which indicate the accuracy of the measurements, are given by substituting the value of $\mathbf{z}$ from (35) into (32), thus

$$\mathbf{v} = \{\mathbf{a(a_t\,a)^{-1}a_t - I}\}\mathbf{M}. \tag{36}$$

In the example of thermal expansion the equation (28) takes the place of $\mathbf{M} = \mathbf{az}$, and so the best value of $\boldsymbol{\alpha}$ is

$$\boldsymbol{\alpha} = \mathbf{RA}, \quad \text{where} \quad \mathbf{R} = (\boldsymbol{\Theta_t\,\Theta})^{-1}\boldsymbol{\Theta_t}.$$

Numerical example. An octagonal disk perpendicular to Ox_2 was cut; the expansions per ° C, measured between opposite facets, were

$$\theta_i = 0° \quad 45° \quad 90° \quad 135°$$
$$10^6 \times A_i = 32 \cdot 0 \quad 16 \cdot 0 \quad 15 \cdot 0 \quad 31 \cdot 5$$

(the first three of these values were used in Exercise 6.5, p. 109). For this case

$$\boldsymbol{\Theta} = \begin{pmatrix} 0 & 0 & 1 \\ \tfrac{1}{2} & 1 & \tfrac{1}{2} \\ 1 & 0 & 0 \\ \tfrac{1}{2} & -1 & \tfrac{1}{2} \end{pmatrix}.$$

Therefore,

$$\boldsymbol{\Theta_t\,\Theta} = \begin{pmatrix} 0 & \tfrac{1}{2} & 1 & \tfrac{1}{2} \\ 0 & 1 & 0 & -1 \\ 1 & \tfrac{1}{2} & 0 & \tfrac{1}{2} \end{pmatrix} \begin{pmatrix} 0 & 0 & 1 \\ \tfrac{1}{2} & 1 & \tfrac{1}{2} \\ 1 & 0 & 0 \\ \tfrac{1}{2} & -1 & \tfrac{1}{2} \end{pmatrix} = \begin{pmatrix} \tfrac{3}{2} & 0 & \tfrac{1}{2} \\ 0 & 2 & 0 \\ \tfrac{1}{2} & 0 & \tfrac{3}{2} \end{pmatrix}$$

$(\boldsymbol{\Theta_t\,\Theta}$ must be symmetrical). Hence,

$$(\boldsymbol{\Theta_t\,\Theta})^{-1} = \frac{1}{\Delta}\left(\begin{array}{ccc} \begin{vmatrix} 2 & 0 \\ 0 & \tfrac{3}{2} \end{vmatrix} & -\begin{vmatrix} 0 & \tfrac{1}{2} \\ 0 & \tfrac{3}{2} \end{vmatrix} & \begin{vmatrix} 0 & \tfrac{1}{2} \\ 2 & 0 \end{vmatrix} \\[2ex] -\begin{vmatrix} 0 & 0 \\ \tfrac{1}{2} & \tfrac{3}{2} \end{vmatrix} & \begin{vmatrix} \tfrac{3}{2} & \tfrac{1}{2} \\ \tfrac{1}{2} & \tfrac{3}{2} \end{vmatrix} & -\begin{vmatrix} \tfrac{3}{2} & \tfrac{1}{2} \\ 0 & 0 \end{vmatrix} \\[2ex] \begin{vmatrix} 0 & 2 \\ \tfrac{1}{2} & 0 \end{vmatrix} & -\begin{vmatrix} \tfrac{3}{2} & 0 \\ \tfrac{1}{2} & 0 \end{vmatrix} & \begin{vmatrix} \tfrac{3}{2} & 0 \\ 0 & 2 \end{vmatrix} \end{array} \right),$$

where Δ is the determinant of $\Theta_t\,\Theta$. In this case $\Delta = 4$. Thus

$$(\Theta_t\,\Theta)^{-1} = \tfrac{1}{4}\begin{pmatrix} 3 & 0 & -1 \\ 0 & 2 & 0 \\ -1 & 0 & 3 \end{pmatrix},$$

so that the equation $R = (\Theta_t\,\Theta)^{-1}\Theta_t$ is

$$R = \tfrac{1}{4}\begin{pmatrix} -1 & 1 & 3 & 1 \\ 0 & 2 & 0 & -2 \\ 3 & 1 & -1 & 1 \end{pmatrix}.$$

Finally, α is given as $\alpha = RA$, or

$$10^6 \times \alpha = \tfrac{1}{4}\begin{pmatrix} -1 & 1 & 3 & 1 \\ 0 & 2 & 0 & -2 \\ 3 & 1 & -1 & 1 \end{pmatrix}\begin{pmatrix} 32\cdot0 \\ 16\cdot0 \\ 15\cdot0 \\ 31\cdot5 \end{pmatrix} = \begin{pmatrix} 15\cdot13 \\ -7\cdot75 \\ 32\cdot13 \end{pmatrix} = 10^6 \times \begin{pmatrix} \alpha_{11} \\ \alpha_{31} \\ \alpha_{33} \end{pmatrix}.$$

The value of $\mathbf{v}$ found from (36) comes out as

$$\begin{pmatrix} 0\cdot13 \\ -0\cdot13 \\ 0\cdot13 \\ -0\cdot13 \end{pmatrix} \times 10^{-6}.$$

Thus, the best values of α_{11}, α_{31}, α_{33} are given by multiplying the array R by the matrix of expansion measurements. If more than four measurements were desired, other arrays similar to R could be calculated for use on 10-sided, 12-sided,... polygonal disks.

The next step is to find the values of the principal expansion coefficients α_1 and α_3 and the orientation of the principal axes. The Mohr circle construction (Ch. II, § 4) provides the most convenient way of doing this. In Fig. 9.1, which is drawn to scale, the principal axes are labelled X_1, X_3. Since α_{31} is negative the representative point P is placed below the axis. Evidently

$$\tan 2\phi = \frac{2|\alpha_{31}|}{\alpha_{33}-\alpha_{11}},$$

which, with the above values, gives

$$\phi = 21\cdot18°.$$

The radius r of the circle is given by

$$r^2 = \tfrac{1}{4}(\alpha_{33}-\alpha_{11})^2 + \alpha_{31}^2,$$

from which we find

$$r = 11\cdot50 \times 10^{-6} \text{ per } °\text{ C.}$$

α_1 and α_3 are then calculated from the equations

$$\left.\begin{aligned}\alpha_1 &= \tfrac{1}{2}(\alpha_{11}+\alpha_{33})-r \\ \alpha_3 &= \tfrac{1}{2}(\alpha_{11}+\alpha_{33})+r\end{aligned}\right\},$$

as

$$\left.\begin{aligned}\alpha_1 &= 12\cdot13\times10^{-6} \text{ per } °\text{C} \\ \alpha_3 &= 35\cdot13\times10^{-6} \text{ per } °\text{C}\end{aligned}\right\}.$$

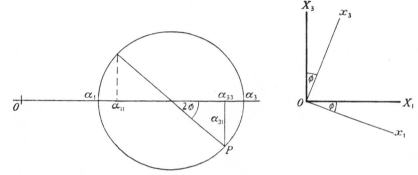

FIG. 9.1. The Mohr circle construction for finding the principal expansion coefficients and the orientation of the principal axes.

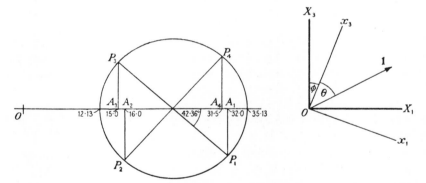

FIG. 9.2. The Mohr circle construction showing the four measured expansions used in the calculation of § 7.1. The values shown are those of $10^6\times\alpha$ (α in ° C^{-1}).

Fig. 9.2 shows on the same Mohr circle diagram the four measured expansions used in the calculation. The orientations $\theta = 0$, 45°, 90°, 135° are shown respectively by the points P_1, P_2, P_3, P_4 and the measured expansions by A_1, A_2, A_3, A_4.

7.2. Principal coefficients and directions for a triclinic crystal. (i) *Determination of coefficients referred to arbitrary axes.* As another and more general illustration of the above methods we may consider measurements of thermal expansion on a triclinic crystal. Since nothing is known beforehand about the orientation of the principal axes of expansion in the crystal, arbitrary axes x_1, x_2, x_3 are chosen,† and a cube is cut with edges parallel to these axes. The expansion coefficients are then measured along the cube edges and the four body diagonals. The direction cosines are

$$(1, 0, 0), \quad (0, 1, 0), \quad (0, 0, 1),$$

$$\tfrac{1}{3}\sqrt{3}(-1, 1, 1), \quad \tfrac{1}{3}\sqrt{3}(1, -1, 1), \quad \tfrac{1}{3}\sqrt{3}(1, 1, -1), \quad \tfrac{1}{3}\sqrt{3}(1, 1, 1).$$

Using the expression [from Ch. I, equation (33)]

$$A_r = l_i l_j \, \alpha_{ij}$$
$$= l_1^2 \alpha_{11} + l_2^2 \alpha_{22} + l_3^2 \alpha_{33} + 2 l_2 l_3 \alpha_{23} + 2 l_3 l_1 \alpha_{31} + 2 l_1 l_2 \alpha_{12},$$

where A_r is the expansion in the l_i direction, we have, in matrix notation,

$$\mathbf{A} = \boldsymbol{\Theta}\boldsymbol{\alpha},$$

where

$$\mathbf{A} = \begin{pmatrix} A_1 \\ A_2 \\ A_3 \\ A_4 \\ A_5 \\ A_6 \\ A_7 \end{pmatrix}, \quad \boldsymbol{\Theta} = \begin{pmatrix} 1 & 0 & 0 & 0 & 0 & 0 \\ 0 & 1 & 0 & 0 & 0 & 0 \\ 0 & 0 & 1 & 0 & 0 & 0 \\ \tfrac{1}{3} & \tfrac{1}{3} & \tfrac{1}{3} & \tfrac{2}{3} & -\tfrac{2}{3} & -\tfrac{2}{3} \\ \tfrac{1}{3} & \tfrac{1}{3} & \tfrac{1}{3} & -\tfrac{2}{3} & \tfrac{2}{3} & -\tfrac{2}{3} \\ \tfrac{1}{3} & \tfrac{1}{3} & \tfrac{1}{3} & -\tfrac{2}{3} & -\tfrac{2}{3} & \tfrac{2}{3} \\ \tfrac{1}{3} & \tfrac{1}{3} & \tfrac{1}{3} & \tfrac{2}{3} & \tfrac{2}{3} & \tfrac{2}{3} \end{pmatrix}, \quad \boldsymbol{\alpha} = \begin{pmatrix} \alpha_{11} \\ \alpha_{22} \\ \alpha_{33} \\ \alpha_{23} \\ \alpha_{31} \\ \alpha_{12} \end{pmatrix}.$$

(Notice that we have not introduced the factors of 2 into the last three components of the $\boldsymbol{\alpha}$ matrix, but have left them in the $\boldsymbol{\Theta}$ matrix. Either procedure is admissible.) Since there are seven measurements and only six unknowns, $\boldsymbol{\Theta}$ is not square, and the least squares method of solution must be adopted.

From equation (35) the best value of $\boldsymbol{\alpha}$ is given by

$$\boldsymbol{\alpha} = (\boldsymbol{\Theta}_t \boldsymbol{\Theta})^{-1} \boldsymbol{\Theta}_t \mathbf{A}.$$

Suppose the measured expansions were

A_1	A_2	A_3	A_4	A_5	A_6	A_7	
10	6	3	3	13/3	17/3	37/3	$\times 10^{-5}$ °C.

† A conventional setting of the x_1, x_2, x_3 axes with respect to the crystallographic axes is given in *Standards on piezoelectric crystals* (1949).

Then we find

$$10^5 \times \boldsymbol{\alpha} = \begin{pmatrix} \frac{17}{21} & -\frac{4}{21} & -\frac{4}{21} & \frac{1}{7} & \frac{1}{7} & \frac{1}{7} & \frac{1}{7} \\ -\frac{4}{21} & \frac{17}{21} & -\frac{4}{21} & \frac{1}{7} & \frac{1}{7} & \frac{1}{7} & \frac{1}{7} \\ -\frac{4}{21} & -\frac{4}{21} & \frac{17}{21} & \frac{1}{7} & \frac{1}{7} & \frac{1}{7} & \frac{1}{7} \\ 0 & 0 & 0 & \frac{3}{8} & -\frac{3}{8} & -\frac{3}{8} & \frac{3}{8} \\ 0 & 0 & 0 & -\frac{3}{8} & \frac{3}{8} & -\frac{3}{8} & \frac{3}{8} \\ 0 & 0 & 0 & -\frac{3}{8} & -\frac{3}{8} & \frac{3}{8} & \frac{3}{8} \end{pmatrix} \begin{pmatrix} 10 \\ 6 \\ 3 \\ 3 \\ \frac{13}{3} \\ \frac{17}{3} \\ \frac{37}{3} \end{pmatrix} = \begin{pmatrix} 10 \\ 6 \\ 3 \\ 2 \\ 3 \\ 4 \end{pmatrix} = 10^5 \times \begin{pmatrix} \alpha_{11} \\ \alpha_{22} \\ \alpha_{33} \\ \alpha_{23} \\ \alpha_{31} \\ \alpha_{12} \end{pmatrix},$$

or, as a 3×3 matrix,

$$10^5 \times \boldsymbol{\alpha} = \begin{pmatrix} 10 & 4 & 3 \\ 4 & 6 & 2 \\ 3 & 2 & 3 \end{pmatrix}. \tag{37}$$

The above measurements were invented so that they all agree per-

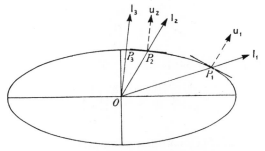

Fɪɢ. 9.3. The successive approximation method of finding the principal axes of a second-rank tensor.

fectly. To show the effect of imperfect measurement we may replace $13/3$ by 4 in the **A** matrix. The result is

$$(\boldsymbol{\Theta}_t \boldsymbol{\Theta})^{-1}\boldsymbol{\Theta}_t \begin{pmatrix} 10 \\ 6 \\ 3 \\ 3 \\ 4 \\ \frac{17}{3} \\ \frac{37}{3} \end{pmatrix} = \begin{pmatrix} 9\cdot95 \\ 5\cdot95 \\ 2\cdot95 \\ 2\cdot12 \\ 2\cdot87 \\ 4\cdot12 \end{pmatrix}.$$

(ii) *Finding the principal coefficients and axes.* Let us assume now that the best value of the $\boldsymbol{\alpha}$ matrix is that given in (37). To refer this matrix to its principal axes involves the solution of a cubic equation (Ch. II, § 3). In practice, a method of successive approximation is useful. Fig. 9.3 shows the principle of the method, which is as follows.

An arbitrary unit vector represented by the single column matrix $\mathbf{l_1}$

is chosen, and the displacement matrix u_1 for unit rise of temperature is calculated from the equation

$$u_1 = \alpha l_1.$$

We know from the property of the representation quadric that the direction of u_1 is that of the normal at the point P_1 where l_1 cuts the surface of the quadric. A new unit vector l_2 is then taken parallel to u_1 and the corresponding matrix u_2 is computed. A third unit vector l_3 parallel to u_2 is taken, and so the process is repeated. If the quadric is an ellipsoid the vectors converge on to the shortest axis, as Fig. 9.3 illustrates. In general, they converge on to that axis of the quadric which corresponds to the numerically greatest principal coefficient.

In the numerical example we start by choosing

$$l_1 = \begin{pmatrix} 1 \\ 0 \\ 0 \end{pmatrix},$$

having noticed the large value of α_{11}. Then

$$10^5 \times u_1 = \begin{pmatrix} 10 & 4 & 3 \\ 4 & 6 & 2 \\ 3 & 2 & 3 \end{pmatrix} \begin{pmatrix} 1 \\ 0 \\ 0 \end{pmatrix} = \begin{pmatrix} 10 \\ 4 \\ 3 \end{pmatrix}.$$

Since only the ratios of the components of the u matrices are significant, it is not necessary to reduce the length of the l vectors to unity each time. We therefore continue multiplication as follows:

$$\begin{pmatrix} 10 & 4 & 3 \\ 4 & 6 & 2 \\ 3 & 2 & 3 \end{pmatrix} \begin{pmatrix} 10 \\ 4 \\ 3 \end{pmatrix} = \begin{pmatrix} 125 \\ 70 \\ 47 \end{pmatrix};$$

and, similarly, we obtain in succession

$$\begin{pmatrix} 1671 \\ 1014 \\ 656 \end{pmatrix} \rightarrow \begin{pmatrix} 22734 \\ 14080 \\ 9009 \end{pmatrix} \rightarrow \begin{pmatrix} 310687 \\ 193434 \\ 123389 \end{pmatrix} \rightarrow \ldots .$$

To show the convergence we list the successive values of l, all reduced to unit length:

$$\begin{pmatrix} 1 \\ 0 \\ 0 \end{pmatrix} \rightarrow \begin{pmatrix} 0{\cdot}895 \\ 0{\cdot}358 \\ 0{\cdot}268 \end{pmatrix} \rightarrow \begin{pmatrix} 0{\cdot}828 \\ 0{\cdot}464 \\ 0{\cdot}312 \end{pmatrix} \rightarrow \begin{pmatrix} 0{\cdot}810 \\ 0{\cdot}492 \\ 0{\cdot}318 \end{pmatrix} \rightarrow \begin{pmatrix} 0{\cdot}806 \\ 0{\cdot}499 \\ 0{\cdot}322 \end{pmatrix} \rightarrow \begin{pmatrix} 0{\cdot}803 \\ 0{\cdot}500 \\ 0{\cdot}319 \end{pmatrix} \rightarrow \ldots .$$

This set of direction cosines, being those of the shortest axis of the quadric, will correspond to the greatest principal expansion coefficient.

To find the least principal expansion coefficient we perform the same operations on the α^{-1} matrix. We find

$$10^{-5}\times\alpha^{-1} = \tfrac{1}{86}\begin{pmatrix} 14 & -6 & -10 \\ -6 & 21 & -8 \\ -10 & -8 & 44 \end{pmatrix}.$$

This time the large value of $(\alpha^{-1})_{33}$ leads us to start with $\mathsf{I}_1 = \begin{pmatrix} 0 \\ 0 \\ 1 \end{pmatrix}.$

Then successive multiplication gives

$$\begin{pmatrix} 14 & -6 & -10 \\ -6 & 21 & -8 \\ -10 & -8 & 44 \end{pmatrix}\begin{pmatrix} 0 \\ 0 \\ 1 \end{pmatrix} = \begin{pmatrix} -10 \\ -8 \\ 44 \end{pmatrix} \rightarrow \begin{pmatrix} -532 \\ -460 \\ 2{,}100 \end{pmatrix} \rightarrow \begin{pmatrix} -25{,}788 \\ -23{,}268 \\ 101{,}400 \end{pmatrix};$$

dividing by 10^6 and proceeding, we further obtain

$$\begin{pmatrix} -1\cdot235 \\ -1\cdot146 \\ 4\cdot906 \end{pmatrix} \rightarrow \begin{pmatrix} -59\cdot47 \\ -55\cdot90 \\ 237\cdot38 \end{pmatrix} \rightarrow \ldots .$$

The corresponding direction cosines are

$$\begin{pmatrix} 0 \\ 0 \\ 1 \end{pmatrix} \rightarrow \begin{pmatrix} -0\cdot218 \\ -0\cdot175 \\ 0\cdot960 \end{pmatrix} \rightarrow \begin{pmatrix} -0\cdot240 \\ -0\cdot208 \\ 0\cdot949 \end{pmatrix} \rightarrow \begin{pmatrix} -0\cdot241 \\ -0\cdot217 \\ 0\cdot946 \end{pmatrix} \rightarrow \begin{pmatrix} -0\cdot238 \\ -0\cdot220 \\ 0\cdot946 \end{pmatrix} \rightarrow \begin{pmatrix} -0\cdot237 \\ -0\cdot223 \\ 0\cdot945 \end{pmatrix} \rightarrow \ldots .$$

By forming the vector product of the direction cosines corresponding to the greatest and least coefficients we find the intermediate axis to be

$$\begin{pmatrix} -0\cdot544 \\ 0\cdot834 \\ 0\cdot061 \end{pmatrix}.$$

The expansion in the $\begin{pmatrix} 0\cdot803 \\ 0\cdot500 \\ 0\cdot319 \end{pmatrix}$ direction is

$$10^{-5}\times\begin{pmatrix} 10 & 4 & 3 \\ 4 & 6 & 2 \\ 3 & 2 & 3 \end{pmatrix}\begin{pmatrix} 0\cdot803 \\ 0\cdot500 \\ 0\cdot319 \end{pmatrix} = 10^{-5}\times\begin{pmatrix} 10\cdot99 \\ 6\cdot85 \\ 4\cdot37 \end{pmatrix}$$

and the length of this vector is $13\cdot67\times10^{-5}$. Similarly, the expansions in the other two principal directions are computed to be $1\cdot776\times10^{-5}$ and $3\cdot527\times10^{-5}$. As a check we note that, with these values,

$$\alpha_1+\alpha_2+\alpha_3 = 18\cdot973\times10^{-5},$$

which may be compared with $\alpha_{11}+\alpha_{22}+\alpha_{33} = 19\times10^{-5}$ from the original matrix. Thus, our final result for the magnitudes and directions of the principal expansions is:

$$\alpha_1 = 13 \cdot 7 \times 10^{-5} \text{ per } ° \text{ C in the direction } (0 \cdot 80,\ 0 \cdot 50,\ 0 \cdot 32),$$
$$\alpha_2 = 3 \cdot 53 \times 10^{-5} \quad ,, \quad ,, \quad ,, \quad (-0 \cdot 54,\ 0 \cdot 83,\ 0 \cdot 06),$$
$$\alpha_3 = 1 \cdot 78 \times 10^{-5} \quad ,, \quad ,, \quad ,, \quad (-0 \cdot 24,\ -0 \cdot 22,\ 0 \cdot 95).$$

For a fuller account of numerical methods of finding principal coefficients and axes, see the book by Hartree (1952), pp. 178–89.

EXERCISE 9.5. Verify that the principal expansions found above are the roots of the secular equation (Ch. II, § 3)

$$|\alpha_{ij} - \lambda \delta_{ij}| = 0.$$

SUMMARY

Matrix algebra (§ 2). *Addition and subtraction* (§ 2.2). Two matrices may be added or subtracted when each has the same number of rows and the same number of columns. The sum of two matrices **A** and **B**, denoted **A**+**B**, is a matrix **C** whose elements are given by

$$C_{ij} = A_{ij} + B_{ij}.$$

The rule for subtraction is similar.

Multiplication (§ 2.1). If **A** is an $(m \times n)$ matrix and **B** is an $(n \times p)$ matrix, the product **AB** is an $(m \times p)$ matrix. The elements of **AB** are given by

$$(AB)_{ik} = A_{ij} B_{jk}.$$

Linear transformations (§ 2.1). Let **x** be an $(m \times 1)$ matrix and **y** be an $(n \times 1)$ matrix connected together by the equation

$$\mathbf{x} = \alpha \mathbf{y},$$

where **α** is an $(m \times n)$ matrix. Let **y**, in turn, be connected with a $(p \times 1)$ matrix **z** by

$$\mathbf{y} = \beta \mathbf{z},$$

where **β** is an $(n \times p)$ matrix. Then **x** is connected with **z** by

$$\mathbf{x} = \alpha \beta \mathbf{z}.$$

General rule. In general, matrices may be added, subtracted and multiplied, but not divided, as though they were ordinary numbers; but, usually, $\mathbf{AB} \neq \mathbf{BA}$.

(§ 4.1) The *transpose* $\mathbf{A_t}$ of a matrix **A** is obtained by interchanging rows and columns.

(§ 4.2) The *reciprocal* α^{-1} of a square, non-singular matrix **α** is such that

$$\alpha \alpha^{-1} = \alpha^{-1} \alpha = I,$$

where **I** is a unit matrix (having 1's on the leading diagonal and zeros elsewhere). Explicitly, the i, jth element of α^{-1} is A_{ji}/Δ, where A_{ji} is the cofactor of α_{ji} and Δ is the determinant $|\alpha_{ij}|$.

Crystal properties (§ 3).

Pyroelectric effect: $\quad\quad\quad \Delta \mathbf{P} = \mathbf{p} \Delta T,$

where $\Delta \mathbf{P}$ and **p** are (3×1) matrices and ΔT is a single number (1×1).

Thermal expansion: $\quad\quad\quad \boldsymbol{\epsilon} = \boldsymbol{\alpha} \Delta T,$

where **ε** and **α** are (6×1) matrices.

Dielectric relations: $\quad \mathbf{D} = \varkappa \mathbf{E}, \quad \mathbf{E} = \beta \mathbf{D}, \quad \beta = \varkappa^{-1},$

where **D** and **E** are (3×1) and **ϰ** and **β** are (3×3) matrices.

Magnetic relation: $\qquad\qquad$ **B** $= \mu$**H,**

where **B** and **H** are (3×1) and μ is a (3×3) matrix.

Direct piezoelectric effect: $\qquad$ **P** $=$ **dσ,**

where **P** is (3×1), **d** is (3×6) and σ is (6×1).

Converse piezoelectric effect: $\qquad$ $\epsilon =$ **d$_t$ E,**

where ϵ is (6×1), **d$_t$** is (6×3) and **E** is (3×1).

Elasticity: $\qquad\qquad$ $\epsilon =$ **sσ,** $\quad \sigma =$ **cϵ,** $\quad$ **c** $=$ **s**$^{-1}$,

where ϵ is (6×1), **c** and **s** are (6×6), and σ is (6×1).

(§ 5) Let an arbitrary direction be represented by the (3×1) matrix **l**. Then, in this direction, the magnitude S of a symmetrical second-rank tensor property given by the (3×3) matrix **S** is

$$S = \mathbf{l}_t\,\mathbf{S}\mathbf{l}.$$

(§ 6) A *transformation of axes* may be represented by the (3×3) matrix **a** $= (a_{ij})$. Then, for the coordinates [(3×1) matrices],

$$\mathbf{x}' = \mathbf{a}\mathbf{x};$$

for the components of a vector [(3×1) matrices],

$$\mathbf{p}' = \mathbf{a}\mathbf{p};$$

and, for the components of a tensor [κ_{ij}] represented by the (3×3) matrix **$\varkappa$,**

$$\mathbf{\varkappa}' = \mathbf{a}\mathbf{\varkappa}\mathbf{a}_t.$$

Example of least squares method (§ 7). Suppose p measurements of thermal expansion on a crystal in different directions are written as a single column $(p \times 1)$ matrix **A**. Then they are related to the thermal expansion coefficients, written as a (6×1) matrix **α**, by the equation

$$\mathbf{A} = \Theta\mathbf{\alpha},$$

where Θ is a $(p \times 6)$ matrix constructed from the direction cosines of the directions of measurement. If $p = 6$ the equation is readily solved as

$$\mathbf{\alpha} = \Theta^{-1}\,\mathbf{A}.$$

If, however, $p > 6$, we have more measurements than unknowns, and they have to be combined to give the best value of **α**. The principle of least squares leads to the result that the best value of **α** is

$$\mathbf{\alpha} = \mathbf{R}\mathbf{A}, \quad \text{where} \quad \mathbf{R} = (\Theta_t\,\Theta)^{-1}\Theta_t.$$

Successive approximation method of finding principal axes [§ 7.2 (ii)]. Suppose the coefficients of a quadric are given as a (3×3) matrix **α**, and we wish to find the principal axes. Draw an arbitrary radius vector represented by a (3×1) matrix **l**. Then the radius-normal property allows us to calculate the normal **u** at its extremity:

$$\mathbf{u} = \mathbf{\alpha}\mathbf{l}.$$

Now take a new radius vector parallel to **u** and calculate the corresponding normal. The process is repeated as often as necessary and, when the quadric is an ellipsoid, the vectors converge on to the shortest axis. The greatest axis is found by repeating the procedure for the inverse quadric given by **α**$^{-1}$; the intermediate axis is then easily written down.

X

THERMODYNAMICS OF EQUILIBRIUM PROPERTIES OF CRYSTALS†

THE preceding chapters have dealt with the thermal, electrical and mechanical properties of a crystal, each property being treated in isolation from the others. In fact the properties are related, and this chapter discusses the connexions between them. A unified treatment of the properties is possible because they all have this feature in common—that they may be described by reference to equilibrium states. That is to say, the properties may be measured with the crystal in equilibrium with its surroundings, so that neither the state of the crystal nor that of its surroundings changes with time; alternatively we may say that the properties are describable by reference to changes which are thermodynamically reversible. We shall later, in Chapters XI and XII, discuss transport properties, where this is not the case.

1. The thermal, electrical and mechanical properties of a crystal

The relations between the properties we shall be concerned with are illustrated by the diagram in Figs. 10.1 a and b.‡ In the three outer corners are temperature T, electric field E_i, and stress σ_{ij}, which may all be thought of as 'forces' applied to the crystal. In the three corresponding inner corners appear entropy per unit volume S, electric displacement D_i, and strain ϵ_{ij} respectively, which are the direct results of these 'forces'. The lines joining these pairs of corners denote what may be called the three *principal effects*:

(i) In a reversible change, and considering unit volume, an increase of temperature produces a change of entropy dS:

$$dS = (C/T)\,dT, \tag{1}$$

where C (a scalar) is the heat capacity per unit volume, and T is the absolute temperature.

---

† This chapter may be omitted on a first reading.

‡ This is a development of a diagram given by Heckmann (1925).

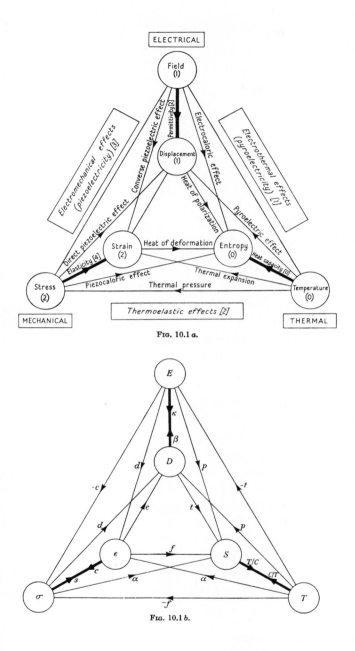

FIG. 10.1 a.

FIG. 10.1 b.

Fig. 10.1. The relations between the thermal, electrical and mechanical properties of a crystal, showing (a) the names of the properties and the variables, and (b) the corresponding symbols. The tensor rank of the variables is shown in round brackets and the tensor rank of the properties in square brackets.

(ii) A small change of electric field dE_i produces a change of electric displacement dD_i according to the equation

$$dD_i = \kappa_{ij} dE_j, \tag{2}$$

where κ_{ij} is the permittivity tensor (Ch. IV).

(iii) A small change of stress $d\sigma_{kl}$ produces a change of strain $d\epsilon_{ij}$ according to the equation

$$d\epsilon_{ij} = s_{ijkl} d\sigma_{kl}, \tag{3}$$

where s_{ijkl} are the elastic compliances (Ch. VIII).

The diagram also illustrates what we shall call the *coupled effects*. These are denoted by lines joining pairs of points which are not both at the same corner. Consider, for example, the two diagonal lines at the bottom of the diagram. One shows *thermal expansion*, the strain produced by a change of temperature, while the other shows the *piezo-caloric effect*, that is, the entropy (heat) produced by a stress. The two horizontal lines at the bottom show that entropy (heat) is produced by strain as *heat of deformation*, and that if the temperature of a crystal is changed without change of shape being allowed to take place a *thermal pressure* is set up. These four coupled effects all connect scalars with second-rank tensors, and are therefore themselves to be specified by second-rank tensors. Thus, for thermal expansion (Ch. VI) the relation is

$$d\epsilon_{ij} = \alpha_{ij} dT. \tag{4}$$

The coupled effects on the left of the diagram have to do with piezo-electricity (Ch. VII). The direct effect is given in differential form by

$$dP_i = d_{ijk} d\sigma_{jk}. \tag{5}$$

Since $D_i = \kappa_0 E_i + P_i,$ [Ch. IV, equation (1)]

we have $dP_i = dD_i - \kappa_0 dE_i.$

Therefore, if the electric field in the crystal is held constant, $dP_i = dD_i$, and we may write (5) as

$$dD_i = d_{ijk} d\sigma_{jk} \quad \text{(E constant)}. \tag{6}$$

Either equation (5) or (6) may be used to express the direct effect. The direct and converse piezoelectric effects thus appear as the diagonals on the left of the diagram; they are seen to be two out of the four possible connexions we could choose to explore between electric field, electric displacement, stress, and strain in piezoelectric crystals. The two other connexions are between strain and displacement and between stress and field. Since these coupled effects all relate a first-rank tensor

(E_i or D_i) with a second-rank tensor (σ_{ij} or ϵ_{ij}), they are themselves given by third-rank tensors.

The coupled effects on the right of the diagram are concerned with pyroelectricity (Ch. IV, § 7). They all connect a vector (D_i or E_i) with a scalar (S or T), and are therefore expressed by first-rank tensors. The equation for the pyroelectric effect may be written

$$dP_i = p_i\, dT. \tag{7}$$

If we now specify that the temperature change is to be carried out with the electric field in the crystal held constant, we have

$$dD_i = p_i\, dT \quad \text{(E constant).} \tag{8}$$

The pyroelectric effect (with the electric field held constant) may thus be regarded either as an electric polarization or as an electric displacement caused by a temperature change. One of the two diagonals on the right of the diagram shows the pyroelectric effect and the other shows the *electrocaloric effect*, the entropy (heat) produced by an electric field. The other connexions show that entropy (heat) is caused by electric displacement as *heat of polarization*, and that there is also a connexion between the temperature and the field analogous to thermal pressure. This completes the description of the diagram of Figs. 10.1 a and b.

Now all these crystal properties are inter-related. In discussing the connexions between them it is necessary to state precisely under what conditions they are to be measured. For example, the elastic compliances may be measured under isothermal or under adiabatic conditions, and different values are obtained experimentally in the two cases; a difference of 1 per cent. is a representative figure. Again, the specific heat of a crystal would normally be measured at constant (zero) stress, but it would be equally acceptable from a theoretical point of view (although not from an experimental one) to measure it at constant strain, and if this were done a slightly different value would be found.

The phenomena of primary and secondary pyroelectricity (Ch. IV, § 7) provide an example of strong interaction between different crystal properties. Reference to Fig. 10.1 b shows that when the temperature is changed an electric displacement is set up by the pyroelectric effect along the direct path $T \to D$; but other paths are possible as well: $T \to \epsilon \to D$, for instance, which would give a displacement by a combination of thermal expansion and a piezoelectric effect. Whether this path is open or not depends upon whether the crystal is allowed to change its shape.

It is clear from these examples that the thermal, electrical and mechanical properties of a crystal must be considered as a whole. To understand their interactions the thermodynamics of the processes must be explored. We shall find that this leads to a precise formulation of the properties, and yields a number of new relations between the coefficients defining the properties.

2. Thermodynamics of thermoelastic behaviour

The essential method of the thermodynamic treatment can be seen by considering a simple system. We choose a crystal that shows no

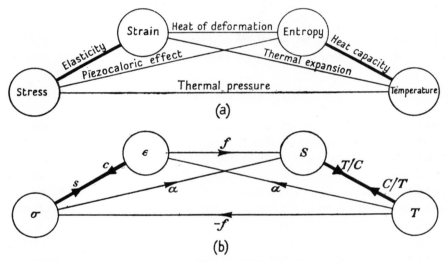

FIG. 10.2. The parts of Figs. 10.1 a and b which are relevant to thermoelastic behaviour.

piezo or pyroelectric effects—a crystal possessing a centre of symmetry would suit the purpose—and we analyse its thermoelastic behaviour. Only the lower part of the diagram in Figs. 10.1 a and b is now relevant; it is shown separately in Figs. 10.2 a and b.

We take the nine stress components σ_{ij} and the temperature T as the independent variables; that is to say, the state of the crystal, and in particular the strain components ϵ_{ij} and the entropy S, are determined when the ten quantities σ_{ij} and T are given. It would be quite possible, as an alternative, to take the ten quantities (ϵ_{ij}, S) as the independent variables, or (σ_{ij}, S), or (ϵ_{ij}, T). Note, however, that the variables (S, T) alone, being only two in number, would not be a possible choice. Thus, we have:

$$\text{independent variables (10)} \quad \sigma_{ij}, T;$$

$$\text{dependent variables} \quad \epsilon_{ij}, S, \text{etc.}$$

Accordingly, considering unit volume, we may write:

$$d\epsilon_{ij} = \overbrace{\left(\frac{\partial \epsilon_{ij}}{\partial \sigma_{kl}}\right)_T}^{\text{elasticity}} d\sigma_{kl} \quad + \quad \overbrace{\left(\frac{\partial \epsilon_{ij}}{\partial T}\right)_\sigma}^{\text{thermal expansion}} dT \tag{9}$$

$$dS = \underbrace{\left(\frac{\partial S}{\partial \sigma_{kl}}\right)_T}_{\text{piezocaloric effect}} d\sigma_{kl} \quad + \quad \underbrace{\left(\frac{\partial S}{\partial T}\right)_\sigma}_{\text{heat capacity}} dT \tag{10}$$

In these equations the summation convention (Ch. I, § 1.1) applies. Thus the first term on the right of equation (9) stands for the sum of nine terms:

$$\left(\frac{\partial \epsilon_{ij}}{\partial \sigma_{kl}}\right)_T d\sigma_{kl} = \left(\frac{\partial \epsilon_{ij}}{\partial \sigma_{11}}\right) d\sigma_{11} + \left(\frac{\partial \epsilon_{ij}}{\partial \sigma_{12}}\right) d\sigma_{12} + \ldots,$$

where the differentiation with respect to each stress component is carried out *keeping all the other stress components*, and the temperature, *constant*. The σ in the symbol $(\partial \epsilon_{ij}/\partial T)_\sigma$ indicates that in differentiation with respect to T all the σ_{ij} are to be kept constant. Since i and j can take values of 1 to 3, equation (9) is short for nine equations; in all, equations (9) and (10) stand for ten equations each with ten terms on the right-hand side.

The four derivatives in (9) and (10) have physical meanings. Putting $dT = 0$ in equation (9) and comparing with equation (3) shows that $(\partial \epsilon_{ij}/\partial \sigma_{kl})_T$ are the isothermal elastic compliances, which may be written s_{ijkl}^T; henceforth superscripts will always indicate the quantity that is to be kept constant. Putting $d\sigma_{kl} = 0$ shows that $(\partial \epsilon_{ij}/\partial T)_\sigma$ are the coefficients of thermal expansion α_{ij} measured at constant stress (compare equation (4)). [There is no need to write α_{ij}^σ for this quantity, because the other coefficient $(\partial \epsilon_{ij}/\partial T)_S$ which might appear at first sight to be a possibility is in fact meaningless; it would imply that each ϵ_{ij} is a function of only two variables T and S, instead of a function of ten variables.] $(\partial S/\partial \sigma_{kl})_T$ measures the increase in entropy caused by applying a stress isothermally. Multiplied by T it gives the heat produced when a crystal is stressed isothermally, the piezocaloric effect. $T(\partial S/\partial T)_\sigma$ measures the heat developed when the temperature is changed but the stress is held constant; it is the heat capacity per unit volume at constant stress C^σ (compare the heat capacity at constant pressure in a gas).

Owing to the interplay between the four effects just described, the four differential coefficients in equations (9) and (10) are not independent. To find the relations between them we consider the energy of the system. Considering unit volume, we know from the first law of thermodynamics that, if a small amount of heat dQ flows into the crystal and a small amount of work dW is done on the crystal by external forces, the increase in the internal energy dU is a perfect differential and is given by

$$dU = dW + dQ. \tag{11}$$

The work done per unit volume in changing the strains by small amounts $d\epsilon_{ij}$ is given (Ch. VIII, § 3) by

$$dW = \sigma_{ij} d\epsilon_{ij}. \tag{12}$$

By the second law of thermodynamics, for a reversible change,

$$dQ = T\, dS. \tag{13}$$

Hence (11) becomes $dU = \sigma_{ij} d\epsilon_{ij} + T\, dS. \tag{14}$

Instead of the function U, which would be convenient for independent variables (ϵ_{ij}, S), we have now to consider the function Φ defined† by

$$\Phi = U - \sigma_{ij}\epsilon_{ij} - TS. \tag{15}$$

The term $\sigma_{ij}\epsilon_{ij}$ stands for the sum of nine separate terms. Differentiation gives $d\Phi = dU - \sigma_{ij} d\epsilon_{ij} - \epsilon_{ij} d\sigma_{ij} - T\, dS - S\, dT,$

which, on combining with (14), leads to

$$d\Phi = -\epsilon_{ij} d\sigma_{ij} - S\, dT. \tag{16}$$

Now in the definition of Φ, equation (15), all the quantities are functions of (σ_{ij}, T), which we agreed to use to define the state of the crystal. Hence Φ itself is a function of (σ_{ij}, T), and therefore, from equation (16),

$$\left(\frac{\partial \Phi}{\partial \sigma_{ij}}\right)_T = -\epsilon_{ij}, \qquad \left(\frac{\partial \Phi}{\partial T}\right)_\sigma = -S. \tag{17}$$

Differentiating the first of these equations with respect to T, and the

† Since only *changes* in U and S are defined by equations (11) and (13), these quantities each contain an arbitrary constant representing the internal energy and the entropy, respectively, in some fixed zero state (which could be specified in terms of the stress and the temperature). In the same way we have not yet defined the state from which ϵ_{ij} is to be measured. To give definite meanings now to U, S and ϵ_{ij} we define the datum state to be the same for all three quantities; specifically, we define U, S and ϵ_{ij} to be simultaneously zero when the crystal is in some fixed condition. Equation (15) then shows that Φ is also zero in this condition.

second with respect to σ_{ij} (keeping T and all the σ_{ij}, except for one, constant), we have the important *thermodynamic relation*

$$-\frac{\partial^2 \Phi}{\partial \sigma_{ij} \partial T} = \left(\frac{\partial \epsilon_{ij}}{\partial T}\right)_\sigma = \left(\frac{\partial S}{\partial \sigma_{ij}}\right)_T. \tag{18}$$

Reference to equations (9) and (10) shows that we have here established a relation between the coefficients of thermal expansion and the coefficients giving the piezocaloric effect. This is the essential result of the thermodynamical analysis. It shows that *the matrix of coefficients on the right-hand side of equations* (9) *and* (10) *is symmetrical.*

Equations (9) and (10) apply whether the four crystal properties are linear or not. However, the properties will always be linear over a sufficiently small range of the variables and hence, for small stresses and temperature changes (that is, for *first-order effects*), the equations may be integrated to give

$$\left. \begin{array}{l} \epsilon_{ij} = s^T_{ijkl}\,\sigma_{kl} + \alpha_{ij}\Delta T \\ \Delta S = \alpha_{ij}\,\sigma_{ij} + (C^\sigma/T)\,\Delta T \end{array} \right\}, \tag{19}$$

where the coefficients are constants. [ϵ_{ij} is taken to be zero when $\sigma_{kl} = \Delta T = 0$.] In the two-suffix notation developed in Chapters VII, VIII and IX equations (19) take the shorter form

$$\left. \begin{array}{l} \epsilon_i = s^T_{ij}\,\sigma_j + \alpha_i\Delta T \\ \Delta S = \alpha_j\,\sigma_j + (C^\sigma/T)\,\Delta T \end{array} \right\}, \tag{20}$$

the α_i's being the matrix elements of equation (16) in Chapter IX. The most compact way of writing the relations is to use single letters for the matrices (Ch. IX). Then we have simply

$$\left. \begin{array}{l} \boldsymbol{\epsilon} = \mathbf{s}^T\boldsymbol{\sigma} + \boldsymbol{\alpha}\Delta T \\ \Delta S = \boldsymbol{\alpha_t}\boldsymbol{\sigma} + (C^\sigma/T)\,\Delta T \end{array} \right\}. \tag{21}$$

Notice that it is $\boldsymbol{\alpha_t}$ (a single-row matrix) not $\boldsymbol{\alpha}$, that appears in the second equation.

2.1. Heat of deformation and thermal pressure. We have now discussed the four effects represented by the four differential coefficients in equations (9) and (10). Since we agreed to take (σ_{ij}, T) as independent variables, the four effects each connect a quantity at the top of Figs. 10.2 *a* and *b* with a quantity at the bottom. There are, however, two other connexions that might be investigated: between ϵ_{ij} and S, and between σ_{ij} and T. The first effect concerns the heat (entropy) developed by a strain; for an isothermal process this heat is given by the coefficient

$T(\partial S/\partial \epsilon_{ij})_T$. The second effect concerns the stress† produced by a temperature change, which, for an experiment in which the crystal is clamped so that there is no strain, would be given by the coefficient $(\partial \sigma_{ij}/\partial T)_\epsilon$. Both these are coupled effects, and just as there is a relation between the thermal expansion coefficients and the piezocaloric coefficients, so there is a relation between the coefficients for the heat of deformation and thermal pressure. The relation may be obtained by considering the function (the Helmholtz free energy)

$$\Psi = U - TS. \tag{22}$$

Differentiating and using equation (14) gives

$$d\Psi = \sigma_{ij}\,d\epsilon_{ij} - S\,dT. \tag{23}$$

The condition that $d\Psi$ is a perfect differential [compare equations (16) and (18)] is

$$\left(\frac{\partial S}{\partial \epsilon_{ij}}\right)_T = -\left(\frac{\partial \sigma_{ij}}{\partial T}\right)_\epsilon = f_{ij}, \quad \text{say,} \tag{24}$$

which is the required thermodynamic relation.

The scheme of coefficients for the properties in Fig. 10.2 a is therefore as shown in Fig. 10.2 b (subscripts and superscripts being omitted). The coefficients in Fig. 10.2 b are not all independent and, in fact, the three tensors s_{ijkl}^T, α_{ij} and C^σ/T which occur in equations (19) are sufficient to define all the first-order thermoelastic effects.

EXERCISE 10.1. Prove that

$$(\partial T/\partial \epsilon_{ij})_S = (\partial \sigma_{ij}/\partial S)_\epsilon, \quad \text{and that} \quad (\partial \epsilon_{ij}/\partial S)_\sigma = -(\partial T/\partial \sigma_{ij})_S.$$

† Some writers (for example, Cady 1946, p. 185) consider that, when a crystal is strained either piezoelectrically or by heating, the strain is caused by an 'internal stress'. To see how this concept arises let a free crystal be heated so that it expands in a certain direction. The applied stress, and therefore the stress in the crystal as we have defined it, is zero, although the strain is not. Now the expansion may be thought of as being produced not by temperature but by a 'body' tensile stress—although such a stress does not really exist. This fictitious stress, which is called the 'internal stress', is exactly equal and opposite to the real stress (a pressure) that would be set up in the crystal if the temperature were applied and the crystal were clamped. We avoid using the idea of fictitious stresses in the present book because it does not seem particularly helpful, and, indeed, when developed it tends rather to confuse than to clarify. In the homogeneous states of stress considered in this chapter the stress within a crystal is given directly by the forces applied to its surfaces (apart from the very small electrostrictive stresses caused by surface charges); if these forces are tensile the stress is tensile, if the forces are compressive the stress is compressive, and if the boundaries of the crystal are free there can be, from equilibrium considerations, no stress in the crystal.

A fictitious internal stress in the above sense is of course quite different from the real internal (locked-up) stresses that are produced when a crystal is non-uniformly heated or deformed plastically.

2.2. Relation between adiabatic and isothermal elastic compliances.

An expression for the difference between the elastic compliances measured adiabatically and isothermally may be obtained as follows. Eliminate dT between equations (9) and (10) and put $dS = 0$. Then we have

$$d\epsilon_{ij} = \left(\frac{\partial \epsilon_{ij}}{\partial \sigma_{kl}}\right)_T d\sigma_{kl} - \left(\frac{\partial \epsilon_{ij}}{\partial T}\right)_\sigma \left(\frac{\partial S}{\partial \sigma_{kl}}\right)_T d\sigma_{kl} \Big/ \left(\frac{\partial S}{\partial T}\right)_\sigma \quad (S \text{ constant}).$$

Dividing by $d\sigma_{kl}$ and using relation (18) we obtain

$$\left(\frac{\partial \epsilon_{ij}}{\partial \sigma_{kl}}\right)_S - \left(\frac{\partial \epsilon_{ij}}{\partial \sigma_{kl}}\right)_T = -\left(\frac{\partial \epsilon_{ij}}{\partial T}\right)_\sigma \left(\frac{\partial \epsilon_{kl}}{\partial T}\right)_\sigma \left(\frac{\partial T}{\partial S}\right)_\sigma. \tag{25}$$

This equation may also be written

$$s^S_{ijkl} - s^T_{ijkl} = -\alpha_{ij}\alpha_{kl}(T/C^\sigma). \tag{26}$$

Since crystals usually, but not invariably (Ch. VI), have positive coefficients of expansion, and since C^σ is positive, the right-hand side of (26) is usually negative. The adiabatic compliances are then smaller than the isothermal ones. We may interpret this fact as follows. In an adiabatic experiment a *tensile* stress produces a *fall* in temperature; this causes a *decrease* in strain. The total strain produced by the stress is then smaller than it would be in an isothermal experiment, and the adiabatic compliance is correspondingly smaller.

An expression analogous to (26) for the difference between the heat capacities at constant strain and constant stress may be found in a similar way.

3. Thermodynamics of thermal, electrical and elastic properties

The thermodynamical methods of the last section can be extended to the more general situation illustrated in Figs. 10.1 a and b where, in addition to interaction between thermal and elastic effects, electrical effects have to be reckoned with.[†]

In what follows we normally choose (σ_{ij}, E_i, T), the outer quantities in Fig. 10.1 b, as the independent variables, with $(\epsilon_{ij}, D_i, S, \text{ etc.})$ as the dependent variables. This is a convenient choice but not the only possible one; for example, for some purposes it would be more natural to think of the strains of the crystal lattice as prescribed rather than the stresses. Other possible choices of independent variables would be (ϵ_{ij}, E_i, T), (σ_{ij}, E_i, S) or (σ_{ij}, D_i, T), etc., in which one quantity is chosen from each corner of the diagram (Fig. 10.1 b). In each case there are

† For the further extension to magnetic effects see the book by Cady (1946), p. 42. The triangular diagram of Fig. 10.1 a and b becomes a tetrahedron.

13 variables. On the other hand, a choice such as (E_i, D_i, S) would give only seven variables and would therefore not be possible; for this reason one has to beware of writing meaningless coefficients such as $(\partial\epsilon_{ij}/\partial E_k)_{D,S}$.

With (σ_{ij}, E_i, T) as independent variables the differentials of ϵ_{ij}, D_i and S may be written as:

$$d\epsilon_{ij} = \underset{\text{elasticity}}{\left(\frac{\partial\epsilon_{ij}}{\partial\sigma_{kl}}\right)_{E,T}} d\sigma_{kl} + \underset{\substack{\text{converse}\\\text{piezoelectricity}}}{\left(\frac{\partial\epsilon_{ij}}{\partial E_k}\right)_{\sigma,T}} dE_k + \underset{\substack{\text{thermal}\\\text{expansion}}}{\left(\frac{\partial\epsilon_{ij}}{\partial T}\right)_{\sigma,E}} dT \qquad (27)$$

$$dD_i = \underset{\substack{\text{direct}\\\text{piezoelectricity}}}{\left(\frac{\partial D_i}{\partial\sigma_{jk}}\right)_{E,T}} d\sigma_{jk} + \underset{\text{permittivity}}{\left(\frac{\partial D_i}{dE_j}\right)_{\sigma,T}} dE_j + \underset{\text{pyroelectricity}}{\left(\frac{\partial D_i}{\partial T}\right)_{\sigma,E}} dT \qquad (28)$$

$$dS = \underset{\substack{\text{piezocaloric}\\\text{effect}}}{\left(\frac{\partial S}{\partial\sigma_{ij}}\right)_{E,T}} d\sigma_{ij} + \underset{\substack{\text{electrocaloric}\\\text{effect}}}{\left(\frac{\partial S}{\partial E_i}\right)_{\sigma,T}} dE_i + \underset{\substack{\text{heat}\\\text{capacity}}}{\left(\frac{\partial S}{\partial T}\right)_{\sigma,E}} dT \qquad (29)$$

There are here $9+3+1 = 13$ equations, each with $9+3+1 = 13$ terms on the right-hand side. Each of the differential coefficients is a measure of the physical effect which is indicated above it. It will be seen that the coefficients on the leading diagonal measure the principal effects while the others measure the coupled effects. Proceeding as in § 2 we now establish the thermodynamic relations between the coefficients in these three equations.

An extra term has to be added to equation (12) to represent the electrical work done on the system in a small change. In Chapter IV, § 4, this was shown to be $E_i\,dD_i$.† Since all the changes are reversible, the first and second laws of thermodynamics then give

$$dU = \sigma_{ij}\,d\epsilon_{ij} + E_i\,dD_i + T\,dS. \qquad (30)$$

If the function Φ is now defined by

$$\Phi = U - \sigma_{ij}\,\epsilon_{ij} - E_i\,D_i - TS, \qquad (31)$$

we have, on differentiating and combining with (30),

$$d\Phi = -\epsilon_{ij}\,d\sigma_{ij} - D_i\,dE_i - S\,dT. \qquad (32)$$

In (31) all the quantities on the right are functions of (σ_{ij}, E_i, T), which

---

† The expression used in Chapter IV was $vE_i\,dD_i$, where v was the volume of the crystal, and it was stated that the field was to be confined to the crystal. If we wish to remove this last stipulation we have to proceed as in Appendix F.

define the state of the system. Hence Φ is also a function of (σ_{ij}, E_i, T) and we may write

$$d\Phi = \left(\frac{\partial\Phi}{\partial\sigma_{ij}}\right)_{E,T} d\sigma_{ij} + \left(\frac{\partial\Phi}{\partial E_i}\right)_{\sigma,T} dE_i + \left(\frac{\partial\Phi}{\partial T}\right)_{\sigma,E} dT. \tag{33}$$

By comparing coefficients in equations (32) and (33) we obtain

$$\left(\frac{\partial\Phi}{\partial\sigma_{ij}}\right)_{E,T} = -\epsilon_{ij}, \qquad \left(\frac{\partial\Phi}{\partial E_i}\right)_{\sigma,T} = -D_i, \qquad \left(\frac{\partial\Phi}{\partial T}\right)_{\sigma,E} = -S. \tag{34}$$

Differentiating the first of these equations with respect to E_k, and the second with respect to σ_{ij} after changing the suffixes from i to k, then gives

$$-\left(\frac{\partial^2\Phi}{\partial\sigma_{ij}\,\partial E_k}\right)_T = \left(\frac{\partial\epsilon_{ij}}{\partial E_k}\right)_{\sigma,T} = \left(\frac{\partial D_k}{\partial\sigma_{ij}}\right)_{E,T} = d_{kij}^T. \tag{35}$$

In a similar way,

$$-\left(\frac{\partial^2\Phi}{\partial\sigma_{ij}\,\partial T}\right)_E = \left(\frac{\partial\epsilon_{ij}}{\partial T}\right)_{\sigma,E} = \left(\frac{\partial S}{\partial\sigma_{ij}}\right)_{E,T} = \alpha_{ij}^E, \tag{36}$$

and

$$-\left(\frac{\partial^2\Phi}{\partial E_i\,\partial T}\right)_\sigma = \left(\frac{\partial D_i}{\partial T}\right)_{\sigma,E} = \left(\frac{\partial S}{\partial E_i}\right)_{\sigma,T} = p_i^\sigma. \tag{37}$$

The three relations (35), (36), (37) establish the fact that the matrix of coefficients on the right of equations (27), (28) and (29) is symmetrical about the leading diagonal. Specifically, they show that:

(i) *the coefficients for the converse piezoelectric effect are numerically equal to the coefficients for the direct effect;*

(ii) *the coefficients for thermal expansion are the same as those for the piezocaloric effect;*

(iii) *the coefficients for the pyroelectric effect are the same as those for the electrocaloric effect.*

These equalities are indicated in Fig. 10.1 b, where each connexion is labelled with the coefficient that measures it. In this figure subscripts and superscripts are omitted.

In integrated form equations (27), (28), (29) may now be written as†

$$\left.\begin{array}{l} \epsilon_{ij} = s_{ijkl}^{E,T}\,\sigma_{kl} + d_{kij}^T E_k + \quad\ \alpha_{ij}^E\,\Delta T \\[4pt] D_i = d_{ijk}^T\,\sigma_{jk} + \kappa_{ij}^{\sigma,T} E_j + \quad p_i^\sigma\,\Delta T \\[4pt] \Delta S = \quad \alpha_{ij}^E\,\sigma_{ij} + \quad p_i^\sigma\,E_i + (C^{\sigma,E}/T)\Delta T \end{array}\right\}. \tag{38}$$

† Some reduction in the number of suffixes can be obtained if the foregoing thermo-dynamical treatment is conducted entirely in the matrix notation rather than in tensor notation. The tensor notation has been chosen because it shows the tensor rank of a property, and because it is a general notation that can be extended, when required, to tensors of any rank ; by contrast, the matrix notation is unsuitable for dealing with the tensors of higher rank than four that are needed to describe second-order effects (Ch. XIII, § 3).

ϵ_{ij} and D_i in these equations are to be reckoned relative to their values for zero stress and field and before any change of temperature.

Résumé of the thermodynamical argument. We give now a summary of the foregoing argument stripped of all subscripts and superscripts, so that the essentials may stand out clearly. It will be clear that the resulting notation is not an entirely precise one but this is the price we have to pay for brevity.

Take σ (9 components), E (3 components), and T as independent variables defining the thermodynamic state. Then ϵ (9 components), D (3 components), and S are functions of σ, E, T. Hence

$$
\left.
\begin{aligned}
d\epsilon &= \frac{\partial \epsilon}{\partial \sigma}\, d\sigma + \frac{\partial \epsilon}{\partial E}\, dE + \frac{\partial \epsilon}{\partial T}\, dT \\[6pt]
dD &= \frac{\partial D}{\partial \sigma}\, d\sigma + \frac{\partial D}{\partial E}\, dE + \frac{\partial D}{\partial T}\, dT \\[6pt]
dS &= \frac{\partial S}{\partial \sigma}\, d\sigma + \frac{\partial S}{\partial E}\, dE + \frac{\partial S}{\partial T}\, dT
\end{aligned}
\right\}, \tag{39}
$$

and the partial derivatives are identifiable as the elastic compliances, the piezoelectric coefficients, the coefficients of thermal expansion, and so on. From the first and second laws of thermodynamics we have

$$dU = \sigma\, d\epsilon + E\, dD + T\, dS. \tag{40}$$

We define a function,

$$\Phi = U - \sigma\epsilon - ED - TS.$$

Then, by differentiating and using (40),

$$d\Phi = -\epsilon\, d\sigma - D\, dE - S\, dT. \tag{41}$$

Since Φ is a function of state it follows that

$$\frac{\partial \Phi}{\partial \sigma} = -\epsilon, \quad \frac{\partial \Phi}{\partial E} = -D, \quad \frac{\partial \Phi}{\partial T} = -S, \tag{42}$$

and hence, by further differentiations,

$$\frac{\partial \epsilon}{\partial E} = \frac{\partial D}{\partial \sigma}, \quad \frac{\partial \epsilon}{\partial T} = \frac{\partial S}{\partial \sigma}, \quad \frac{\partial D}{\partial T} = \frac{\partial S}{\partial E}; \tag{43}$$

so that the matrix of coefficients on the right-hand side of equations (39) is symmetrical.† The direct and converse piezoelectric effects are therefore given by the same coefficients; the same is true for thermal

† Shortly, the Jacobian matrix of ϵ, D, S with respect to σ, E, T [the matrix of coefficients on the right-hand side of (39)] is equal to minus the Hessian of Φ with respect to σ, E, T, and this is symmetrical.

expansion and the piezocaloric effect, and also for the pyroelectric and the electrocaloric effects.

Equations (38) may now be written out in full in the matrix notation with suffixes as follows:

$$
\begin{aligned}
\epsilon_1 &= s_{11}^{E,T}\sigma_1 + s_{12}\sigma_2 + s_{13}\sigma_3 + s_{14}\sigma_4 + s_{15}\sigma_5 + s_{16}\sigma_6 \;+\; d_{11}^{T}E_1 + d_{21}E_2 + d_{31}E_3 \;+\; \alpha_1^{E}\Delta T \\
\epsilon_2 &= s_{12}\sigma_1 + s_{22}\sigma_2 + s_{23}\sigma_3 + s_{24}\sigma_4 + s_{25}\sigma_5 + s_{26}\sigma_6 \;+\; d_{12}E_1 + d_{22}E_2 + d_{32}E_3 \;+\; \alpha_2\Delta T \\
\epsilon_3 &= s_{13}\sigma_1 + s_{23}\sigma_2 + s_{33}\sigma_3 + s_{34}\sigma_4 + s_{35}\sigma_5 + s_{36}\sigma_6 \;+\; d_{13}E_1 + d_{23}E_2 + d_{33}E_3 \;+\; \alpha_3\Delta T \\
\epsilon_4 &= s_{14}\sigma_1 + s_{24}\sigma_2 + s_{34}\sigma_3 + s_{44}\sigma_4 + s_{45}\sigma_5 + s_{46}\sigma_6 \;+\; d_{14}E_1 + d_{24}E_2 + d_{34}E_3 \;+\; \alpha_4\Delta T \\
\epsilon_5 &= s_{15}\sigma_1 + s_{25}\sigma_2 + s_{35}\sigma_3 + s_{45}\sigma_4 + s_{55}\sigma_5 + s_{56}\sigma_6 \;+\; d_{15}E_1 + d_{25}E_2 + d_{35}E_3 \;+\; \alpha_5\Delta T \\
\epsilon_6 &= s_{16}\sigma_1 + s_{26}\sigma_2 + s_{36}\sigma_3 + s_{46}\sigma_4 + s_{56}\sigma_5 + s_{66}\sigma_6 \;+\; d_{16}E_1 + d_{26}E_2 + d_{36}E_3 \;+\; \alpha_6\Delta T \\[6pt]
D_1 &= d_{11}^{T}\sigma_1 + d_{12}\sigma_2 + d_{13}\sigma_3 + d_{14}\sigma_4 + d_{15}\sigma_5 + d_{16}\sigma_6 \;+\; \kappa_{11}^{\sigma,T}E_1 + \kappa_{12}E_2 + \kappa_{13}E_3 \;+\; p_1^{\sigma}\Delta T \\
D_2 &= d_{21}\sigma_1 + d_{22}\sigma_2 + d_{23}\sigma_3 + d_{24}\sigma_4 + d_{25}\sigma_5 + d_{26}\sigma_6 \;+\; \kappa_{12}E_1 + \kappa_{22}E_2 + \kappa_{23}E_3 \;+\; p_2\Delta T \\
D_3 &= d_{31}\sigma_1 + d_{32}\sigma_2 + d_{33}\sigma_3 + d_{34}\sigma_4 + d_{35}\sigma_5 + d_{36}\sigma_6 \;+\; \kappa_{13}E_1 + \kappa_{23}E_2 + \kappa_{33}E_3 \;+\; p_3\Delta T \\[6pt]
\Delta S &= \alpha_1^{E}\sigma_1 + \alpha_2\sigma_2 + \alpha_3\sigma_3 + \alpha_4\sigma_4 + \alpha_5\sigma_5 + \alpha_6\sigma_6 \;+\; p_1^{\sigma}E_1 + p_2E_2 + p_3E_3 + \\
&\qquad\qquad\qquad\qquad\qquad\qquad\qquad\qquad\qquad\qquad\qquad\qquad\qquad (C^{\sigma,E}/T)\Delta T
\end{aligned}
$$

$$\tag{44}$$

For brevity, only typical superscripts are inserted. The $\boldsymbol{\alpha}$ matrix has the form shown in Chapter IX, equation (16). The preceding chapters have described the special forms taken by the various matrices in the 32 crystal classes. These results are collected together in Appendix E, where the complete matrix of coefficients on the right-hand side of equations (44) is illustrated for each crystal class.

When the matrices in equations (44) are denoted by single letters we finally have the following compact expression:

$$
\left.
\begin{aligned}
\boldsymbol{\epsilon} &= \mathbf{s}^{E,T}\boldsymbol{\sigma} + \mathbf{d}_t^{T}\mathbf{E} + \quad\;\; \boldsymbol{\alpha}^{E}\Delta T \\
\mathbf{D} &= \mathbf{d}^{T}\boldsymbol{\sigma} + \mathbf{x}^{\sigma,T}\mathbf{E} + \quad\;\; \mathbf{p}^{\sigma}\Delta T \\
\Delta S &= \boldsymbol{\alpha}_t^{E}\boldsymbol{\sigma} + \mathbf{p}_t^{\sigma}\mathbf{E} + (C^{\sigma,E}/T)\Delta T
\end{aligned}
\right\}.
$$

$$\tag{45}$$

It will be observed that the method used to prove the thermodynamical relations (35), (36) and (37) is essentially the same as that used to prove the reciprocal relations $\kappa_{ij} = \kappa_{ji}$ in Chapter IV, § 4, and $s_{ij} = s_{ji}$ in Chapter VIII, § 3: namely, to form the second partial derivatives of a thermodynamic potential (Φ, Ψ and U are all thermodynamic potentials) with respect to two of the independent variables and to interchange the order of differentiation. Thus, *this one principle suffices to prove the symmetry about the leading diagonal of the whole matrix of coefficients on the right-hand side of equations* (44).

The three direct connexions in Fig. 10.1 b between S, D and ϵ and the three connexions between T, E and σ remain to be discussed. By considering the appropriate thermodynamical potentials the following relations can be found between the six coefficients in question:

$$\left(\frac{\partial S}{\partial D_i}\right)_{T,\sigma} = -\left(\frac{\partial E_i}{\partial T}\right)_{D,\sigma} = t_i^\sigma, \quad \text{say,} \tag{46}$$

$$\left(\frac{\partial D_i}{\partial \epsilon_{jk}}\right)_{E,T} = -\left(\frac{\partial \sigma_{jk}}{\partial E_i}\right)_{\epsilon,T} = e_{ijk}^T, \quad \text{say,} \tag{47}$$

$$\left(\frac{\partial S}{\partial \epsilon_{ij}}\right)_{T,E} = -\left(\frac{\partial \sigma_{ij}}{\partial T}\right)_{\epsilon,E} = f_{ij}^E, \quad \text{say.} \tag{48}$$

These three relations are indicated in Fig. 10.1 b by labelling the three inner sides t, e and f and the three outer sides $-t$, $-e$ and $-f$.

The coefficients on the right-hand side of equations (44) are sufficient to define all the first-order effects. Other first derivatives, such as those in (46), (47) and (48) can all be regarded as dependent upon those in (44). For the piezoelectric effects, for example, it is readily proved, by writing down the partial derivatives, that

$$d_{ijk} = e_{ilm} s_{lmjk}^E, \qquad e_{ijk} = d_{ilm} c_{lmjk}^E,$$

either for T or S constant throughout.

4. Relations between coefficients measured under different conditions

We have seen in § 2.2 that the elastic compliances are different under isothermal and adiabatic conditions. We now have to consider the similar differences between the various coefficients that will occur in the general situation of Figs. 10.1 a and b, which include electrical effects.

To define the elastic compliances completely it is necessary to specify not only the thermal conditions, which might be S constant or T constant, but also the electrical conditions; for example $\mathbf{E}$ might be kept constant, or, alternatively, $\mathbf{D}$ might be kept constant. In a similar way the heat capacity of a crystal depends upon both the mechanical and the electrical conditions under which the measurement is made; it depends upon how E_i, D_i, σ_{ij} and ϵ_{ij} are allowed to vary. The conditions we shall be most concerned with are as follows.

T constant (isothermal change). This is the condition in an experiment carried out slowly enough for the crystal to be in thermal equilibrium with its surroundings at all times.

S constant (adiabatic change). Here no heat is allowed to flow into or out of the crystal. This situation is realized when a crystal is in elastic vibration, where the changes occur sufficiently fast for there to be very little heat exchange between the different parts of the crystal.

E *constant.* **E** can be kept zero by holding the whole of the surface of the crystal at the same electrical potential. When **E** = 0 the crystal is said to be *electrically free*, by analogy with the mechanically free state described below.

D *constant.* If **D** is constant in a crystal, $d\mathbf{D} = \kappa_0\,d\mathbf{E} + d\mathbf{P} = 0$; any change in **P** due to piezoelectric or pyroelectric effects must therefore

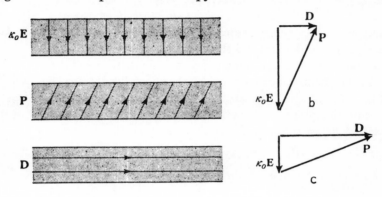

FIG. 10.3. (*a*) The vectors $\kappa_0\mathbf{E}$, **P** and **D** in an isolated thin slab of crystal, far from its edges, (*b*) the relation between the vectors, and (*c*) the effect of a change in **P**.

be balanced by an equal and opposite change in $\kappa_0\mathbf{E}$. This is difficult to achieve in the general case by any simple experimental arrangement. It is, nevertheless, instructive to consider the case of a thin slab of crystal which is completely isolated. We shall only discuss the conditions in the central section of the slab far from its edges (Fig. 10.3 *a*), although it is not difficult to extend the discussion in a qualitative way to cover the edge regions as well. Outside the crystal **D** is zero. Since the normal component of **D** is continuous across the boundary, the only components of **D** that can exist in the crystal are parallel to its surfaces (lower diagram). **E**, on the other hand, must be perpendicular to the surfaces owing to the continuity of its tangential component (which is zero) across the boundary. **P** will, in general, lie in some oblique direction.† The relation between **D**, **E** and **P** is shown in

† In a static experiment compensating charges will, in practice, settle on the surface and will screen the polarization charges. We assume in the theoretical discussion that the insulation is perfect so that this does not happen. The effect does not arise in a dynamic experiment in which **P** is reversed so frequently that there is no time for charges to accumulate.

Fig. 10.3 b. It can be seen from this diagram that, if the conditions were such that **P** was always perpendicular to the surfaces, **D** would always be zero; **E**, which is a depolarizing field caused by the surface polarization charges, would then always be such as to cancel **P** completely $(\kappa_0 \mathbf{E} = -\mathbf{P})$. Only in this case would **D** be constant (zero) in an isolated plate. To see that **D** is not constant in the general case we may consider the effect of a small change, in magnitude and direction, in **P**. **D** and $\kappa_0 \mathbf{E}$ are constrained to be respectively always parallel and perpendicular to the surfaces of the slab; if, therefore, we pass from the configuration of Fig. 10.3 b to that of Fig. 10.3 c, **D** and $\kappa_0 \mathbf{E}$ will change their magnitudes, but will remain fixed in direction. What remains fixed (and equal to zero) during the change is thus the normal component of **D** and the transverse components of **E**.

In this way we see that, while making the surface of the crystal an equipotential ensures that **E** is zero, isolating the crystal does not ensure that **D** is zero. In general, a change in an isolated plate of large extent, far from its edges, is carried out neither with **D** constant nor with **E** constant, but with the normal component of **D** and the transverse components of **E** constant. In general, when **D** = 0 the crystal is said to be *electrically clamped*.

σ *constant*.† The (*mechanically*) *free state* ($\sigma_{ij} = 0$) can be approximated by mounting the crystal in a way that allows strains to take place as freely as possible.

ϵ *constant*.† For ϵ_{ij} to be constant a system of mechanical stresses must be applied which is such that *all* components of strain are held constant. The crystal may be thought of as being surrounded by, and having its surfaces firmly attached to, a medium with infinite elastic stiffness. When $\epsilon_{ij} = 0$ the crystal is said to be (*mechanically*) *clamped*.

4.1. Principal effects. The following expressions show how the conditions of measurement affect the coefficients describing the principal effects. The first of them has already been obtained in § 2.2. The derivation of the other expressions is similar, but requires certain further manipulations of the suffixes; to illustrate these manipulations we work out equation (51) in Appendix G. A study of Fig. 10.1 b will make it clear that all the expressions have essentially the same form. Any expression involving the quantities at one corner of the triangle evidently has a symmetrical counterpart for each of the other two corners. Thus, when one of the expressions has been obtained, the

---

† We write σ and ϵ for σ_{ij} and ϵ_{ij} for brevity in contexts such as this.

others may be written down, except for the detailed placing of the suffixes, simply by inspecting Fig. 10.1 b.

(i) *Isothermal and adiabatic elastic compliances (at constant field).*

$$s^S_{ijkl} - s^T_{ijkl} = -\alpha_{ij}\alpha_{kl}(T/C^\sigma) \quad \textbf{(E} \text{ constant).} \tag{49}$$

The term on the right may be interpreted as a thermoelastic contribution to the compliance.

(ii) *Electrically clamped and free elastic compliances (isothermal).*

$$s^D_{ijkl} - s^E_{ijkl} = -d_{mij}d_{nkl}\beta^\sigma_{mn} \quad (T \text{ constant}) \tag{50}$$

(the piezoelectric contribution to the compliance).

(iii) *Clamped and free permittivities (isothermal).*

$$\kappa^\epsilon_{ij} - \kappa^\sigma_{ij} = -d_{ikl}d_{jmn}c^E_{klmn} \quad (T \text{ constant}) \tag{51}$$

(the piezoelectric contribution to the permittivity).

(iv) *Isothermal and adiabatic permittivities (at constant stress).*

$$\kappa^S_{ij} - \kappa^T_{ij} = -p_i p_j(T/C^E) \quad (\sigma \text{ constant}) \tag{52}$$

(the pyroelectric contribution to the permittivity).

(v) *Electrically clamped, and free, heat capacities (at constant stress).*

$$C^D - C^E = -T p_i p_j \beta^T_{ij} \quad (\sigma \text{ constant}) \tag{53}$$

(the pyroelectric contribution to the heat capacity).

(vi) *Clamped and free heat capacities (at constant field).*

$$C^\epsilon - C^\sigma = -T\alpha_{ij}\alpha_{kl}c^T_{ijkl} \quad \textbf{(E} \text{ constant)} \tag{54}$$

(the thermoelastic contribution to the heat capacity).

(vii) *Elastic compliances at constant normal displacement (isothermal).* As an example of an intermediate type of condition we may consider the isothermal elastic compliances measured on a plate which is of large extent compared to its thickness and which is electrically isolated. As we have seen, this arrangement keeps constant the component of **D** normal to the plate and the components of **E** transverse to the plate. If Ox_1 is the direction normal to the plate, the measured elastic compliances are $(\partial\epsilon_{ij}/\partial\sigma_{kl})_{D_1,E_2,E_3} = s^*_{ijkl}$, say ($T$ constant). It may be shown that the measured compliances are related to s^E_{ijkl} by

$$s^*_{ijkl} - s^E_{ijkl} = -d_{1ij}d_{1kl}/\kappa^\sigma_{11} \quad (T \text{ constant}). \tag{55}$$

4.2. Coupled effects. Similar expressions to those in § 4.1 may be derived for the coefficients describing coupled effects. Again, all the expressions have a common form which is readily apparent on reference to Fig. 10.1 b.

(i) *Pyroelectric and electrocaloric coefficients at constant stress and strain* (see § 4.4).

$$p_i^\epsilon - p_i^\sigma = -\alpha_{jk}^E c_{jklm}^{E,T} d_{ilm}^T. \tag{56}$$

(ii) *Electrically clamped, and free, thermal expansion and piezocaloric coefficients.*

$$\alpha_{ij}^D - \alpha_{ij}^E = -d_{kij}^T \beta_{kl}^{\sigma,T} p_l^\sigma. \tag{57}$$

(iii) *Isothermal and adiabatic direct and converse piezoelectric coefficients.*

$$d_{ijk}^S - d_{ijk}^T = -p_i^\sigma (T/C^{\sigma,E}) \alpha_{jk}^E. \tag{58}$$

4.3. The magnitude of the effects. The magnitude of all the possible interactions between the crystal properties we are considering is given when the magnitudes of the partial differential coefficients in equations (27), (28), (29) are given. Some numerical values have been quoted in previous chapters. The following schematic array summarizes the orders of magnitude of the coefficients at room temperature for non-metallic and non-ferroelectric crystals, when they are not forced to be zero for reasons of symmetry:

	σ	E	T			σ	E	T
ϵ	s	d	α		ϵ	10^{-11}	3×10^{-12}	10^{-5}
D	d	κ	p		D	3×10^{-12}	10^{-10}	3×10^{-6}
S	α	p	C/T		S	10^{-5}	3×10^{-6}	10^4

(Values in rationalized m.k.s. units)

These values are necessarily very approximate. They do not take account of the spread in the values of the different components of each tensor; nor do they indicate the considerable spread in the magnitudes of the coefficients for different crystals. [For ferroelectric crystals (Ch. IV, § 8) the orders of magnitude of some of the coefficients are quite different from those shown above, and, in addition, some coefficients show a marked temperature dependence.]

For the purpose of order of magnitude calculations the tensors may be treated as scalars. The relative difference between the adiabatic and the isothermal elastic compliances, $(s^{S,E} - s^{T,E})/s^{T,E}$, is, from equation (49), of order of magnitude $\dfrac{\alpha^2}{(C/T)s}$. This is a dimensionless ratio which measures the strength of the thermoelastic interaction;† when the numerical values quoted above are inserted, the ratio is found to

† The square roots of ratios such as this are known as *thermoelastic, electrothermal and electromechanical coupling factors.*

be equal to 10^{-3}. This and other relative differences are given in Table 11. The last column of the table shows that the differences between the coefficients measured under different conditions are, in general, small corrections of up to 1 per cent. (although it must be remembered that the numbers given are the products and quotients of coefficients already only roughly specified). It will be noticed, however, that the

TABLE 11

Relative differences between properties measured under different conditions

Equation	Difference	Order of magnitude of relative difference	
(49)	$s^{S,E} - s^{T,E}$	$\dfrac{\alpha^2}{(C/T)s}$	10^{-3}
(50)	$s^{D,T} - s^{E,T}$	$\dfrac{d^2}{\kappa s}$	10^{-2}
(51)	$\kappa^{\epsilon,T} - \kappa^{\sigma,T}$	$\dfrac{d^2}{\kappa s}$	10^{-2}
(52)	$\kappa^{S,\sigma} - \kappa^{T,\sigma}$	$\dfrac{p^2}{(C/T)\kappa}$	10^{-5}
(53)	$C^{D,\sigma} - C^{E,\sigma}$	$\dfrac{p^2}{(C/T)\kappa}$	10^{-5}
(54)	$C^{\epsilon,E} - C^{\sigma,E}$	$\dfrac{\alpha^2}{(C/T)s}$	10^{-3}
(56)	$p^\epsilon - p^\sigma$	$\dfrac{d\alpha}{sp}$	1
(57)	$\alpha^D - \alpha^E$	$\dfrac{dp}{\kappa\alpha}$	10^{-2}
(58)	$d^S - d^T$	$\dfrac{p\alpha}{(C/T)d}$	10^{-3}

relative difference between the pyroelectric effect at constant stress and constant strain is not a small one; it is of the order of 100 per cent. This special position of pyroelectricity is the reason for the attention that is paid to *secondary pyroelectricity*, which we discuss in the next section. It should not be thought that the singling out of pyroelectricity for special attention means that it is different in principle from the other coupled effects. A similar discussion could equally well be given of 'secondary' thermal expansion, or 'secondary' piezoelectricity, if the magnitude of these effects warranted it.

4.4. Primary and secondary pyroelectricity. Throughout this section E is supposed constant. Since $D = f_1(\epsilon, T)$ and $\epsilon = f_2(\sigma, T)$, we may write

$$dD = \left(\frac{\partial D}{\partial \epsilon}\right)_T d\epsilon + \left(\frac{\partial D}{\partial T}\right)_\epsilon dT, \qquad d\epsilon = \left(\frac{\partial \epsilon}{\partial \sigma}\right)_T d\sigma + \left(\frac{\partial \epsilon}{\partial T}\right)_\sigma dT;$$

the suffixes are omitted, but could readily be inserted if required for the argument. Substituting the value of $d\epsilon$ from the second equation into the first, putting $d\sigma = 0$ and dividing by dT, we obtain

$$\overset{\text{primary}}{} \quad \overset{\text{secondary}}{}$$

$$\left(\frac{\partial D}{\partial T}\right)_\sigma = \left(\frac{\partial D}{\partial T}\right)_\epsilon + \left(\frac{\partial D}{\partial \epsilon}\right)_T \left(\frac{\partial \epsilon}{\partial T}\right)_\sigma \quad \text{(E constant).} \qquad (59)$$

The term $(\partial D/\partial T)_\sigma$ in (59) represents the pyroelectric effect at constant

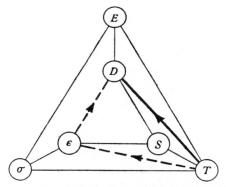

Fig. 10.4. Primary and secondary pyroelectricity. The full line illustrates the primary effect (with the strain ϵ constant); the broken line illustrates the secondary effect which can occur when the crystal is free to deform.

stress. It is the effect that would be measured if the crystal were free to change its shape. The equation shows that the measured effect can be divided into two parts, represented by the two terms on the right. There is first the *primary* (or 'true') *pyroelectric effect* given by $(\partial D/\partial T)_\epsilon$. This measures the displacement (polarization) that would be produced by a temperature change if the shape and volume of the crystal were held fixed. The second term measures the *secondary pyroelectric effect*, or 'false pyroelectricity of the first kind': that is, the additional pyroelectricity produced because the crystal is free to deform. This additional pyroelectricity is seen to arise from the path $T \to \epsilon \to D$ (Fig. 10.4). Thus, when the crystal is free to deform, thermal expansion causes a strain, which in turn, by the piezoelectric path $\epsilon \to D$, produces a displacement.

If $(\partial D/\partial\epsilon)_T$ in (59) is expressed as a product of two derivatives, we have

$$\underset{\text{primary}}{\left(\frac{\partial D}{\partial T}\right)_\sigma} = \underset{}{\left(\frac{\partial D}{\partial T}\right)_\epsilon} + \underset{\text{secondary}}{\left(\frac{\partial D}{\partial \sigma}\right)_T\left(\frac{\partial \sigma}{\partial \epsilon}\right)_T\left(\frac{\partial \epsilon}{\partial T}\right)_\sigma} \qquad \text{(E constant)}. \qquad (60)$$

This equation, which is equivalent to (56), gives the secondary effect in terms of the piezoelectric coefficients, the elastic stiffnesses, and the thermal expansion coefficients (path $T \to \epsilon \to \sigma \to D$ in Fig. 10.4).

In practice it happens that the secondary effect is numerically greater than the primary, and it may even be that the primary effect is too small to be observed in any crystal. $(\partial D/\partial T)_\sigma$ is what is normally measured, and to deduce $(\partial D/\partial T)_\epsilon$ from it requires a knowledge of the piezoelectric, elastic and thermal expansion coefficients. Since $(\partial D/\partial T)_\epsilon$ is relatively so small, a small error in the determination of $(\partial D/\partial T)_\sigma$ may make the difference between finding a finite primary effect and none at all.†

The relation of secondary pyroelectricity to the crystal symmetry is a question that deserves some attention. All crystals show the three principal effects indicated at the corners of Figs. 10.1 a and b. All crystals, too, show thermal expansion and thermoelastic effects. However, we have seen that pyroelectricity (primary and secondary) and piezoelectricity can only exist in crystals belonging to certain classes. Now the symmetry of some crystals allows them to show piezoelectricity but not pyroelectricity, and this raises the question: what then becomes of the secondary pyroelectricity that arises through thermal expansion and piezoelectricity? By the crystal symmetry it cannot exist, although the symmetry nevertheless allows both thermal expansion and piezo-electricity. α-quartz (class 32) is an example of such a crystal. It possesses three diad axes, and pressure along any one of these causes piezoelectricity to appear. When it is heated it expands, and equal strains are set up along each of the three diad axes. These strains would give rise to piezoelectricity were it not for the fact that they are mutually inclined at 120° to one another, so that the effects all cancel. There is therefore no secondary pyroelectric effect. It may be noted incidentally that, since the symmetry of quartz forbids it to be pyro-electric, it will not show any 'secondary piezoelectricity' arising from thermoelastic and pyroelectric effects [equation (58)].

The same considerations apply to the cubic form of zinc sulphide (zinc blende), class $\bar{4}3m$, which possesses four polar triad axes. It is

† For further details see Cady's *Piezoelectricity* (1946), chapter xx.

piezoelectric but not pyroelectric. On the other hand, the hexagonal form (wurtzite), class *6mm*, in which the hexad axis is a unique direction, is both piezoelectric and pyroelectric.

In all the phenomena dealt with in this chapter the state of the crystal is assumed to be the same at all points. *Tertiary pyroelectricity*, or 'false pyroelectricity of the second kind', is essentially the result of a non-uniform state of affairs. Uneven heating causes temperature gradients which, by thermal expansion, cause non-uniform stresses and strains. These in turn can produce a polarization by piezoelectric effects, which, unless the experiments are carefully performed, can be mistaken for primary or secondary pyroelectricity.

SUMMARY

FIGS. 10.1 *a* and *b* summarize the relations between the thermal, electrical and mechanical properties of a crystal. If the quantities at the three outer corners are taken as the independent variables, the first-order (linear) effects may be written in matrix notation as

$$\left.\begin{array}{l} \boldsymbol{\epsilon} = \mathbf{s}^{E,T}\boldsymbol{\sigma} + \quad \mathbf{d}_t^T \mathbf{E} + \qquad \boldsymbol{\alpha}^E \Delta T \\ \mathbf{D} = \quad \mathbf{d}^T\boldsymbol{\sigma} + \mathbf{x}^{\sigma,T}\mathbf{E} + \qquad \mathbf{p}^\sigma \Delta T \\ \Delta S = \quad \boldsymbol{\alpha}_t^E\boldsymbol{\sigma} + \quad \mathbf{p}_t^\sigma \mathbf{E} + (C^{\sigma,E}/T)\Delta T \end{array}\right\}. \qquad (45)$$

The terms on the leading diagonal represent the *principal effects*, namely, elasticity, permittivity and heat capacity. The off-diagonal terms represent the *coupled effects*, namely, the direct and converse piezoelectric effect, thermal expansion and the piezocaloric effect, and pyroelectricity and the electrocaloric effect. The symmetry of the complete matrix of coefficients on the right-hand side of these equations follows from the thermodynamical argument that is given in § 3 and summarized on pp. 181–2.

The diagram of Figs. 10.1 *a* and *b* emphasizes the formal symmetry between the properties. Thus, a calculation of the difference between the isothermal and adiabatic elastic compliances (at constant field) proceeds in an exactly analogous way to a calculation of the difference between the heat capacities at constant stress and constant strain (at constant field). A calculation of the difference between the isothermal permittivities at constant stress and constant strain is likewise similar. Expressions for such differences are given as equations (49) to (58). Equations (49) to (54) express, in order, the thermoelastic and the piezoelectric contributions to the elastic compliances, the piezoelectric and pyroelectric contributions to the permittivities, and the pyroelectric and the thermoelastic contributions to the heat capacity. Equations (56) to (58) express the secondary pyroelectric effect arising from thermal expansion and piezoelectricity, the secondary thermal expansion arising from pyroelectricity and piezoelectricity, and the secondary piezoelectricity arising from the piezocaloric effect and pyroelectricity. These secondary effects are usually 1 per cent. or less of the main effects. However, secondary pyroelectricity is not a small effect and represents a major part of the pyroelectric phenomenon usually measured.

PART 3

TRANSPORT PROPERTIES

Chapters XI and XII describe three further crystal properties represented by tensors, namely, thermal and electrical conductivity, and thermo-electricity. Since they are concerned with transport processes and thermodynamically irreversible phenomena, these properties do not fit into the scheme of reversible effects set up in Chapter X. It is therefore more convenient and logical to treat them separately here.

THERMAL AND ELECTRICAL CONDUCTIVITY

1. The thermal conductivity and resistivity tensors

(i) *Conductivity*. When a difference of temperature is maintained between different parts of a solid there is, in general, a flow of heat. If h_1, h_2, h_3 are the quantities of heat traversing, in unit time, unit areas perpendicular to Ox_1, Ox_2, Ox_3, respectively, it is easy to show that they are the components of a vector $\mathbf{h}$ (in the sense defined in Ch. I, § 3). $\mathbf{h}$ is in the direction of heat flow and, if a unit cross-section is taken perpendicular to $\mathbf{h}$, the rate of flow of heat across it is h. In an isotropic solid the heat conduction obeys the law

$$\mathbf{h} = -k \operatorname{grad} T, \tag{1}$$

where k is a positive coefficient, the *thermal conductivity*. The heat thus flows in the direction of the greatest fall in temperature, and the rate of flow is directly proportional to the temperature gradient. In suffix notation,

$$h_i = -k \frac{\partial T}{\partial x_i}. \tag{2}$$

In a crystal $\mathbf{h}$ is not, in general, parallel to $\operatorname{grad} T$ and equation (2) is replaced by

$$h_i = -k_{ij} \frac{\partial T}{\partial x_j}. \tag{3}$$

Each component of $\mathbf{h}$ now depends on all three components of temperature gradient, and not only on one of them as in an isotropic medium. Since the coefficients k_{ij} relate two vectors, they form a second-rank tensor, the *thermal conductivity tensor*. Each of the k_{ij} can be given a physical meaning. For example, if unit temperature gradient is set up along Ox_2 the heat flow† along this axis is $-k_{22}$, while $-k_{12}$ and $-k_{32}$ measure the two transverse heat flows parallel to Ox_1 and Ox_3.

Now it may be proved, as we shall discuss in §§ 5 and 6, that

$$k_{ij} = k_{ji}. \tag{4}$$

$[k_{ij}]$ is thus a symmetrical tensor and may be written

$$\begin{bmatrix} k_{11} & k_{12} & k_{31} \\ k_{12} & k_{22} & k_{23} \\ k_{31} & k_{23} & k_{33} \end{bmatrix},$$

† By 'heat flow' in contexts like this we shall always mean 'rate of flow of heat per unit cross-section', i.e. time rate per unit cross-section.

or, referred to its principal axes,

$$\begin{bmatrix} k_1 & 0 & 0 \\ 0 & k_2 & 0 \\ 0 & 0 & k_3 \end{bmatrix}.$$

Referred to the principal axes, equations (3) become simply

$$h_1 = -k_1 \frac{\partial T}{\partial x_1}, \qquad h_2 = -k_2 \frac{\partial T}{\partial x_2}, \qquad h_3 = -k_3 \frac{\partial T}{\partial x_3}. \tag{5}$$

The representation quadric for thermal conductivity is

$$k_{ij} x_i x_j = 1, \tag{6}$$

or, referred to the principal axes,

$$k_1 x_1^2 + k_2 x_2^2 + k_3 x_3^2 = 1. \tag{7}$$

Experimentally, k_1, k_2 and k_3 are always found to be positive, and so (7) is always an ellipsoid, the *thermal conductivity ellipsoid*. Its shape and orientation must conform to the crystal symmetry (Ch. I, § 5.1). A few numerical values are given in Table 12.

TABLE 12

Thermal conductivities of crystals

Values of the principal conductivities in m.k.s. units [joule/(m sec °C)]

Crystal	System	Temperature °C	k_1, k_2	k_3
Quartz . .	trigonal	30	6·5	11·3
Calcite . .	,,	30	4·18	4·98
Bismuth . .	,,	18	9·24	6·65
Graphite . .	hexagonal	30	355	89
Aluminium. .	cubic	30	208	
Copper . .	,,	0	410	

Note. $k_1 = k_2$ for all the crystals listed, but this result would not, of course, be true for crystals of the optically biaxial systems. Values are taken from the *International Critical Tables* (1929), **5**, 231.

(ii) *Resistivity.* If the three equations (3) were solved for the $\partial T / \partial x_j$ the result would be

$$\frac{\partial T}{\partial x_i} = -r_{ij} h_j, \tag{8}$$

where the r_{ij} are functions of the k_{ij}. The r_{ij} relate two vectors and hence form a second-rank tensor, the *thermal resistivity tensor*. Written as a matrix $\mathbf{r} = (r_{ij})$ the resistivity is evidently the reciprocal of the conductivity matrix $\mathbf{k} = (k_{ij})$ (Ch. IX, § 4.2):

$$\mathbf{r} = \mathbf{k}^{-1}. \tag{9}$$

[It is hardly necessary to remark that the relation between the individual components of $[k_{ij}]$ and $[r_{ij}]$ is not, in general, one of simple reciprocity. Thus, for example,

$$r_{12} \neq (k_{12})^{-1}.$$

k_{12} and r_{12} refer to quite different experiments. k_{12} gives the result of measuring the x_1 component of heat flow in an experiment in which the temperature gradient is along x_2; thus

$$h_1 = -k_{12}\frac{dT}{dx_2}.$$

r_{12}, on the other hand, refers to an experiment in which the x_1 component of temperature gradient is measured while the total flow of heat is along x_2; thus

$$\frac{\partial T}{\partial x_1} = -r_{12}h_2.]$$

Since $k_{ij} = k_{ji}$, it is easily proved from the relation

$$r_{ij} = \frac{(-1)^{i+j}\Delta_{ij}}{|k_{ij}|},$$

where Δ_{ij} is the determinant which remains after striking out the ith row and the jth column of $|k_{ij}|$ (Ch. IX, § 4.2), that

$$r_{ij} = r_{ji}. \tag{10}$$

When the symmetrical tensor $[r_{ij}]$ is referred to its principal axes, whose directions are the same as those of the conductivity tensor, we have

$$\frac{\partial T}{\partial x_1} = -r_1 h_1, \qquad \frac{\partial T}{\partial x_2} = -r_2 h_2, \qquad \frac{\partial T}{\partial x_3} = -r_3 h_3,$$

and the relations

$$r_1 = 1/k_1, \qquad r_2 = 1/k_2, \qquad r_3 = 1/k_3.$$

The resistivity ellipsoid, whose equation referred to the principal axes is

$$r_1 x_1^2 + r_2 x_2^2 + r_3 x_3^2 = 1,$$

thus has semi-axes of lengths inversely proportional to those of the conductivity ellipsoid.

2. Two special cases of steady heat flow

It is instructive to consider the geometry of two special cases of steady-state heat flow. By a steady state is meant one in which the heat flow and the temperature at any point do not change with time.

(i) *Heat flow across a flat plate.* The first case is illustrated in Fig. 11.1 a; the two surfaces of a large flat plate of crystal are in contact with two good conductors which maintain the surfaces at different temperatures. Since the crystal is of large extent compared with its thickness, the isothermal surfaces must run parallel to the surfaces of the crystal, apart from end effects. The temperature gradient is therefore prescribed to be perpendicular to the plate, and the heat flow must, in general, lie in some other direction. Figs. 11.1 b and c show, in a

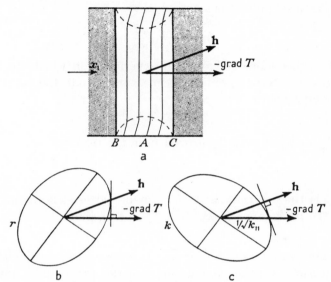

FIG. 11.1. Heat flow across a flat plate between good conductors. The directions of $-\mathrm{grad}\, T$ and **h** in relation to (a) the plate, (b) the resistivity ellipsoid, and (c) the conductivity ellipsoid.

two-dimensional representation, how **h** and $-\mathrm{grad}\, T$ are arranged in relation to the resistivity ellipsoid and the conductivity ellipsoid respectively. The situation is similar to that of an anisotropic dielectric between parallel conducting plates, shown in Fig. 4.3, p. 73.

To express the result analytically we note that, since it is the direction of $\mathrm{grad}\, T$ rather than **h** that is prescribed, it is better to work with conductivities than with resistivities; we write

$$h_1 = -k_{11}\frac{dT}{dx_1}, \qquad h_2 = -k_{12}\frac{dT}{dx_1}, \qquad h_3 = -k_{31}\frac{dT}{dx_1},$$

where x_1 is taken perpendicular to the plates. Now the rate of flow of heat across the plate is given not by h but by the component h_1. h_2 and h_3 merely denote transverse heat flows (we discuss below what becomes of h_2 and h_3 at the edges). k_{11} is therefore the quantity most readily measured in the experiment. $1/\sqrt{k_{11}}$ is shown in Fig. 11.1 c as the length of the radius vector in the x_1 direction.

(ii) *Heat flow down a long rod.* Let the crystal shown in the form of a long rod in Fig. 11.2 a be of the same substance and in the same orientation as the crystal in Fig. 11.1 a. If a temperature difference is now maintained between the two ends of the rod, the conductivity of the crystal is supposed to be so much greater than that of its surroundings that the direction of heat flow must be parallel to its axis. The

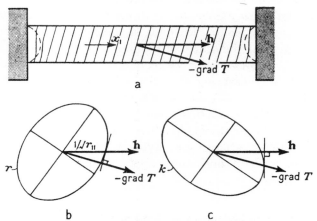

Fig. 11.2. Heat flow down a long rod. The directions of $-\mathrm{grad}\ T$ and **h** in relation to (a) the rod, (b) the resistivity ellipsoid, and (c) the conductivity ellipsoid.

temperature gradient must therefore, in general, be in an oblique direction. The arrangement of **h** and $-\mathrm{grad}\ T$ in relation to the representation ellipsoids is shown in Figs. 11.2 b and c.

Since the direction of **h** is prescribed (parallel to x_1) it is now better to work with resistivities. Thus,

$$\frac{\partial T}{\partial x_1} = -r_{11}h_1, \qquad \frac{\partial T}{\partial x_2} = -r_{12}h_1, \qquad \frac{\partial T}{\partial x_3} = -r_{31}h_1.$$

The temperature gradient down the length of the rod, which is what would be most easily measured in the experiment, is $\partial T/\partial x_1$. This is connected with the heat flow down the rod by r_{11}. The value of $1/\sqrt{r_{11}}$ is indicated in Fig. 11.2 b as the length of the radius vector in the x_1 direction.

(iii) *End effects.* At the edges of the plate in Fig. 11.1 a the heat flow must be parallel to x_1, which in turn means that the isothermals can no longer be parallel to the plate surfaces. In fact the arrangement of **h** and grad T at a point such as A in Fig. 11.1 a is the same as in Fig. 11.2. B and C are singular points. There is thus an end effect over the region roughly outlined by the broken line.

If the ends of the rod in Fig. 11.2 a were rectangular and were placed in contact with good conductors, as shown, there would be a similar end effect. At the ends the isothermals are constrained to be transverse to the rod axis, and the direction of **h** is accordingly that shown in Fig. 11.1 a. The end effects might be removed by shaping the edges of the plate in Fig. 11.1 a parallel to the direction taken by **h** at the centre of the plate; and by making the ends of the rod in Fig. 11.2 a parallel to the isothermals at the centre of the rod.

(iv) *General.* The two cases we have discussed provide the basis of most methods of measuring the thermal conductivities of crystals. The quantity most directly measurable with the flat plate arrangement is evidently k_{11}, the *conductivity* normal to the plate, while with the long rod it is r_{11}, the *resistivity* along the rod. The situation is analogous to that in elasticity, where by applying a known stress one measures a compliance, while if one could apply a known strain one would measure a stiffness.

The principal conductivities and resistivities of a crystal, and the orientation of the principal axes, may be calculated from a number of measurements made in different directions. The calculation is essentially the same as that for any other second-rank tensor property; some examples have been given in the exercises on p. 109 and in Chapter IX, § 7.

EXERCISE 11.1. Suppose the crystal in Fig. 11.2 a is quartz. Use the conductivities given in Table 12 to calculate the orientation of the crystal for which the angle θ between the vectors $(-\text{grad } T)$ and **h**, far from the ends of the rod, is a maximum. Calculate also the value of θ_{max}.

3. Steady-state heat flow in general

Let us now consider the general equations that govern the conduction of heat in a crystal when a steady state has been set up. Suppose, for generality, that there are sources and sinks of heat distributed continuously throughout the crystal, so that the net rate of production of heat per unit volume of the crystal is $\dot{q}$. $\dot{q}$ may be positive or negative. In a steady state the difference between the heat leaving and entering any volume must equal the heat created within it. We therefore have the conservation equation

$$\text{div}\,\mathbf{h} = \dot{q} \quad \text{or} \quad \frac{\partial h_i}{\partial x_i} = \dot{q}.$$

Substituting for h_i from equation (3) gives

$$k_{ij}\frac{\partial^2 T}{\partial x_i\,\partial x_j} = -\dot{q}, \tag{11}$$

a differential equation for the temperature T which is identical in form with the equation satisfied by the electrostatic potential ϕ in a crystal containing a continuous distribution of charge (Ch. IV, equation (17)). Using the fact that $k_{ij} = k_{ji}$, and transforming to the principal axes of $[k_{ij}]$, we obtain

$$k_1\frac{\partial^2 T}{\partial x_1^2} + k_2\frac{\partial^2 T}{\partial x_2^2} + k_3\frac{\partial^2 T}{\partial x_3^2} = -\dot{q} \tag{12}$$

as the equation to be satisfied by T.

The essential problem is: knowing the temperatures on certain boundaries and knowing the distribution of sources of heat within the crystal, if any, to solve equation (12) and thereby calculate the temperature distribution.

If the medium were isotropic with conductivity k, equation (12) would take the simpler form

$$k\left(\frac{\partial^2 T}{\partial x_1^2} + \frac{\partial^2 T}{\partial x_2^2} + \frac{\partial^2 T}{\partial x_3^2}\right) = -\dot{q}, \tag{13}$$

which is Poisson's equation. Many solutions of this equation for various boundary conditions and distributions of sources are known—from electrostatics, for example. We shall now describe a simple substitution whereby the solution to any steady heat flow problem in a crystal can be derived from a corresponding solution in an isotropic medium.

Let us transform equation (12) by the equations

$$x_1 = (k_1\,k_2\,k_3)^{-\frac{1}{3}}k_1^{\frac{1}{2}}X_1, \qquad x_2 = (k_1\,k_2\,k_3)^{-\frac{1}{3}}k_2^{\frac{1}{2}}X_2,$$
$$x_3 = (k_1\,k_2\,k_3)^{-\frac{1}{3}}k_3^{\frac{1}{2}}X_3. \tag{14}$$

This transformation may be pictured as a homogeneous strain. However, the volume of a small element remains fixed during the transformation, for

$$dX_1\,dX_2\,dX_3 = dx_1\,dx_2\,dx_3;$$

hence, the source strength $\dot{q}$ per unit volume is unchanged. Accordingly, equation (12) becomes

$$(k_1\,k_2\,k_3)^{\frac{1}{3}}\left(\frac{\partial^2 T}{\partial X_1^2} + \frac{\partial^2 T}{\partial X_2^2} + \frac{\partial^2 T}{\partial X_3^2}\right) = -\dot{q}.$$

But this is the equation for the temperature distribution produced by sources $\dot{q}$ in an isotropic medium of conductivity $(k_1 k_2 k_3)^{\frac{1}{3}}$.

The above result allows the following procedure for solving a problem in an anisotropic medium. Start with the solution to an appropriate heat-flow problem in an isotropic medium of conductivity $(k_1 k_2 k_3)^{\frac{1}{3}}$. We form a picture of the solution by imagining the sources, the boundaries, the isothermals, and the lines of heat flow as drawn in space. Now distort the picture according to the transformation (14). Then the new picture of sources, boundaries, isothermals and lines of heat flow represents the solution to a problem in an anisotropic medium with principal conductivities k_1, k_2, k_3. During the distortion we stipulate that the sources retain their strengths and the isothermals retain their temperatures.

The reader should satisfy himself that this procedure is valid. That the transformation works for the isothermals follows from what we have done above. That it works also for the heat flow vectors requires additional justification. It is readily proved that, if the components of heat flow at X_1, X_2, X_3 in the isotropic problem are H_1, H_2, H_3 and the components at x_1, x_2, x_3 in the corresponding anisotropic problem are h_1, h_2, h_3, then

$$\frac{h_1}{H_1} = \frac{x_1}{X_1}, \quad \text{etc.}$$

That is, the heat flow components transform like the coordinates, which is the result we want. Notice that a similar result would not apply to the temperature gradient vectors, for they are in both cases perpendicular to the isothermals.

The transformation procedure is illustrated by the example given in the next section § 3.1.

3.1. Heat flow from a point source. The above method may be used to calculate the temperature distribution around a point source of heat situated in an infinite crystal. But first we may notice that the form of the isothermal surfaces in this problem may be deduced very directly by using the radius-normal property of the representation quadric. We anticipate the result that heat flows out from the point source in straight lines (§ 5); if, therefore, the resistivity ellipsoid is constructed centred on the point (Fig. 11.3), the radius vector is always the direction of **h** and the outward normal where it cuts the surface is the direction of $-\text{grad } T$. Hence, the ellipsoid itself is similar to an isothermal surface. *The isothermal surfaces surrounding a point source of heat in an infinite crystal are all similar in shape and orientation to the resistivity ellipsoid for the crystal.*

Coming now to the temperature distribution, we begin by solving the corresponding problem in an isotropic medium. Let the point source be situated at the origin and emitting an amount of heat $\dot{Q}$ per unit

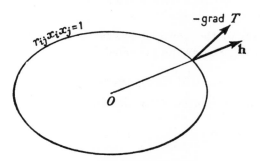

FIG. 11.3. The isothermal surfaces surrounding a point source of heat in an infinite crystal are all similar in shape and orientation to the resistivity ellipsoid for the crystal.

time. At all points of the medium except at the source $\dot{q} = 0$, and therefore, from equation (13), using X_1, X_2, X_3 in place of x_1, x_2, x_3,

$$\frac{\partial^2 T}{\partial X_1^2} + \frac{\partial^2 T}{\partial X_2^2} + \frac{\partial^2 T}{\partial X_3^2} = 0$$

(Laplace's equation). It is well known, and may readily be verified, that a solution of this equation is

$$T = \frac{A}{R} + T_\infty, \tag{15}$$

where A is a constant, $R = (X_1^2 + X_2^2 + X_3^2)^{\frac{1}{2}}$, and T_∞ is the temperature far from the source. This solution is chosen because we know that the isothermals are spheres. According to (15), T is infinite at the origin, but the difficulty may be simply removed by imagining the heat to be supplied not from a point but from a small finite spherical source.

The value of A is fixed by calculating the rate of heat flow across the surface of a sphere of radius R, and equating it to $\dot{Q}$. Thus, the radial temperature gradient is $-A/R^2$, the heat flow per unit area of the surface is then kA/R^2, and therefore

$$\frac{kA}{R^2} \cdot 4\pi R^2 = \dot{Q}.$$

Hence
$$A = \frac{\dot{Q}}{4\pi} \cdot k^{-1}.$$

Substituting this value of A into equation (15), we find the following expression for the temperature distribution in the isotropic case:

$$T - T_\infty = \frac{Q}{4\pi} \cdot k^{-1} (X_1^2 + X_2^2 + X_3^2)^{-\frac{1}{2}}.$$

The transition to the anisotropic case is made as described in § 3. Choosing Ox_1, Ox_2, Ox_3 as the principal axes, putting

$$k = (k_1 k_2 k_3)^{\frac{1}{3}},$$

and making the substitution (14), we obtain immediately

$$T - T_\infty = \frac{Q}{4\pi} (k_1 k_2 k_3)^{-\frac{1}{3}} \left(\frac{x_1^2}{k_1} + \frac{x_2^2}{k_2} + \frac{x_3^2}{k_3} \right)^{-\frac{1}{2}} \tag{16}$$

as the required distribution.

Along any given radius $(T - T_\infty)$ evidently falls off inversely as the distance from the source. It will be observed that our previous result for the shape of the isothermals is verified, for the spherical isothermals of the isotropic problem evidently distort into ellipsoids similar in shape to the resistivity ellipsoid. Notice also that, in accordance with what was said at the end of § 3, the lines of heat flow remain straight, while the lines of maximum temperature gradient do not.

4. Electrical conductivity

The formal analysis of the conduction of electricity in anisotropic crystals is similar to that of the conduction of heat. The fundamental equation is the generalized form of Ohm's Law:

$$j_i = -\sigma_{ik} \frac{\partial \phi}{\partial x_k} = \sigma_{ik} E_k, \tag{17}$$

where j_i is the current density, σ_{ik} is the electrical conductivity tensor, ϕ is the potential and E_k is the electric field intensity. Alternatively, the field components may be written in terms of the current density components by using the electrical resistivity tensor ρ_{ik}, thus

$$E_i = \rho_{ik} j_k. \tag{18}$$

The resistivity matrix is the reciprocal of the conductivity matrix. Numerical values of the electrical resistivities of some metallic crystals are given in Table 13.

The rate of *Joule heating* of a conductor is expressed by the scalar product of the current density and the field. In a crystal, therefore, the heat produced in unit time and unit volume is

$$j_i E_i = \rho_{ik} j_i j_k = \rho j^2, \tag{19}$$

where j is the magnitude of the current density and ρ is the resistivity in the direction of the current.

TABLE 13

Electrical resistivities of metallic crystals

Values of principal resistivities at 20° C (unit = 10^{-8} ohm–m)

Crystal	System	ρ_1, ρ_2	ρ_3
Tin	tetragonal	9·9	14·3
Bismuth . . .	trigonal	109	138
Cadmium . . .	hexagonal	6·80	8·30
Zinc	,,	5·91	6·13
Tungsten . . .	cubic	5·48	
Copper . . .	,,	1·51	

Values taken from *International Critical Tables* (1929), **6**, 135.

5.† The reciprocal relation $k_{ij} = k_{ji}$

During the foregoing discussion of heat conduction in anisotropic crystals we assumed the validity of the relation

$$k_{ij} = k_{ji}. \tag{4}$$

That this assumption is justifiable is by no means obvious. It asserts, for example, that, if a temperature gradient in the x_1 direction produces a certain heat flow in the x_2 direction (given by k_{21}), the same temperature gradient applied in the x_2 direction would give a heat flow of precisely the same magnitude in the x_1 direction (given by k_{12}). To see what is involved in the assumption it is helpful first to discover what the consequences would be if (4) did not hold.

First split $[k_{ij}]$ into a symmetrical and an antisymmetrical part (p. 97). Then rotate the axes of reference to coincide with the principal axes of the symmetrical part. The other part remains antisymmetrical during the transformation (Ch. I, § 3.2), and so the tensor takes the new form

$$\begin{bmatrix} k_{11} & 0 & 0 \\ 0 & k_{22} & 0 \\ 0 & 0 & k_{33} \end{bmatrix} + \begin{bmatrix} 0 & k_{12} & -k_{31} \\ -k_{12} & 0 & k_{23} \\ k_{31} & -k_{23} & 0 \end{bmatrix} = \begin{bmatrix} k_{11} & k_{12} & -k_{31} \\ -k_{12} & k_{22} & k_{23} \\ k_{31} & -k_{23} & k_{33} \end{bmatrix}.$$

The presence of symmetry in the crystal will tend to impose restrictions

† The rest of this chapter is more difficult and may be omitted on a first reading.

on this scheme of coefficients. By way of illustration we take the case of a crystal having a 2-, 3-, 4-, or 6-fold axis (either rotation or inverse) parallel to x_3. If a temperature gradient is established parallel to x_3, the heat flow must also be parallel to x_3, by symmetry. Hence

$$k_{23} = k_{31} = 0,$$

and the tensor reduces to

$$\begin{bmatrix} k_{11} & k_{12} & 0 \\ -k_{12} & k_{22} & 0 \\ 0 & 0 & k_{33} \end{bmatrix}. \tag{20}$$

It is also evident from symmetry considerations that the presence of

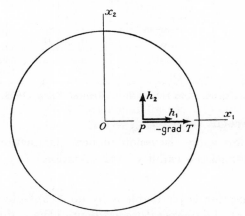

Fig. 11.4. Illustrating spiral heat flow where
$$k_{ij} \neq k_{ji}.$$

a 3-, 4-, or 6-fold axis (either rotation or inverse) ensures that $k_{11} = k_{22}$; hence for these symmetries we have

$$\begin{bmatrix} k_{11} & k_{12} & 0 \\ -k_{12} & k_{11} & 0 \\ 0 & 0 & k_{33} \end{bmatrix}. \tag{21}$$

It may be proved in a straightforward way that the tensor in this form is quite unaltered by *any* rotation of axes about x_3.

If planes of symmetry parallel to x_3 are present, it is readily proved that the further restriction

$$k_{12} = 0 \tag{22}$$

is imposed.

Now let us imagine a circular plate (Fig. 11.4) of a crystal belonging to a class in which $[k_{ij}]$ has the form (21); the plate is cut perpendicular to the principal symmetry axis x_3 and is heated at the centre O. To ensure that the outer edge of the plate shall be an isothermal we surround the plate with a circular box of highly conducting material. Now, as we have just seen, the conductivity tensor, and therefore the equations governing the flow of heat, have rotational symmetry about x_3. The geometry of the arrangement is also rotationally symmetrical. It follows that the isothermals on the flat surface of the plate are circles.

At the point P on the x_1 axis the heat flows are

$$h_1 = -k_{11}\frac{\partial T}{\partial x_1}, \qquad h_2 = -k_{21}\frac{\partial T}{\partial x_1} = k_{12}\frac{\partial T}{\partial x_1}, \qquad h_3 = 0.$$

h therefore makes an angle θ with Ox_1, where

$$\tan \theta = \frac{h_2}{h_1} = -\frac{k_{12}}{k_{11}},$$

and, owing to the rotational symmetry, **h** must make this same angle with the radius vector everywhere on the plate. The heat therefore flows outwards in equiangular spirals.

A number of attempts have been made to detect a spiral heat flow in crystals, but they have all failed (Soret 1893, 1894; Voigt 1903). Accordingly, the experimenters drew the general conclusion that, at least within the error of the experiments, heat flow from an isolated point takes place in straight lines and that relation (4) holds.

6. Thermodynamical arguments. Onsager's Principle

We did not have to appeal directly to experiment in establishing the symmetry of tensors like the permittivity, the permeability, and the elastic stiffness. The symmetry of these tensors followed because we were able to write down an expression for the energy (change) of a crystal as a function of the electric and magnetic fields and the components of strain. We used the fact that the energy was a function only of these variables (provided they were not too large). Such a procedure cannot be used in the present instance because no such energy expression exists; we are dealing, in fact, with an essentially different kind of crystal property. The other crystal properties were described by reference to equilibrium states; for instance, when we were describing elasticity we strained the crystal, and in doing so we moved as slowly as we liked from one equilibrium state to another; in the language of thermodynamics, the changes were reversible. Thermal conduction, on the other hand, cannot be described by reference to an equilibrium state; heat flows through the crystal at a definite rate, and the thermodynamical laws for reversible processes do not apply. It is true that the crystal may be in a steady state—its state may not change with time—but it cannot be considered to be in equilibrium because the systems with which it interacts (the heat reservoirs, or, if the heat is being produced electrically, the electric batteries, and so on) are undergoing continuous changes. The thermodynamical theory appropriate to this situation was first formulated by Onsager (1931 a, b).[†] The older thermodynamics has been more accurately described as 'thermostatics'; in contrast, Onsager's theory is essentially the thermo-*dynamics* of irreversible processes.

We shall now describe what is known as *Onsager's Principle*. This is a general principle which applies to all transport phenomena—under which term may be included the conduction of heat, the conduction of electricity, and the transport of matter that takes place by diffusion.

The equation governing the flow of an electric current in an isotropic material,
$$\mathbf{j} = -\sigma\,\mathrm{grad}\,\phi,$$
may be written
$$\mathbf{j}_1 = L_1\mathbf{X}_1, \tag{23}$$
where $\mathbf{j}_1 = \mathbf{j}$ and $\mathbf{X}_1 = -\mathrm{grad}\,\phi$. $\mathbf{j}_1$ represents a *flux* (of charge) and $\mathbf{X}_1$ represents a *force*. In the same way the flux of heat is proportional to the force on the heat according to the equation
$$\mathbf{j}_2 = L_2\mathbf{X}_2, \tag{24}$$
where the flux $\mathbf{j}_2$ is the heat flow, formerly designated by $\mathbf{h}$, and the force $\mathbf{X}_2 = -1/T\,\mathrm{grad}\,T$. The reason for the factor $1/T$ will be referred to later (T is the absolute temperature).

Equations (23) and (24) would be correct as they stand if the two processes of transport of electricity and of heat took place independently of one another. But, in general, this condition is not satisfied; for when the two processes occur simultaneously in a circuit composed of two different metals they interfere, as is shown by the appearance of the various thermoelectric effects. In general it is necessary to use the more comprehensive equations
$$\left.\begin{aligned}\mathbf{j}_1 &= L_{11}\mathbf{X}_1 + L_{12}\mathbf{X}_2\\ \mathbf{j}_2 &= L_{21}\mathbf{X}_1 + L_{22}\mathbf{X}_2\end{aligned}\right\}, \tag{25}$$
in which each of the fluxes is linearly related to *both* the forces instead of only to one of them. That the fluxes are linearly related to the first powers of the forces is an assumption of the theory; in general we only expect it to hold when the forces are sufficiently small. Onsager's

† For a full account with more recent developments see the book by de Groot (1951).

Principle states that in such a case, provided the fluxes and forces are correctly chosen (see below),

$$L_{12} = L_{21}. \tag{26}$$

It may be verified that, with our choice of definitions for the $\mathbf{j}$'s and $\mathbf{X}$'s, the dimensions of L_{12} and L_{21} are the same.

In more complicated situations, where there are more than two fluxes and associated forces, equations (25) are generalized to

$$\mathbf{j}_i = L_{ij}\mathbf{X}_j \quad (i,j = 1, 2, ..., n), \tag{27}$$

and Onsager's Principle asserts that

$$L_{ij} = L_{ji}. \tag{28}$$

It is important to notice that the $\mathbf{j}$'s and $\mathbf{X}$'s in (25) and (27) must be correctly chosen before Onsager's Principle can be applied. By 'correctly chosen' we mean that they must obey certain rules. It will be clear that in (27), for example, we must know first of all which forces go with which fluxes, in order to know which are the leading diagonal terms and which are the cross terms. In the case under discussion, equations (25), it is fairly obvious that we must link the heat flow with the temperature gradient and the electric current with the potential gradient, but a general rule to cover all cases is evidently required. A succinct account of the rules is given by de Groot (1951), pp. 5–7. There is a certain freedom of choice of the $\mathbf{j}$'s and $\mathbf{X}$'s within the rules, and for all such choices the Onsager relation (28) holds. The factor $1/T$ in the definition of $\mathbf{X}_2$ in (25) arises from an application of the rules.

We shall be returning to the discussion of equations (25) when we come to examine thermoelectric effects in the next chapter. We give the equations here because they provide a simple illustration of the application of Onsager's Principle. In this chapter we are more directly concerned with thermal conductivity in anisotropic crystals, to which Onsager's Principle also applies. The fluxes in question are the three components of heat flow parallel to the three axes of reference. They do interfere with one another, in general; this is clearly shown by the fact that a temperature gradient parallel to x_1 produces not only a heat flow along x_1 but also heat flows along x_2 and x_3. Onsager's Principle applied to this case (somewhat naïvely, as we shall see below) gives the reciprocal relation

$$k_{ij} = k_{ji}. \tag{4}$$

It will be realized that we have not really proved (4), but we have made it depend on the wider assertion represented by Onsager's Principle.

Casimir's objection. Casimir (1945) has drawn attention to certain difficulties in applying Onsager's Principle to equations (3). In the first place, the heat flows do not satisfy one of the rules for the proper choice of fluxes, which requires that they shall be the time derivatives of the variables which define the thermodynamical state. In the second place, the three heat flows are not directly observable physical quantities; for in any experiment it is, strictly, only the divergence of the heat flow (the difference between the heat entering and leaving a volume) that we measure. This last fact means that we can add to $[k_{ij}]$ any tensor that does not alter $\mathrm{div}\,\mathbf{h}$. Thus we may add to $[k_{ij}]$ a tensor $[k_{ij}^0]$ which is such that†

$$\frac{\partial}{\partial x_i}\left(k_{ij}^0 \frac{\partial T}{\partial x_j}\right) = 0,$$

or

$$\frac{\partial k_{ij}^0}{\partial x_i}\frac{\partial T}{\partial x_j} + k_{ij}^0 \frac{\partial^2 T}{\partial x_i \partial x_j} = 0.$$

This equation is satisfied for all temperature distributions provided

$$k_{ij}^0 = -k_{ji}^0, \tag{29}$$

and

$$\frac{\partial k_{ij}^0}{\partial x_i} = 0. \tag{30}$$

There is thus an uncertainty in $[k_{ij}]$, because the addition of an anti-symmetrical tensor $[k_{ij}^0]$ satisfying (30) gives no change in $\mathrm{div}\,\mathbf{h}$ and therefore no observable consequences. In view of this we cannot expect to be able to prove by any method that $[k_{ij}]$ is necessarily symmetrical. In agreement with this conclusion, Casimir shows that when the Onsager theory is correctly applied the result is not

$$k_{ij} - k_{ji} = 0,$$

but rather

$$\frac{\partial(k_{ij} - k_{ji})}{\partial x_i} = 0. \tag{31}$$

Without additional assumptions we can go no further than (31). However, we have just seen that an antisymmetrical part of $[k_{ij}]$ gives no observable physical consequences provided it satisfies (30); Onsager's Principle shows us, by equation (31), that the antisymmetrical part of $[k_{ij}]$ does indeed satisfy (30). The conclusion is that **an**

† We have to allow $[k_{ij}]$ and $[k_{ij}^0]$ to be functions of position because we wish later to have equations which apply even when we cross the boundary of our crystal.

antisymmetrical part of $[k_{ij}]$ is not observable. In view of this there is no reason why we should not put it equal to zero† and write

$$k_{ij} = k_{ji}. \tag{4}$$

We are not forced to do so but it is permissible.

One of the consequences of not accepting (4) would be that we should have to assume that the conductivity of a vacuum is not zero. To see this, take a flat surface of a crystal perpendicular to x_3. In the crystal, if (4) is not accepted, we have

$$k_{31} - k_{13} \neq 0. \tag{32}$$

Onsager's Principle, equation (31), shows that

$$\frac{\partial(k_{31} - k_{13})}{\partial x_3} = 0,$$

the differentiation being across the crystal boundary. Therefore, (32) still holds in the vacuum, and hence k_{31} and k_{13} cannot both be zero in the vacuum. (We cannot of course *prove* that there is no heat flow in a vacuum because we have no way of detecting heat flows that do not have a divergence.) Conversely, if $[k_{ij}]$ is *assumed* symmetrical in a vacuum,‡ it *must* also be symmetrical in the crystal; and, *a fortiori*, if $[k_{ij}]$ is assumed to be *zero* in a vacuum, it must be symmetrical in the crystal.

We found that in a circular crystal plate heated at the centre an antisymmetrical part of $[k_{ij}]$ would lead to heat flow along spirals. The heat flow would thus consist of two parts, a purely radial flow given by the strength of the source and a purely circular flow. This second part has no divergence; one might be tempted to say that it is there-fore in principle unobservable, and that attempts to find it by experi-ment, such as those of Soret and Voigt, were misguided and must have been based on faulty reasoning. Such a conclusion would not be quite correct. What the experiments really showed was that, *if $[k_{ij}]$ is assumed zero in a vacuum*, then, within experimental error, $[k_{ij}]$ is symmetrical in the crystals investigated. Onsager's Principle, by pro-viding equation (31), has now given the theoretical reason for the experimental findings.

6.1. Theoretical basis of Onsager's Principle. We have not so far spoken of any proof of Onsager's Principle. The theoretical basis

† We adopted a similar procedure with regard to the symmetry of the piezoelectric tensors (Ch. VII, § 1).

‡ This could be regarded as following from Neumann's Principle (p. 20).

for its truth lies in statistical mechanics. It is beyond the scope of this book to go into it here and we confine ourselves to a few general remarks.

Relations (28) may be shown to be a consequence of the assumption of 'microscopic reversibility': that if the velocities of all the particles present in a system are reversed simultaneously the particles will re-trace their former paths, reversing the entire succession of configurations—or, alternatively expressed, the mechanical equations of motion of individual particles are symmetrical with respect to past and future, that is, they are invariant under the transformation $t \to -t$. This is an assumption about the laws that govern processes on an atomic scale. It is true for the classical and also for the quantum-mechanical laws of motion, in the absence of magnetic fields.† It is interesting to notice that in a magnetic field we do not have microscopic reversibility for the path of a charged particle (because of the form of the expression for the Lorentz force) unless we reverse the field as well as the velocity. This leads to the relations

$$L_{ij}(\mathbf{H}) = L_{ji}(-\mathbf{H}) \tag{33}$$

in place of (28). In other words, L_{ij} is the same function of $\mathbf{H}$ as L_{ji} is of $-\mathbf{H}$.

The proof of his Principle given by Onsager (1931 a, b) is general and does not depend on the particular details of the transport processes; the result is therefore valid for all irreversible transport processes. Onsager's Principle in irreversible thermodynamics is the counterpart of the second law in thermodynamics (thermostatics). The second law may either be regarded as an axiom, or it may be derived from the principles of statistical mechanics; the status of Onsager's Principle is precisely the same.

SUMMARY

THE law of conduction of heat in crystals is

$$h_i = -k_{ij}(\partial T/\partial x_j), \tag{3}$$

where h_i is the rate of flow of heat per unit cross-section, $\partial T/\partial x_j$ is the temperature gradient, and the k_{ij} are the coefficients of thermal conductivity. The k_{ij} form a second-rank tensor. Equation (3) may be solved for the $\partial T/\partial x_j$ to give

$$\partial T/\partial x_i = -r_{ij}h_j, \tag{8}$$

† And in the absence of Coriolis forces (Onsager 1931 a, b).

where the r_{ij} are the coefficients of thermal resistivity. The resistivity matrix $\mathbf{r} = (r_{ij})$ is the reciprocal of the conductivity matrix $\mathbf{k} = (k_{ij})$:

$$\mathbf{r} = \mathbf{k}^{-1}. \tag{9}$$

It may be proved (see below) that $k_{ij} = k_{ji}$. Hence $r_{ij} = r_{ji}$, and both the conductivity and resistivity tensors may be referred to their common principal axes. The principal conductivities and resistivities are reciprocally related: $r_1 = 1/k_1$, $r_2 = 1/k_2$, $r_3 = 1/k_3$. The conductivity ellipsoid

$$k_1 x_1^2 + k_2 x_2^2 + k_3 x_3^2 = 1, \tag{7}$$

and the resistivity ellipsoid

$$r_1 x_1^2 + r_2 x_2^2 + r_3 x_3^2 = 1,$$

must conform to the symmetry of the crystal in accordance with Neumann's Principle.

When heat flows across a flat plate whose surfaces are isothermals there is, in general, a flow of heat parallel to the surfaces, in addition to the flow normal to the plate. The ratio of the heat flow normal to the plate to the temperature gradient is the conductivity normal to the plate. On the other hand, when heat flows down a long rod there is, in general, a transverse as well as a longitudinal temperature gradient. The ratio of the longitudinal temperature gradient to the heat flow is the resistivity along the rod.

For the steady-state flow of heat in a homogeneous crystal equation (3) may be combined with the conservation equation

$$\operatorname{div} \mathbf{h} = \dot{q} \quad \text{or} \quad \partial h_i / \partial x_i = \dot{q},$$

where $\dot{q}$ is the rate of production of heat per unit volume, to give

$$k_{ij} \frac{\partial^2 T}{\partial x_i \partial x_j} = -\dot{q}; \tag{11}$$

or, referred to the principal axes of $[k_{ij}]$,

$$k_1 \frac{\partial^2 T}{\partial x_1^2} + k_2 \frac{\partial^2 T}{\partial x_2^2} + k_3 \frac{\partial^2 T}{\partial x_3^2} = -\dot{q}. \tag{12}$$

By means of a simple transformation (14) the solution to any steady heat-flow problem in an anisotropic medium may be obtained directly from the solution to a corresponding isotropic problem.

The isothermal surfaces surrounding a point source of heat in an infinite crystal are all similar in shape and orientation to the resistivity ellipsoid for the crystal. This result follows directly from the radius-normal property of the representation quadric. Along any given direction from the point source the excess temperature $(T - T_\infty)$ falls off inversely as the distance from the source.

The formal equations governing the conduction of electricity in crystals are similar to those for the conduction of heat:

$$j_i = -\sigma_{ik}(\partial \phi / \partial x_k) = \sigma_{ik} E_k, \qquad E_i = \rho_{ik} j_k, \tag{17, 18}$$

where j_i is the current density, ϕ is the potential, E_k is the electric field, and σ_{ik} and ρ_{ik}, respectively, are the electrical conductivity and resistivity tensors. The rate of Joule heating per unit volume is

$$j_i E_i = \rho_{ik} j_i j_k. \tag{19}$$

The relation $k_{ij} = k_{ji}$. If $k_{ij} \neq k_{ji}$ heat flows out from an isolated point source in spirals. When experiments failed to detect such a spiral flow it was concluded that $k_{ij} = k_{ji}$. This relation is now recognized as a special instance of the application of Onsager's Principle. Onsager's Principle asserts that, if appropriately chosen fluxes $\mathbf{j}_i$ and forces $\mathbf{X}_i$ are associated by the linear equations

$$\mathbf{j}_i = L_{ij}\mathbf{X}_j \quad (i,j = 1, 2, ..., n), \tag{27}$$

then $L_{ij} = L_{ji}$ [or, if magnetic fields are present, $L_{ij}(\mathbf{H}) = L_{ji}(-\mathbf{H})$].

Straightforwardly applied, this gives $k_{ij} = k_{ji}$ immediately. However, a more searching analysis shows that, before one can deduce $k_{ij} = k_{ji}$ rigorously from Onsager's Principle, it is necessary to assume that $[k_{ij}]$ is symmetrical—or, in particular, zero—in a vacuum.

Onsager's Principle in irreversible thermodynamics is the counterpart of the second law in thermodynamics (thermostatics). They may both be derived from the principles of statistical mechanics.

XII

THERMOELECTRICITY†

THE conduction of heat and the conduction of electricity in crystals were treated in the last chapter as two separate processes. This was possible because we were concerned with situations where only one of the processes occurred at a time. It was mentioned, however, that when both processes occurred together they interfered with one another, the results of this interference being observed as the phenomena of thermo-electricity. In the present chapter we formulate the basic equations which govern thermoelectricity in crystals and we show how the equations lead to the various observed effects. It is first necessary to discuss thermoelectricity in isotropic conductors.

1. Thermoelectric effects in isotropic conductors

There are three thermoelectric effects in isotropic conductors:

(i) *The thermoelectric e.m.f.* (*Seebeck effect*). If a circuit is made of two different metals, a and b, and the junctions are maintained at different temperatures, an e.m.f. is set up in the circuit. If a condenser is inserted in the conductor a, for instance (Fig. 12.1), it becomes charged.

(ii) *The Peltier heat*. When current is allowed to flow across a junction between two different metals it is found that heat must be continuously added or subtracted at the junction in order to maintain its temperature constant. The heat is proportional to the current flowing and changes sign when the current is reversed. We write

$$\dot{Q}_{ab} = \Pi_{ab} J, \tag{1}$$

where $\dot{Q}_{ab}$ is the rate at which heat is absorbed at the junction when a current J passes from metal a to metal b, and Π_{ab} is the Peltier coefficient, which depends on the nature of the conductors and the temperature.

(iii) *The Thomson heat*. When a current flows in a wire, of homogeneous material and of constant cross-section but with a non-uniform temperature, heat must be supplied to keep the temperature distribution steady. The heat that must be supplied in unit time to an element

† This chapter, which is more difficult than the others, may be omitted on a first reading.

of the wire in which the temperature rise, in the direction of the current,
is dT is

$$dQ = \tau J\, dT,\tag{2}$$

where τ is the *Thomson coefficient.*

1.1. Derivation of the Thomson relations. There are two relations
between the magnitudes of the thermoelectric effects (i), (ii), (iii) above,
known as the *Thomson relations.* We now derive them, using the
methods of irreversible thermodynamics outlined in the last chapter.†
Consider the circuit of Fig. 12.1. Suppose the junctions are at

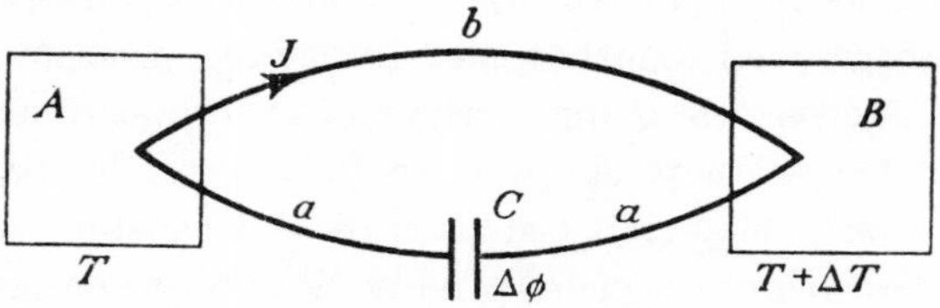

FIG. 12.1. Hypothetical circuit for obtaining the
Thomson relations.

temperatures T and $T+\Delta T$ in two large heat reservoirs A and B, and
that a potential difference $\Delta\phi$ is established across the condenser C,
the side nearer to B being at the higher potential. The condenser is
supposed to have no heat capacity and the wires a and b are supposed
to be thermally (and electrically) insulated. In Chapter XI, § 6, we wrote
the equations connecting the flow of electricity j_1, and of heat j_2, with
the 'forces' $X_1 = -\operatorname{grad}\phi$, and $X_2 = -1/T\operatorname{grad} T$, as

$$\left.\begin{aligned}
j_1 &= L_{11}X_1 + L_{12}X_2\\
j_2 &= L_{21}X_1 + L_{22}X_2
\end{aligned}\right\},\tag{3}$$

with the Onsager relation

$$L_{12} = L_{21}.\tag{4}$$

For the circuit here the equations take the form

$$\left.\begin{aligned}
J &= -L_{11}\Delta\phi - L_{12}\frac{\Delta T}{T}\\[2mm]
H &= -L_{21}\Delta\phi - L_{22}\frac{\Delta T}{T}
\end{aligned}\right\},\tag{5}\;(6)$$

where H is the rate at which heat flows down the wires from A to B
and J is the current flowing round the circuit in the sense indicated
in Fig. 12.1. (For $\Delta T = 0$ and $\Delta\phi$ positive, J is *negative*; and, for

† The following treatment, due to Polder, is taken, with some changes, from de Groot
(1951), § 57.

$\Delta\phi = 0$ and ΔT positive, H is *negative*.) Heat will, of course, be produced in the wires themselves both by Joule heating and by the Thomson effect. The rate of production of Joule heat is proportional to J^2, which, from (5), is proportional to ΔT^2; the rate of production of Thomson heat is proportional to $J\,\Delta T$ or, from (5), proportional to ΔT^2. The rate of heat flow itself is, from (6), proportional to ΔT. Hence, to the first order, the Joule and Thomson heats may be neglected here, and it is fair to speak of a *uniform* heat flow between the reservoirs. It will be appreciated that the whole theory is in any case only concerned with first-order terms, and that in equations (5) and (6), for example, terms of higher order than ΔT and $\Delta\phi$ are omitted.

Let us first notice that if $\Delta T = 0$ we have

$$J = -L_{11}\,\Delta\phi, \tag{7}$$

and that therefore L_{11} is the electrical conductance of the circuit connecting the condenser plates, in the isothermal state. The heat conductance λ of the path between the reservoirs, with zero current ($J = 0$), is found by eliminating $\Delta\phi$ between (5) and (6); this gives

$$H = -\lambda\,\Delta T, \tag{8}$$

with $\qquad\qquad \lambda = \{(L_{11}\,L_{22} - L_{12}\,L_{21})/L_{11}\}(1/T).$

Starting from equations (5) and (6), it is now a simple matter to deduce the existence of the thermoelectric phenomena (i), (ii), (iii), referred to above, and the relations between them. Suppose first that ΔT is fixed, and consider the steady state where no current flows ($J = 0$). Then, from (5),

$$\frac{\Delta\phi}{\Delta T} = -\frac{L_{12}}{L_{11}\,T}. \tag{9}$$

$\Delta\phi$ is the thermoelectric e.m.f. The equation gives the *thermoelectric power* of the couple, that is, the potential difference per unit temperature difference, with zero current flowing.

Now let us put $\Delta T = 0$. A potential difference will now cause both electric current and a flow of heat, Peltier heat, between the reservoirs. Their ratio is

$$\frac{H}{J} = \frac{L_{21}}{L_{11}} = \Pi_{ab}, \tag{10}$$

the Peltier coefficient giving the heat absorbed by the junction at A.

The Onsager relation (4), with (9) and (10), shows immediately that

$$\frac{d\phi}{dT} = -\frac{\Pi_{ab}}{T}, \tag{11}$$

which is the second of the Thomson relations; it connects the thermo-electric power with the Peltier coefficient. It is interesting to recall that Thomson first derived this relation by applying the laws of thermo-dynamics (thermostatics). The second law of thermostatics (written as an equality) cannot be applied to the thermoelectric effects because of the irreversible processes of heat conduction and electrical conduction that inevitably accompany them. Thomson proceeded on the assumption that the Thomson and Peltier heats could be considered as re-versible and applied the second law to them alone. In this way he derived equation (11) (see, for example, Epstein 1937, pp. 361–4), which was later completely confirmed by experiment. The splitting up of the phenomenon into an irreversible and a reversible part cannot be justi-fied on thermodynamical grounds and, as Thomson clearly realized, represents an additional assumption. The necessity for the assumption remained an objection to the theory until the advent of the new thermo-dynamics. In the new approach, which entirely avoids the difficulty and in which the phenomena are treated as one whole, it will be seen that equation (11) is essentially a consequence of Onsager's Principle.

Thomson's first relation follows directly from the first law of thermo-statics,

$$\Pi_{ab} J - (\Pi_{ab} + \Delta\Pi_{ab})J + (\tau_b - \tau_a)J\,\Delta T = J\,\Delta\phi,$$

or,

$$-\Delta\Pi_{ab} + (\tau_b - \tau_a)\Delta T = \Delta\phi, \tag{12}$$

where τ_a and τ_b are the Thomson heats. When ϕ is eliminated between the Thomson relations, (11) and (12), we have

$$\tau_b - \tau_a = \frac{d\Pi_{ab}}{dT} - \frac{\Pi_{ab}}{T}, \tag{13}$$

as the connexion between the Thomson and Peltier heats; by eliminating Π_{ab} we find

$$\tau_a - \tau_b = T\frac{d^2\phi}{dT^2}, \tag{14}$$

as the connexion between the Thomson heats and the derivative of the thermoelectric power.

It is useful to remember that, at a given temperature, the Peltier heat is proportional to $d\phi/dT$, by (11), while the difference of the Thomson heats is proportional to $d^2\phi/dT^2$, by (14).

2. Thermoelectric effects in isotropic continuous media

§ 1.1 dealt with thermoelectric effects in a circuit composed of two homogeneous wires of different metals. Before passing to crystals we need to set up a theory of the thermoelectric effect in a single continuous isotropic medium (de Groot 1951, Callen 1948). We allow the properties of the medium to change with position, as is clearly necessary if a Peltier effect is to be included. The Peltier effect treated by the theory will then be distributed throughout the medium. The ordinary Peltier effect at an interface is quite simply included in the theory by treating both metals as if they formed one medium, but with a discontinuous change of properties at the interface.

We shall formulate the theory in a way that facilitates the generalization to anisotropic media that has to be made later.

2.1. Formulation of the flow equations. We start with the following fundamental equations connecting fluxes with forces:

$$\mathbf{j}^e = -\alpha \operatorname{grad} \bar{\mu} - \beta \frac{\operatorname{grad} T}{T} \qquad (15)$$

$$\mathbf{h} = -\beta \operatorname{grad} \bar{\mu} - \gamma \frac{\operatorname{grad} T}{T} \qquad (16)$$

These equations are generalizations of equations (5) and (6). The electric current, of density $\mathbf{j}$, is assumed to be carried by electrons of charge $-e$. $\mathbf{j}^e = -\mathbf{j}/e$ is the electron current density (the number of electrons passing through unit cross-section normal to their path per unit time). $\mathbf{h}$ is the heat current density, as before. We shall have more to say about the strict definition of $\mathbf{h}$ later. $\bar{\mu}$ is the electrochemical potential of a conduction electron in the metal (see below). α, β and γ are constants. The choice of fluxes, $\mathbf{j}^e$ and $\mathbf{h}$, and corresponding forces, $-\operatorname{grad} \bar{\mu}$ and $-(\operatorname{grad} T)/T$, is one that obeys the rules of irreversible thermodynamics (p. 209),[†] and we are therefore justified in applying Onsager's Principle and putting the coefficient of $-\operatorname{grad} \bar{\mu}$ in (16) equal to β.

By writing $$\bar{\mu} = \mu - e\phi \qquad (17)$$

we may divide $\bar{\mu}$ into a chemical part μ, which depends on the composition and the temperature of the medium, and an electrical part $-e\phi$. It should be noted that, owing to the temperature dependence of μ, $\operatorname{grad} \bar{\mu}$ includes a part that depends on $\operatorname{grad} T$. Thus, the effect of a temperature gradient is not represented solely by the second term in equations (15) and (16). This is inelegant and it would be possible to recast the equations to show two independent terms, one representing the effect of a gradient of composition only, and the other representing

† Although a more rigorous treatment would have to take account of Casimir's objection (p. 210).

the effect of a gradient of temperature only. It is nevertheless better to accept the disadvantage and use the present system, because, once equations (15) and (16) are adopted, the remaining analysis is comparatively simple.

2.2. The rate of evolution of heat. The physical meaning of α in equation (15) may be seen by considering a homogeneous material at a uniform temperature. Then $\operatorname{grad} \mu = 0$, and so $\operatorname{grad} \bar{\mu} = -e \operatorname{grad} \phi$. Hence

$$\mathbf{j}^e = -\mathbf{j}/e = \alpha e \operatorname{grad} \phi,$$

and we see that $\alpha e^2 = \sigma$, the electrical conductivity, or

$$\alpha^{-1} = e^2 \rho, \tag{18}$$

where ρ is the electrical resistivity.

We now calculate from equations (15) and (16) an expression for the rate of production of heat energy. First find an expression for $\operatorname{grad} \bar{\mu}$ from (15) and then substitute in (16). Thus,

$$\operatorname{grad} \bar{\mu} = -\alpha^{-1}\mathbf{j}^e - \alpha^{-1}\beta \frac{\operatorname{grad} T}{T} \tag{19}$$

$$\mathbf{h} = \alpha^{-1}\beta \mathbf{j}^e + (\alpha^{-1}\beta^2 - \gamma) \frac{\operatorname{grad} T}{T} \tag{20}$$

Now write $\alpha^{-1}\beta/T = \Sigma$. We shall see later (p. 222) that $-\Sigma/e$ is the absolute thermoelectric power of the conductor at the particular point and temperature under consideration. The equations now become

$$\operatorname{grad} \bar{\mu} = -\alpha^{-1}\mathbf{j}^e - \Sigma \operatorname{grad} T \tag{21}$$

$$\mathbf{h} = T\Sigma \mathbf{j}^e - k \operatorname{grad} T \tag{22}$$

where $k = -(\alpha^{-1}\beta^2 - \gamma)/T$, the thermal conductivity for zero electric current ($\mathbf{j}^e = 0$). Notice that Σ connects the electrochemical potential gradient with the temperature gradient.

When there is no electric current the flow of energy (rate per unit cross-section), which we denote by $\mathbf{j}^u$, is equal simply to the flow of heat $\mathbf{h}$. When there is a transport of electrons, however, $\mathbf{j}^u$ must include a term due to their movement. It may be shown that, in fact,

$$\mathbf{j}^u = \mathbf{h} + \bar{\mu}\mathbf{j}^e. \tag{23}$$

When an electric current flows, what is meant by the rate of flow of heat is not quite obvious, and, indeed, the definitions adopted by different authors are not the same. For the purposes of this book we ask the reader to accept equation (23) as at least a plausible definition of $\mathbf{h}$ and to accept also that $\mathbf{h}$ defined in this way satisfies equation (16).

With the expression for **h** given by (22), equation (23) for the **energy** flow may be written

$$\mathbf{j}^u = (T\Sigma + \bar{\mu})\mathbf{j}^e - k\,\mathrm{grad}\,T. \tag{24}$$

$\mathbf{j}^u$ thus consists of one part due to the passage of the electric current and another part, with which we are already familiar, due to the temperature gradient.

The rate at which heat is evolved, per unit volume, at any point in the material is the convergence (negative of the divergence) of $\mathbf{j}^u$. Thus

$$-\mathrm{div}\,\mathbf{j}^u = -\mathbf{j}^e.\,\mathrm{grad}(T\Sigma + \bar{\mu}) - (T\Sigma + \bar{\mu})\mathrm{div}\,\mathbf{j}^e + \mathrm{div}(k\,\mathrm{grad}\,T).$$

Since we are only concerned with steady states, $\mathrm{div}\,\mathbf{j}^e = 0$, and the second term on the right vanishes. We therefore have

$$-\mathrm{div}\,\mathbf{j}^u = -\mathbf{j}^e.\,(\Sigma\,\mathrm{grad}\,T + \mathrm{grad}\,\bar{\mu}) - T\mathbf{j}^e.\,\mathrm{grad}\,\Sigma + \mathrm{div}(k\,\mathrm{grad}\,T),$$

or, by using (21) and (18),

$$\overset{\text{Joule}}{-\mathrm{div}\,\mathbf{j}^u = e^2\rho(j^e)^2} \quad \overset{\text{thermoelectric}}{-\quad T\mathbf{j}^e.\,\mathrm{grad}\,\Sigma} \quad \overset{\text{conduction}}{+\quad \mathrm{div}(k\,\mathrm{grad}\,T).} \tag{25}$$

This is the complete expression for the rate of evolution of heat per unit volume in a body when there are simultaneously present, (1) an electric current (of density $-ej^e$), (2) a composition gradient ($\mathrm{grad}\,\Sigma$), and (3) a temperature gradient ($\mathrm{grad}\,T$). The first term on the right-hand side is the Joule heat. The second term represents the thermoelectric heat, which we denote by $\dot{q}$ and which is our main concern:

$$\dot{q} = -T\mathbf{j}^e.\,\mathrm{grad}\,\Sigma. \tag{26}$$

The last term represents simply the convergence of the heat flows given by the simple heat conduction theory in the absence of electric currents.

If the state of the body is steady there can, of course, be no evolution of heat within it; for, if there were, the temperature would rise. Thus, in the steady state, $-\mathrm{div}\,\mathbf{j}^u = 0$, and the three terms on the right in equation (25) must adjust themselves so that they exactly cancel. For instance, an *evolution* of heat in a certain region due to Joule heating and thermoelectric heating would be balanced by the temperature distribution so adjusting itself that the last term gave an exactly equal *subtraction* of heat.

2.3. Derivation of the observed thermoelectric effects from the equations. We started with equations (15) and (16); from them we found equations (21) and (22), and, finally, equation (25) for the rate of evolution of heat. Let us now see how the observed thermoelectric effects—the thermoelectric e.m.f., the Peltier heat, and the Thomson heat—may be derived from the equations.

(i) *The thermoelectric e.m.f.* An expression for the thermoelectric e.m.f. in a couple in terms of the Σ's for the two metals is obtained by integrating equation (21) round the circuit of Fig. 12.1. In the steady state $\mathbf{j}^e = 0$ and (21), which takes the form

$$\operatorname{grad}(\mu - e\phi) = -\Sigma \operatorname{grad} T, \tag{27}$$

integrates to give $\quad \int d\mu - e \int d\phi = -\int \Sigma \, dT.$

If we start from one side of the condenser and integrate round the circuit to the other side the initial and final values of μ are equal, since we are in the same metal at the same temperature, and so the first term gives zero. We are left with

$$e\Delta\phi = (\Sigma^{(b)} - \Sigma^{(a)}) \Delta T,$$

and so the thermoelectric power is

$$\frac{\Delta\phi}{\Delta T} = -(\Sigma^{(a)} - \Sigma^{(b)})/e. \tag{28}$$

The terms $-\Sigma^{(a)}/e$ and $-\Sigma^{(b)}/e$ may be called the *absolute thermoelectric powers* of the two metals, in the sense that their difference for any pair of metals gives the observed thermoelectric power.

(ii) *The Peltier heat.* Since $\operatorname{div} \mathbf{j}^e = 0$, $\mathbf{j}^e$ may be taken inside the differentiation in (26) and the expression for the thermoelectric heat put in the alternative form

$$\dot{q} = -T \operatorname{div}(\Sigma \mathbf{j}^e). \tag{26}'$$

At an interface between two conductors (26)' degenerates to:

$$\frac{\text{rate of evolution of heat}}{\text{per unit area of interface}} = -T\Delta_n(\Sigma \mathbf{j}^e),$$

where $\Delta_n(\Sigma \mathbf{j}^e)$ denotes the discontinuity in the normal component of $\Sigma \mathbf{j}^e$. Since the component of $\mathbf{j}^e$ normal to the interface is continuous, the expression may be written

$$-T(\Sigma^{(a)} - \Sigma^{(b)})j_n^e, \tag{29}$$

where (a) and (b) refer to the two different conductors and where j_n^e is the normal component directed into a. (29) is then a general expression for the Peltier heat at an interface and is valid even when the current does not cross the interface normally. If an electric current of density

j does pass normally from a to b (so that the electron current passes from b to a) we have for the rate of evolution of heat per unit area

$$- T(\Sigma^{(a)} - \Sigma^{(b)}) j/e.$$

Hence, from the definition (1),

$$\Pi_{ab} = T(\Sigma^{(a)} - \Sigma^{(b)})/e. \tag{30}$$

Equation (30) when combined with (28) leads directly to the second Thomson relation (11).

(iii) *The Thomson heat.* To see how the Thomson heat arises from equation (26) we may reproduce the conditions under which the Thomson heat was defined (p. 215)—a wire of homogeneous material and of constant cross-section A, with a non-uniform temperature and carrying an electric current J. The only spatial variation is along the conductor, which we take as the x direction. Let J flow in the direction of increasing x; then $\mathbf{j}^e$ will be in the direction of decreasing x. Hence (26) gives

$$\dot{q} = T j^e \frac{d\Sigma}{dx}.$$

The heat evolved by an element of length dx in unit time is

$$\dot{q} A \, dx = T j^e A \, d\Sigma = T(J/e) \, d\Sigma = -\tau J \, dT,$$

where the Thomson coefficient τ is related to Σ by

$$\tau = -\frac{T}{e} \left(\frac{\partial \Sigma}{\partial T} \right)_x. \tag{31}$$

Equation (31) written down for two conductors a and b and combined with (30) readily leads to equation (13).

(iv) *Peltier and Thomson heats.* Another way of seeing that equation (26) contains both the Peltier and the Thomson heats is to notice that Σ depends on position in two ways: it varies with position as the properties of the medium change (for instance, as we cross an interface to a different metal) and, secondly, it varies with position on account of the non-uniformity of the temperature distribution. grad Σ therefore contains two terms, and we have

$$\dot{q} = -j_i^e \, T \, \frac{\partial \Sigma}{\partial x_i} = \overset{\text{Peltier}}{-j_i^e \, T \left(\frac{\partial \Sigma}{\partial x_i} \right)_T} \overset{\text{Thomson}}{- j_i^e \, T \left(\frac{\partial \Sigma}{\partial T} \right)_x \frac{\partial T}{\partial x_i}}. \tag{32}$$

The first term is the Peltier heat, proportional to the current and existing by virtue of a change in the properties of the medium with position (at constant temperature). The second term is the Thomson heat, proportional to the current and the temperature gradient.

2.4. Order of magnitude of the coefficients. Let us estimate the order of magnitude of the coefficients α, β, γ occurring in equations (15) and (16). Typical values of measurable quantities for a good conductor in m.k.s. units and degrees Centigrade are:

absolute thermoelectric power $= 1\cdot0\times10^{-5}$ volts/° C,

$k = 420$ joules/(m sec ° C),

$\rho = 1\cdot8\times10^{-8}$ ohm m.

With $e = 1\cdot6\times10^{-19}$ m.k.s. units we have

$$\Sigma/e = 1\cdot0\times10^{-5}, \qquad \Sigma = \mathbf{1\cdot6\times10^{-24}}.$$

We also find

$$\alpha^{-1} = e^2\rho = 4\cdot6\times10^{-46}; \quad \alpha = \mathbf{2\cdot2\times10^{45}}.$$

If we take $T = 300°$ K,

$$\beta = T\Sigma\alpha = \mathbf{1\cdot1\times10^{24}}.$$

Since $k = -(\alpha^{-1}\beta^2-\gamma)/T$,

$$\gamma = kT+\alpha^{-1}\beta^2 = (1\cdot3\times10^5)+(5\cdot6\times10^2)$$

and we may take $\gamma = \mathbf{1\cdot3\times10^5}$.

The ratio $\beta^2/(\alpha\gamma)$ is dimensionless, and gives a measure of the strength of the coupling between the thermal and electrical effects. We have here

$$\beta^2/(\alpha\gamma) = \mathbf{4\times10^{-3}}.$$

3. Thermoelectric effects in crystals

3.1. Formulation of the flow equations. Now that we have set up equations (15) and (16) for isotropic conductors, and have seen how they lead to the observed thermoelectric effects, it is quite simple to generalize them into a form appropriate to anisotropic conductors (Domenicali 1953).† The only change is that the constants α, β, γ are now second-rank tensors. To find out what type of thermoelectric behaviour this change implies, we have to manipulate the equations in the same way as before. In doing so it is best to regard α, β, γ, and all the vectors, as matrices (Ch. IX). We then have, in suffix notation,

$$j_i^e = -\alpha_{ik}\frac{\partial\bar\mu}{\partial x_k}-\beta_{ik}\frac{1}{T}\frac{\partial T}{\partial x_k} \right\} \tag{33}$$

$$h_i = -\beta_{ki}\frac{\partial\bar\mu}{\partial x_k}-\gamma_{ik}\frac{1}{T}\frac{\partial T}{\partial x_k} \tag{34}$$

Onsager's Principle (p. 207) demands that the complete matrix of

† The following treatment is formally somewhat different from Domenicali's but is essentially equivalent to it.

coefficients on the right-hand side of these equations should be symmetrical. It follows at once that (α_{ik}) and (γ_{ik}) are symmetrical matrices. It also follows that the nine coefficients of the $(1/T)(\partial T/\partial x_k)$ in (33) are the same as the nine coefficients of the $\partial\bar\mu/\partial x_k$ in (34); but it may be seen by writing out the terms that it is β_{ki} rather than β_{ik} that must appear in (34). Thus we must write, in matrix notation,

$$\mathbf{j}^e = -\boldsymbol{\alpha}\,\mathbf{grad}\,\bar\mu - \boldsymbol{\beta}\,\frac{\mathbf{grad}\,T}{T} \tag{33'}$$

$$\mathbf{h} = -\boldsymbol{\beta_t}\,\mathbf{grad}\,\bar\mu - \boldsymbol{\gamma}\,\frac{\mathbf{grad}\,T}{T} \tag{34'}$$

with $\boldsymbol{\beta_t}$, the transpose of $\boldsymbol{\beta}$, in the second equation. $\boldsymbol{\beta}$ is not symmetrical, in general.

3.2. The rate of evolution of heat. As in the isotropic case, equation (18), we have

$$\boldsymbol{\alpha}^{-1} = e^2\boldsymbol{\rho}, \tag{35}$$

where $\boldsymbol{\rho}$ is the electrical resistivity matrix. The symmetry of $\boldsymbol{\rho}$ then follows from the symmetry of $\boldsymbol{\alpha}$.

Equations (21) and (22) now take the form†

$$\mathbf{grad}\,\bar\mu = -\boldsymbol{\alpha}^{-1}\mathbf{j}^e - \boldsymbol{\Sigma}\,\mathbf{grad}\,T \tag{36}$$

$$\mathbf{h} = T\boldsymbol{\Sigma_t}\,\mathbf{j}^e - \mathbf{k}\,\mathbf{grad}\,T \tag{37}$$

where $\boldsymbol{\Sigma} = (\boldsymbol{\alpha}^{-1}\boldsymbol{\beta})/T$, and $\mathbf{k} = -(\boldsymbol{\beta_t}\boldsymbol{\alpha}^{-1}\boldsymbol{\beta}-\boldsymbol{\gamma})/T$, both $\boldsymbol{\Sigma}$ and $\mathbf{k}$ now being matrices. $\mathbf{k}$ is the thermal conductivity matrix and $\boldsymbol{\Sigma}$ is a matrix characterizing the thermoelectric properties of the crystal. Their components are functions of position and temperature. It is readily proved that the symmetry of $\boldsymbol{\alpha}$ and $\boldsymbol{\gamma}$ implies the symmetry of $\mathbf{k}$. But, in general, $\boldsymbol{\Sigma}$ is not symmetrical.

We note in passing that, by putting $\mathbf{j}^e = 0$ in (36) and splitting up $\bar\mu$ into its electrical and chemical parts according to (17), we obtain an expression for the thermoelectric e.m.f. per unit length:

$$-\mathbf{grad}\,\phi = -\frac{1}{e}(\boldsymbol{\Sigma}\,\mathbf{grad}\,T + \mathbf{grad}\,\mu), \tag{38}$$

or, in suffix notation,

$$-\frac{\partial\phi}{\partial x_i} = -\frac{1}{e}\left(\Sigma_{ik}\frac{\partial T}{\partial x_k}+\frac{\partial\mu}{\partial x_i}\right). \tag{38'}$$

The energy flow is given, as before in equation (23), by

$$\mathbf{j}^u = \mathbf{h}+\bar\mu\mathbf{j}^e, \tag{39}$$

† It may be easily shown that, if **A** and **B** are matrices, $(\mathbf{AB})_t = \mathbf{B_t}\mathbf{A_t}$. Hence
$$\boldsymbol{\Sigma_t} = (1/T)\boldsymbol{\beta_t}(\boldsymbol{\alpha}^{-1})_t = (1/T)\boldsymbol{\beta_t}\,\boldsymbol{\alpha}^{-1},$$
since $\boldsymbol{\alpha}^{-1}$ is symmetrical.

all three terms being single column matrices. The generalization of (24) is therefore

$$\mathbf{j}^u = (T\boldsymbol{\Sigma}_t + \bar{\mu}\mathbf{I})\mathbf{j}^e - \mathbf{k}\,\mathrm{grad}\,T, \tag{40}$$

or, in suffix notation,

$$j_k^u = (T\Sigma_{ik} + \bar{\mu}\delta_{ik})j_i^e - k_{ki}\frac{\partial T}{\partial x_i}. \tag{40}'$$

The rate of evolution of heat is

$$-\frac{\partial j_k^u}{\partial x_k} = -\left(\Sigma_{ik}\frac{\partial T}{\partial x_k} + \frac{\partial \bar{\mu}}{\partial x_k}\delta_{ik}\right)j_i^e - T\frac{\partial}{\partial x_k}(j_i^e \Sigma_{ik}) + \frac{\partial}{\partial x_k}\left(k_{ki}\frac{\partial T}{\partial x_j}\right). \tag{41}$$

Equation (36) shows that the expression between the first pair of brackets is the ith component of the matrix $-\boldsymbol{\alpha}^{-1}\mathbf{j}^e$, namely, $-(\alpha^{-1})_{ik}j_k^e$. The first term on the right in (41) is therefore the Joule heat,

$$e^2 \rho_{ik} j_i^e j_k^e,$$

and we have, finally,

$$\overset{\text{Joule}}{} \qquad \overset{\text{thermoelectric}}{} \qquad \overset{\text{conduction}}{}$$

$$-\frac{\partial j_k^u}{\partial x_k} = e^2 \rho_{ik} j_i^e j_k^e - T\frac{\partial}{\partial x_k}(j_i^e \Sigma_{ik}) + \frac{\partial}{\partial x_k}\left(k_{ki}\frac{\partial T}{\partial x_i}\right). \tag{42}$$

3.3. Discussion of the thermoelectric heating. The second term on the right in (42),

$$\dot{q} = -T\frac{\partial}{\partial x_k}(j_i^e \Sigma_{ik}), \tag{43}$$

represents the heat evolved by thermoelectric effects.†

We see that $\dot{q}/T$ is the convergence of a vector $j_i^e \Sigma_{ik}$, which is in general not parallel to the electron current density vector j_i^e. One may think of $j_i^e \Sigma_{ik}$ as being the density of the flow of entropy carried by the electron current.

At an interface (43) becomes:

$$\genfrac{}{}{0pt}{}{\text{rate of evolution of heat}}{\text{per unit area}} = -T\Delta_n(j_i^e \Sigma_{ik}), \tag{44}$$

where, as before, the symbol $\Delta_n(\quad)$ indicates the discontinuity in the normal component of the vector in the brackets. Alternatively, we may use j_i rather than j_i^e and write the expression for the heat evolution as

$$\Delta_n(j_i \Pi_{ik}), \tag{45}$$

---

† This equation was first derived by Ehrenfest and Rutgers (1929) from the postulate that the thermoelectric effects could be treated as reversible and independent of the irreversible effects that accompanied them. This was before the Onsager theory of irreversible processes had been formulated. The 'pseudo-thermostatic' approach which Ehrenfest and Rutgers had to use is not satisfactory, as we saw on p. 218, although it does give the correct result.

where Π_{ik} is the *Peltier matrix* or *tensor*, defined for each conductor by

$$\Pi_{ik} = T(\Sigma_{ik}/e). \tag{46}$$

The presence of symmetry in the crystal reduces the number of independent components of Σ_{ik} (and Π_{ik}) in a way that can readily be calculated by the usual methods. It has to be remembered that, in general,

$$\Sigma_{ik} \neq \Sigma_{ki}. \tag{47}$$

The forms of $[\Sigma_{ik}]$ in the standard orientations (Appendix B) for the various classes are given in Table 14 in the dot and ring symbolism used before.

TABLE 14

Forms of the thermoelectric tensor $[\Sigma_{ik}]$

By setting up special cases, as was done for isotropic materials, the significance of the various components of Σ_{ik} may be found. The components on the leading diagonal are associated with similar effects to those found in isotropic materials, although now of course the effects depend on direction for all crystals except those of the cubic system. The off-diagonal components of Σ_{ik}, however, imply, through equation (44), the existence of a *transverse Peltier effect* that is peculiar to non-cubic crystals. This may be seen by considering a rod of non-cubic crystal (Fig. 12.2) with its length parallel to x_1 and of rectangular cross-section. Let a uniform electron current density $\mathbf{j}^e = [j^e, 0, 0]$ flow down the rod. The vector $j_i^e \Sigma_{ik}$ will not, in general, be parallel to $\mathbf{j}^e$; in addition to the longitudinal component $j^e \Sigma_{11}$ there will be lateral components $j^e \Sigma_{12}$ and $j^e \Sigma_{13}$. On the upper face normal to x_2 there is a discontinuity in the normal component of $j_i^e \Sigma_{ik}$, namely $-j^e \Sigma_{12}$, and hence a corresponding evolution of heat, given by equation (44), of amount

$$Tj^e \Sigma_{12}.$$

On the opposite face there will be an equal absorption of heat. A similar transverse heating and cooling occurs on the faces perpendicular to x_3.

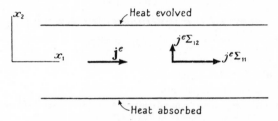

FIG. 12.2. An example of the transverse Peltier effect.

A remarkable result follows if we carry out the differentiation on the right-hand side of equation (43). We have to remember that Σ_{ik} depends on x_k in two ways, just as for an isotropic medium: it varies with position as the properties of the medium change (for instance, as we cross an interface either to another crystal or to the same crystal in a different orientation) and, secondly, it varies with position on account of any non-uniformity in the temperature distribution. It therefore gives rise to two terms when we differentiate, thus:

$$\underset{\text{Peltier}}{\quad} \quad \underset{\text{Thomson}}{\quad} \quad \underset{\text{Bridgman}}{\quad}$$
$$\dot{q} = -j_i^e T \left(\frac{\partial \Sigma_{ik}}{\partial x_k} \right)_T - j_i^e T \left(\frac{\partial \Sigma_{ik}}{\partial T} \right)_x \frac{\partial T}{\partial x_k} - T \Sigma_{ik} \frac{\partial j_i^e}{\partial x_k}. \qquad (48)$$

The first term on the right represents the Peltier heat, proportional to current and existing by virtue of a change in the properties of the medium with position (at constant temperature). The fact that j_i^e and x_k have different subscripts in this term shows that changes in Σ_{ik} that are not in the direction of the current are relevant; they give rise to the transverse Peltier effect. This is in addition to the normal Peltier effect given by the terms with $i = k$.

The second term is the Thomson heat, proportional to the current and the temperature gradient; since temperature gradients not in the direction of the current are relevant now, we have a *transverse Thomson heat*, in addition to the normal Thomson heat. The complete expression for the Thomson heat evolved per unit volume and per unit time may be written

$$-j_i \tau_{ik} \frac{\partial T}{\partial x_k},$$ (49)

where τ_{ik} is the *Thomson heat tensor* given by

$$\tau_{ik} = -\frac{T}{e}\left(\frac{\partial \Sigma_{ik}}{\partial T}\right)_x;$$ (50)

compare equation (31).

The last term in (48) represents an effect we have not yet met in

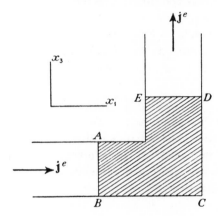

Fig. 12.3. An example of the Bridgman heat. Heat is evolved simply because the electron current j^e has changed direction in the crystal.

this chapter and which, like the transverse Peltier and Thomson heats, is peculiar to crystals. In an isotropic crystal only the diagonal terms of Σ_{ik} survive and this last term is simply

$$-T\Sigma \frac{\partial j_i^e}{\partial x_i};$$

since $\partial j_i^e/\partial x_i = 0$ for steady currents, the term vanishes. In an anisotropic crystal, on the other hand, the term gives rise to an additional production of heat, the *Bridgman heat*. This is essentially connected with any non-uniformity in the distribution of current.

As an example of the Bridgman heat, consider the L-shaped crystal with arms of unit square cross-section shown in Fig. 12.3. The orientation of the crystal is the same in both arms and the whole crystal is at the same temperature. A uniform electron current of density j^e flows along the arm parallel to x_1, and then changes direction and flows parallel to x_3 along the other arm. In order to avoid a transverse Peltier heat we take the crystal to have a symmetry that reduces $[\Sigma_{ik}]$ to the form

$$\begin{bmatrix} \bullet & \cdot & \cdot \\ \cdot & \bullet & \cdot \\ \cdot & \cdot & \bullet \end{bmatrix}$$

and we orient it so that the principal axis of symmetry is parallel to x_3. We apply equation (43) to the shaded volume. The Σ_{ik} are constants and $-\dot{q}/T$ is the divergence of $j_i^e \Sigma_{ik}$ integrated over the volume. By Gauss's Theorem this is the flux of the outward normal components of $j_i^e \Sigma_{ik}$ crossing the sides. The outward normal components crossing AB and DE are respectively $-j^e\Sigma_{11}$ and $j^e\Sigma_{33}$. Near the surfaces of the crystal, the electron current flows parallel to the sides; outside the crystal it is zero. There are therefore no normal components of $j_i^e \Sigma_{ik}$ on these sides.

Hence
$$\dot{q} = Tj^e(\Sigma_{11} - \Sigma_{33}),$$

and we have an expression for the Bridgman heat, due in this case simply to the fact that $\mathbf{j}^e$ has changed direction in the crystal.

In general, any non-uniform distribution of current density in an anisotropic crystal will give non-vanishing $\partial j_i^e/\partial x_k$ terms in (48) and hence a Bridgman heating.

SUMMARY

[1] **Thermoelectric effects in isotropic conductors.** (i) *The thermoelectric e.m.f. of a couple.* If a difference of temperature ΔT is maintained between the two junctions of a thermocouple a thermoelectric e.m.f. $\Delta\phi$ is set up. $d\phi/dT$ is the *thermoelectric power*.

(ii) *The Peltier heat.* The rate of heat absorption at a junction as a current J passes from metal a to metal b is

$$\dot{Q}_{ab} = \Pi_{ab} J, \tag{1}$$

where Π_{ab} is the *Peltier coefficient*.

(iii) *The Thomson heat.* The rate of supply of heat needed to keep the temperature steady in an element of wire in which the temperature rise, in the direction of the current, is dT is

$$d\dot{Q} = \tau J \, dT, \tag{2}$$

where τ is the *Thomson coefficient*.

Thomson's first relation follows from the first law of thermodynamics:

$$-\Delta\Pi_{ab} + (\tau_b - \tau_a)\Delta T = \Delta\phi. \tag{12}$$

Thomson's second relation follows from Onsager's Principle:

$$d\phi/dT = -(\Pi_{ab}/T). \tag{11}$$

Hence we may write for the Peltier heat and the difference of the Thomson heats,

$$\left. \begin{array}{l} \Pi_{ab} = -T(d\phi/dT) \\ \tau_a - \tau_b = T(d^2\phi/dT^2) \end{array} \right\}. \tag{11$'$} \tag{14}$$

[2] **Thermoelectric effects in isotropic continuous media.** By starting with appropriate flow equations connecting the fluxes of electric charge and of heat with the forces acting on them (due to the potential gradient and the temperature gradient), and using Onsager's Principle, we find the following equations.

For the thermoelectric e.m.f. per unit length:

$$-\operatorname{grad}\phi = -(1/e)(\Sigma \operatorname{grad} T + \operatorname{grad}\mu), \qquad \text{(27)$'$ [from (27)]}$$

where Σ is a constant for the particular point of the conductor, and the temperature, under consideration.

For the rate of evolution of heat per unit volume:

$$\begin{array}{ccc} \text{Joule} & \text{thermoelectric} & \text{conduction} \end{array}$$
$$-\operatorname{div}\mathbf{j}^u = e^2\rho j^{e2} \quad - \quad T\operatorname{div}(\Sigma\mathbf{j}^e) \quad + \quad \operatorname{div}(k\operatorname{grad} T). \tag{25}$$
$$\text{[with (26)$'$]}$$

The thermoelectric term, which may also be written as $-T\mathbf{j}^e.\operatorname{grad}\Sigma$, contains both the Peltier and the Thomson heats, thus:

$$\begin{array}{cc} \text{Peltier} & \text{Thomson} \end{array}$$
$$-j_i^e T\frac{\partial\Sigma}{\partial x_i} = -j_i^e T\left(\frac{\partial\Sigma}{\partial x_i}\right)_T - j_i^e T\left(\frac{\partial\Sigma}{\partial T}\right)_x \frac{\partial T}{\partial x_i}. \tag{32}$$

At an interface the rate of evolution of heat per unit area is

$$- T(\Sigma^{(a)} - \Sigma^{(b)})j_n^e, \tag{29}$$

where j_n^e is the normal component directed into a. Thus

$$\Pi_{ab} = T(\Sigma^{(a)} - \Sigma^{(b)})/e. \tag{30}$$

All the observed thermoelectric effects noted in [1] above may be derived from these equations.

[3] **Thermoelectric effects in crystals.** In crystals the scalar Σ is replaced by the tensor Σ_{ik}. The equations corresponding to (27)$'$ and (25) above are

$$-\frac{\partial\phi}{\partial x_i} = -\frac{1}{e}\left(\Sigma_{ik}\frac{\partial T}{\partial x_k} + \frac{\partial\mu}{\partial x_i}\right), \tag{38$'$}$$

and

$$\begin{array}{ccc} \text{Joule} & \text{thermoelectric} & \text{conduction} \end{array}$$
$$-\frac{\partial j_k^u}{\partial x_k} = e^2\rho_{ik}j_i^e j_k^e - T\frac{\partial}{\partial x_k}(j_i^e\Sigma_{ik}) + \frac{\partial}{\partial x_k}\left(k_{ki}\frac{\partial T}{\partial x_i}\right). \tag{42}$$

The thermoelectric term may be expanded as

$$\begin{array}{ccc} \text{Peltier} & \text{Thomson} & \text{Bridgman} \end{array}$$
$$-j_i^e T\left(\frac{\partial\Sigma_{ik}}{\partial x_k}\right)_T - j_i^e T\left(\frac{\partial\Sigma_{ik}}{\partial T}\right)_x \frac{\partial T}{\partial x_k} - T\Sigma_{ik}\frac{\partial j_i^e}{\partial x_k}. \tag{48}$$

At an interface the *Peltier heat* per unit area is

$$- T\Delta_n(j_i^e\Sigma_{ik}) = \Delta_n(j_i\Pi_{ik}), \tag{44), (45}$$

where Δ_n denotes the discontinuity in the normal component, and where

$\Pi_{ik} = T(\Sigma_{ik}/e)$. Π_{ik} is the *Peltier matrix* or *tensor*. In general, there are both longitudinal and transverse Peltier heats.

The *Thomson heat* per unit volume and per unit time may be written

$$-j_i \tau_{ik} \frac{\partial T}{\partial x_k}, \qquad \tau_{ik} = -\frac{T}{e}\left(\frac{\partial \Sigma_{ik}}{\partial T}\right)_x; \qquad (49),\ (50)$$

τ_{ik} is the *Thomson heat tensor*. There are, in general, both longitudinal and transverse Thomson heats.

The *Bridgman heat* per unit volume and per unit time, given by the last term in (48), is an effect peculiar to crystals and arises from any non-uniformity in the current distribution.

All the thermoelectric effects are thus given by the tensor Σ_{ik} and its temperature and space derivatives. In general

$$\Sigma_{ik} \neq \Sigma_{ki}.$$

Cubic crystals behave isotropically. All other crystals behave anisotropically, and, in the most general case, there are nine independent Σ_{ik} coefficients.

PART 4
CRYSTAL OPTICS

XIII

NATURAL AND ARTIFICIAL DOUBLE RE-
FRACTION. SECOND-ORDER EFFECTS

1. Double refraction

In this chapter we shall study some of the phenomena connected with the transmission of light through crystals. We shall be mainly concerned with *birefringence* or *double refraction*: both the natural birefringence which results from the natural crystal anisotropy, and the artificial birefringence that is produced by an electric field (electro-optical effect) or by a stress (photoelastic effect). The question of *rotatory polarization* or *optical activity* is discussed in the next chapter. We shall not deal in this book with the absorption of light by crystals.

It will be assumed that the reader has some knowledge of the elementary treatment of double refraction (Hartshorne and Stuart 1950). §§ 1.1, 1.2, 1.3 summarize the main elementary facts that are of interest for our purpose. The proofs, from Maxwell's equations, of the properties of the indicatrix described in § 1.1 are given in Appendix H; they are not necessary for an understanding of the rest of the chapter.

1.1. The indicatrix. In an *isotropic* medium the dielectric properties at optical frequencies are given by

$$\mathbf{D} = \kappa \mathbf{E}, \quad \text{or} \quad \mathbf{D} = \kappa_0 K \mathbf{E}, \tag{1}$$

where κ is the permittivity of the medium, κ_0 the permittivity of a vacuum, and K the dielectric constant (κ and K are not, in general, the same as the permittivity and dielectric constant measured statically or at low frequencies). Maxwell's equations show that, if the relative magnetic permeability $M = \mu/\mu_0$ (p. 54) is taken as 1, the velocity of propagation of electromagnetic waves through the medium is given by

$$v = c/\sqrt{K}, \tag{2}$$

where c [equal to $1/\sqrt{(\kappa_0 \mu_0)}$] is the velocity in a vacuum (Abraham and Becker 1937, pp. 182–7). The *refractive index n*, defined as c/v, is thus

$$n = \sqrt{K}. \tag{3}$$

In an *anisotropic* medium we recall, from p. 69, that relations (1) have to be replaced by

$$D_i = \kappa_{ij} E_j, \quad \text{or} \quad D_i = \kappa_0 K_{ij} E_j. \tag{4}$$

When relations (4) instead of (1) are used in conjunction with Maxwell's equations, they lead, as we prove in Appendix H, to the conclusion that not one but *two* waves, of different velocity, may in general be propagated through a crystal with a given wave normal; moreover, these waves are plane polarized. The value of c/v for each wave may be called the *refractive index n* for that wave. The refractive indices of the two waves, as functions of the direction of their common wave normal, are obtained by drawing an ellipsoid known as the *indicatrix*.

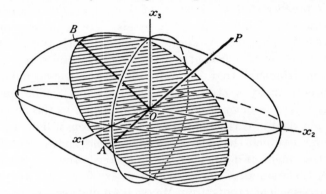

FIG. 13.1. The indicatrix construction giving the two refractive indices and the vibration directions of **D** for the two plane polarized waves associated with a given wave normal.

If x_1, x_2, x_3 are the principal axes of the dielectric constant (or the permittivity) tensor, the indicatrix is defined by the equation

$$\frac{x_1^2}{n_1^2} + \frac{x_2^2}{n_2^2} + \frac{x_3^2}{n_3^2} = 1, \tag{5}$$

where $n_1 = \sqrt{K_1}$, $n_2 = \sqrt{K_2}$, $n_3 = \sqrt{K_3}$. K_1, K_2, K_3 are the principal dielectric constants.

The indicatrix has the following important property (Fig. 13.1). Draw through the origin a straight line OP in an arbitrary direction. Draw the central section of the indicatrix perpendicular to it. This will be an ellipse. Then the two wave-fronts normal to OP that may be propagated through the crystal have refractive indices equal to the semi-axes, OA and OB, of this ellipse. The displacement vector **D** in the plane polarized wave with refractive index equal to OA vibrates parallel to OA. Similarly, the displacement vector in the wave with refractive index equal to OB vibrates parallel to OB. From this it follows, as a special case, that the two possible waves with wave normal x_1 have refractive indices n_2 and n_3; and **D** in the two waves is parallel to x_2 and x_3 respectively. Similarly, the wave normal x_2 corresponds to two waves of refractive indices n_3 and n_1, with $\mathbf{D}$ parallel to x_3 and x_1 respectively. A similar statement applies to the wave normal x_3. For this reason n_1, n_2, n_3 are called the *principal refractive indices*.

Note that refractive index itself is not a tensor, although its variation with direction is determined as we have seen, by the dielectric constant, which is a tensor.

Equation (5) may be alternatively written as

$$B_1 x_1^2 + B_2 x_2^2 + B_3 x_3^2 = 1, \tag{6}$$

where $B_1 = 1/n_1^2 = 1/K_1$, and B_2 and B_3 have similar meanings. B_1, B_2, B_3, being the reciprocals of the prin-cipal dielectric constants, are the *rela-tive dielectric impermeabilities*. We recognize (6) as the representation quadric (p. 16) for the relative di-electric impermeability. This repre-sentation quadric and the indicatrix are therefore identical.

1.2. The effect of crystal sym-metry. So far we have dealt with the general case where the optical proper-ties of the crystal are not restricted by crystal symmetry. When crystal sym-metry is present, the shape and orienta-tion of the indicatrix are subject to the same limitations as the representation quadric for a second-rank tensor pro-perty (Ch. I, § 5.1). It follows that the

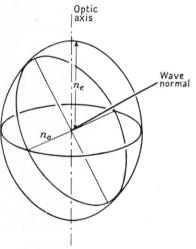

FIG. 13.2. The indicatrix for a (positive) uniaxial crystal.

indicatrix for a *cubic* crystal is a sphere and, since all central sections are circles, there is no double refraction.

For *hexagonal*, *tetragonal* and *trigonal* crystals the indicatrix is neces-sarily an ellipsoid of revolution about the principal symmetry axis (Fig. 13.2). With x_3 as this axis the equation is written

$$\frac{x_1^2}{n_0^2} + \frac{x_2^2}{n_0^2} + \frac{x_3^2}{n_e^2} = 1.$$

The central section perpendicular to the principal axis, and only this central section, is a circle (radius n_0). Hence, only for a wave normal along the principal axis is there no double refraction. The principal axis is called the *optic axis*, and the crystals are said to be *uniaxial*.

n_0 and n_e are called the *ordinary* and *extraordinary* refractive indices respectively. The crystal is said to be *positive* when $(n_e - n_0)$ is positive, and *negative* when $(n_e - n_0)$ is negative.

For the three remaining crystal systems, the *orthorhombic, monoclinic* and *triclinic*, the indicatrix is a triaxial ellipsoid. There are *two* circular sections, as illustrated in Fig. 13.3, and hence two privileged wave normal directions for which there is no double refraction. These two

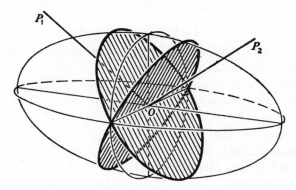

Fig. 13.3. The two circular sections of the indicatrix, and the two primary optic axes OP_1, OP_2, for a biaxial crystal. Both circular sections contain the principal axis of the indicatrix that is intermediate in length between the other two. Their radius is therefore equal to the intermediate principal refractive index.

directions are called the *primary optic axes*, or simply the *optic axes*, and the crystals are said to be *biaxial*.

1.3. The wave surface.† Suppose a point source of light is situated within a crystal. The wave front emitted at any instant forms a continuously expanding surface. We imagine this front fixed at some later instant and we call the surface so formed the *wave surface*.‡

(i) *Uniaxial crystal.* The wave surface for a uniaxial crystal consists of a sphere and an ellipsoid of revolution touching one another along the optic axis, as shown in Fig. 13.4 a and b. This may be proved from the indicatrix construction in the following way. The indicatrix is an ellipsoid of revolution (Fig. 13.2) and the radius of the central circular section is n_0. Therefore, any central section will always have one

---

† This section is not needed for understanding the rest of the chapter, but we shall call on its results in Chapter XIV.

‡ This nomenclature agrees with that of Dana, Hartshorne and Stuart, Houstoun, Rogers and Kerr, Tutton, and Wood. Groth, Johannsen and. Wooster call the surface so defined the *ray surface*, since it is the locus of points reached by the rays at any instant.

principal axis equal to n_0. The other principal axis will be intermediate in length between n_0 and n_e (Fig. 13.2). We know that two waves are

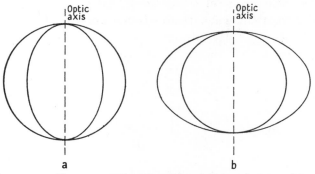

FIG. 13.4. The wave surface for a uniaxial crystal; (a) positive crystal, (b) negative crystal.

propagated outwards from the point source in any direction. The indicatrix construction shows that one of them always has the same refractive index n_0, and therefore the same velocity, regardless of direction; the other has a refractive index that varies from n_0 to n_e according to the direction of the wave normal. The wave surface will therefore consist of two sheets: a sphere (the ordinary wave), and a surface of revolution about the optic axis which touches the sphere at the ends of the optic axis (the extraordinary wave).

The extraordinary wave is shown separately in Fig. 13.5. It is necessary to distinguish between the wave normal and the *ray direction*, for it is clear from the figure that these two directions are not identical; for instance, at the point P the ray direction is OP and does not coincide with the normal to the surface. *All the statements we have made about the indicatrix have referred to the wave normal.*† In Fig. 13.5 the *wave velocity* for the element of wave front at P is

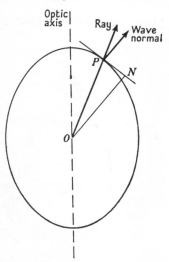

FIG. 13.5. The wave surface for a (positive) uniaxial crystal, showing the extraordinary wave only. The wave velocity v of the element of wave front at P is proportional to ON; the ray velocity is proportional to OP.

† For constructing the ray direction from the indicatrix see the textbook by Wooster (1938). This contains further relevant geometrical constructions and relations that must be omitted here.

proportional to ON, drawn normal to the tangent at P; by contrast the *ray velocity* is proportional to OP. The velocity given by the indicatrix construction is the wave velocity. The indicatrix construction therefore gives the length of ON when its direction is known. It is then a simple

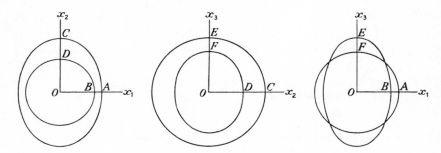

FIG. 13.6. Principal sections of the wave surface for a biaxial crystal. In the case illustrated the optic axes lie in the $x_1 x_3$ plane. For lettering see Fig. 13.7.

piece of analytical geometry, which we shall omit here, to show that the extraordinary wave surface is an ellipsoid of revolution about the optic axis. The lengths of the axes are evidently proportional to $1/n_e$, $1/n_e$, $1/n_0$.

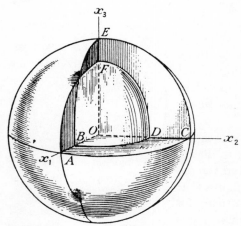

FIG. 13.7. A cut-away view of the wave surface for a biaxial crystal (Hartshorne and Stuart, *Crystals and the polarising microscope*, Arnold, 1950). Compare Fig. 13.6.

(ii) *Biaxial crystal*. The wave surface for a biaxial crystal is not as simple as that for a uniaxial crystal. For wave normals which lie in any of the three principal planes of the indicatrix the situation is very similar to that described above for a uniaxial crystal. The reader may easily show from the indicatrix construction that the wave surface consists of two sheets, and that each principal plane cuts the sheets in a circle and an ellipse. These curves are shown in Fig. 13.6 and the complete surface† is illustrated in Fig. 13.7.

† See Joos (1934) for an analytical treatment of this surface.

2. The electro-optical and photoelastic effects

2.1. Introduction to the electro-optical effect. In Chapter X a general definition of permittivity, at constant stress and temperature, was given as $(\partial D_i/\partial E_j)_{\sigma,T}$. For many purposes it is sufficient to assume that the permittivity coefficients of a crystal at a given temperature are constants and do not depend on the strength of the field. However, the effect we now have to discuss is a direct consequence of the dependence of permittivity on field. Consider a tetragonal or monoclinic crystal in which a field **E** acts parallel to the principal symmetry axis. This example is chosen because we want, for simplicity, to have **D** parallel to **E**, but, for reasons that will appear later, we do not want necessarily to have a centre of symmetry present. D may be expressed as a function of E by

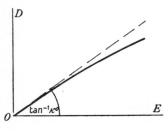

$$D = \kappa^0 E + \alpha E^2 + \beta E^3 + ..., \qquad (7)$$

FIG. 13.8.

where κ^0, α, β are constants. (κ^0 is to be distinguished from κ_0, the permittivity of a vacuum.) In Fig. 13.8 the straight line of slope κ^0 represents the first term; the second and higher-order terms give a deviation from the straight line which approaches zero as $E \to 0$. In practice α is negative, and so a given field E produces a smaller D than would be given by the first term alone.

The permittivity, defined by the slope of the curve, is

$$\kappa = dD/dE = \kappa^0 + 2\alpha E + 3\beta E^2 + ... \qquad (8)$$

At the field strengths that can readily be attained the second and higher-order terms represent extremely small additional contributions to the permittivity, and the second-order effects to which they give rise are usually difficult to detect. However, small changes of permittivity *at optical frequencies* are equivalent to small changes in refractive index, and these can be measured with great precision by using the effects of birefringence and by optical interferometry. A difference in refractive index for two waves, plane polarized at right angles, of one part in 10^6, which corresponds to a difference in permittivity of two parts in 10^6, can be readily measured. Thus, the correction to the permittivity due to the finite field, although not necessarily relatively greater than the corrections to other quantities, such as the elastic constants, is nevertheless accorded prominence because it can be more easily measured. The change in refractive index of a crystal produced by an electric field is known as the *electro-optical effect*.

It has to be remembered that permittivity depends on the frequency of the electric field. Keeping to the special case used above, where **D** was parallel to **E**, the situation is illustrated schematically in Fig. 13.9. Curve A represents D as a function of E for changes carried out slowly. Curve B represents the relation between D and E for changes at optical frequencies. Suppose, then, that a light wave travels through the crystal with **E** parallel to the unique axis. The oscillation due to the light wave is represented by the double-headed arrow through the origin. The refractive index is proportional to the square root of the slope of curve B at the origin. Now suppose that a static electric field E_0 is applied in the direction we are considering, so that D and E due to the static field are represented by the point P. The oscillation due to a light wave passing through the crystal is now shown by the double-headed arrow C drawn through P. Its slope, which gives the new refractive index, will be less than that of curve A at P. The electro-optical effect is thus a measure of the change in the slope of C as the biasing field E_0 is changed. It is not given simply by equation (8), because the biasing field is static, while the field in the derivative dD/dE is a rapidly changing one. However, we may write n^2, which is equal to κ/κ_0, as a power series in the biasing field E_0 and hence, taking square roots, n may be written as

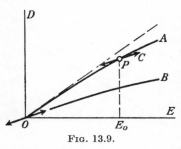

Fig. 13.9.

$$n = n^0 + aE_0 + bE_0^2 + ..., \qquad (9)$$

where a and b are constants and n^0 is the refractive index for $E_0 = 0$.

Now suppose that the crystal has a centre of symmetry. If the field is reversed in direction the physical situation, and therefore the refractive index, is essentially unaltered, but the sign of E_0 in (9) must be reversed. We then have

$$n = n^0 - aE_0 + bE_0^2 - .... \qquad (10)$$

Expressions (9) and (10) for n are only compatible if $a = 0$. We are left with

$$n = n^0 + bE_0^2 + ... \qquad (11)$$

If, on the other hand, there were no symmetry element present which made the two directions of the field equivalent from the point of view of the crystal, then there is no reason why reversing the field should not change the refractive index, and so the first-order term (aE_0) would remain.† The number of constants necessary to represent the electro-optical effect in crystals of different symmetries is discussed below. The present argument shows that if a centre of symmetry is present no first-order effect can exist. In crystals with a centre of symmetry, and in liquids, only the second-order term (bE_0^2) and higher even-order terms can exist, and so, as one might expect, the effects are very small in moderate fields. The effect given by the bE_0^2 term and higher even-order terms, which can occur in all substances, is known as the *Kerr effect*.

2.2. Introduction to the photoelastic effect. The permittivity and dielectric constant, and hence the refractive index, are, in general, functions not only of the applied electric field, but also of the stress on the crystal. The change of refractive index caused by stress is called the *photoelastic effect*. We choose for illustration the same crystal as in the last section. If a uniaxial stress σ is applied to the crystal along the unique axis, in addition to the biasing field E_0, the refractive index for a light wave whose electric vector is in this direction is given by an expression of the form

$$n = n^0 + aE_0 + a'\sigma + bE_0^2 + b'\sigma^2 + b''E_0\sigma + ..., \qquad (12)$$

where a, a', b, b', b'' are constants. The argument given in § 2.1 for the vanishing of the term aE_0 in a centrosymmetrical crystal does not apply to the first-order stress term $a'\sigma$. For, whereas reversing the direction of E_0 in such a crystal does not essentially change the situation, reversing the sign of σ changes the state of the crystal from tension to compression, and so, in general, changes n. This is why even an isotropic material like glass shows a first-order photoelastic effect, but cannot show a first-order electro-optical effect.

2.3. The general case. The refractive index of a crystal is specified by the indicatrix, which, as we have seen in § 1, is an ellipsoid whose coefficients are the components of the relative dielectric impermeability tensor B_{ij} at optical frequencies, namely,

$$B_{ij}x_i x_j = 1. \qquad (13)$$

(By definition, $B_{ij} = \kappa_0\, \partial E_i/\partial D_j$.) Thus, in general, the small change of refractive index produced by electric field and stress is, more precisely, a small change in the shape, size and orientation of the indicatrix.

---

† We call aE_0 the first-order term because it gives a first-order *change* $(n-n^0)$ in refractive index.

This change is most conveniently specified by giving the small changes in the coefficients B_{ij}. If we neglect higher-order terms than the first in the fields and stresses, the changes ΔB_{ij} in the coefficients, under an applied field E_k and an applied stress σ_{kl}, are given by

$$\Delta B_{ij} = z_{ijk} E_k + \pi_{ijkl} \sigma_{kl}, \tag{14}$$

where z_{ijk} is a third-rank tensor whose components give the electro-optical effect, and π_{ijkl} is a fourth-rank tensor giving the photoelastic effect. The z_{ijk} are the *electro-optical coefficients* and the π_{ijkl} the *piezo-optical coefficients*. Typical orders of magnitude, in m.k.s. units, are:

$$z_{ijk} \sim 10^{-12} \text{ metres/volt},$$

$$\pi_{ijkl} \sim 10^{-12} \text{ metres}^2/\text{newton} \ (= 10^{-13} \text{ cm}^2/\text{dyne}).$$

[Equation (14) is closely similar to the equation giving the strains of a crystal produced by a field (converse piezoelectric effect) and by a stress (elasticity):

$$\epsilon_{ij} = d_{kij} E_k + s_{ijkl} \sigma_{kl}, \tag{15}$$

where the d_{kij} are the piezoelectric coefficients and the s_{ijkl} are the compliances. The ϵ_{ij}, like the ΔB_{ij}, are dimensionless. The physical dimensions of the d's and s's are the same as those of the z's and π's respectively. Moreover, the typical orders of magnitude are closely similar; in m.k.s. units, $d_{kij} \sim 3 \times 10^{-12}$, $s_{ijkl} \sim 10^{-11}$.]

The photoelastic effect is sometimes expressed in terms of the strains instead of in terms of the stresses. We have the elastic relations (p. 132), $\sigma_{kl} = c_{klrs} \epsilon_{rs}$. Hence, (14) may be alternatively written,

$$\Delta B_{ij} = z_{ijk} E_k + p_{ijrs} \epsilon_{rs}, \tag{16}$$

where $$p_{ijrs} = \pi_{ijkl} c_{klrs}, \qquad \pi_{ijkl} = p_{ijrs} s_{rskl}. \tag{17}$$

The p_{ijrs} are the *elasto-optical coefficients*. Notice that they are dimensionless.

Since $\pi \sim 10^{-12}$ and $c \sim 10^{11}$, the order of magnitude of the p's is given by $p \sim 10^{-12} \times 10^{11} = 10^{-1}$. Since the B_{ij} are of order of magnitude unity, the ΔB_{ij} are a measure of the relative distortion of the indicatrix. We see, therefore, that this relative distortion is, quite roughly, one-tenth of the strain.

2.4. The primary and secondary electro-optical effect. In discussing the electro-optical effect we have not yet specified the mechanical constraints on the crystal: for example, the crystal might be free so that the stress is zero, or clamped, so that the strain is zero. This is important, because, if the crystal is free, a static electric field will cause a strain by the converse piezoelectric effect, and this in turn will give a change in refractive index by the photoelastic effect. The situation is very like that of primary and secondary pyroelectricity (Ch. X, § 4.4). The electro-optical effect that would be obtained if the crystal were not allowed to strain is called the *primary* (or 'true') *effect*, and the effect due to piezoelectricity and photoelasticity is then called the *secondary* (or 'false') *effect*. The observed effect in a free crystal is the sum of the primary and secondary effects.

The order of magnitude of the secondary effect may be found as follows. A field E produces a piezoelectric strain, $\epsilon = dE$. This produces in turn a change in B given by

$$\Delta B = p\epsilon = (pd)E.$$

According to the rough figures we have been using, the coefficient pd, giving the secondary effect, is of order of magnitude

$$10^{-1} \times 3 \times 10^{-12} = 3 \times 10^{-13}.$$

Thus, the secondary effect is comparable in magnitude to the observed effect in a free crystal, which is given, as we have seen, by $z \sim 10^{-12}$.

In an experiment in which a static field is applied to a free crystal, what is measured is the effect at constant stress—that is, primary plus secondary. The effect at constant strain (primary effect) may be found by applying an alternating electric field of high frequency. The crystal strains are then very small, provided the frequency and the geometry of the arrangement are not such as to set the crystal into resonance. The resulting birefringence, which alternates in sign with the frequency of the field, can be measured (Carpenter 1950).

2.5. The effect of symmetry. (i) *General symmetry arguments.* In visualizing the effect on the indicatrix produced by a field or by a stress the following general principle is helpful: *a crystal under an external influence will exhibit only those symmetry elements that are common to the crystal without the influence and the influence without the crystal.* The truth of the principle is almost self-evident. If we operate with one of their common symmetry elements on the influence (before applying it to the crystal) and on the crystal (without the influence), no change in either will be apparent. It follows that the crystal under the external influence will also possess this symmetry element. If, on the other hand, we operate in the same way with a symmetry element that is not common, there will be a change. It follows that the crystal under the external influence will not, in general, possess this symmetry element.

As a simple application, suppose an electric field is applied to a cubic crystal of class *23* (Fig. 13.10 a) parallel to a triad axis. The triad axis is the only symmetry element that the field shares with the crystal. Hence, the optical properties of the crystal in the field will be those appropriate to the trigonal class *3*, which is uniaxial.

We can immediately see that a uniaxial stress applied parallel to a 3-, 4- or 6-fold axis will give an optically uniaxial crystal with the optic axis parallel to the stress. On the other hand, a uniaxial stress applied

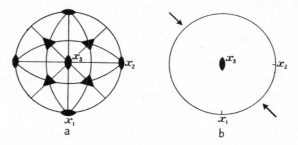

FIG. 13.10. A cubic crystal of class *23*. (*a*) Stereogram of the symmetry elements without an electric field or a stress, (*b*) stereogram of the symmetry elements which are common to both the unstressed crystal and a uniaxial compressive stress parallel to [110].

along a diad axis or a general direction will give an optically biaxial crystal. The same remarks are true when an electric field is applied in place of the uniaxial stress.

An interesting illustration of the principle is the following: a crystal of class *23* under a uniaxial pressure along the [110] direction. Fig. 13.10 a shows the symmetry elements of the unstressed crystal, and Fig. 13.10 b shows that the only symmetry element that the unstressed crystal shares with the stress is a diad axis along x_3. x_3 must therefore be a principal axis of the indicatrix; but the other principal axes lie in the $x_1 x_2$ plane with no particular relationship to the stress direction. This result holds also for class *m3* (Figs. 13.11 a and b). However, in the more symmetrical cubic classes $\bar{4}3m$, *432*, *m3m*, the [110] direction is either a diad axis or lies in a plane of symmetry. Therefore, if a crystal of one of these classes were compressed along [110], one of the indicatrix axes would be parallel to the stress axis.

(ii) *Reduction in the number of independent electro-optical coefficients.* We know that $B_{ij} = B_{ji}$. Therefore, in equation (14), $\Delta B_{ij} = \Delta B_{ji}$ for all E_k and σ_{kl}. It follows that

$$z_{ijk} = z_{jik}. \tag{18}$$

This relation reduces the number of independent coefficients z_{ijk} from 27 to 18, and it makes possible a similar abbreviated notation to that used for the piezoelectric tensors (p. 113). The change is made as follows. The B_{ij} are written with only one suffix, running from 1 to 6:

$$\begin{bmatrix} B_{11} & B_{12} & B_{31} \\ B_{12} & B_{22} & B_{23} \\ B_{31} & B_{23} & B_{33} \end{bmatrix} \rightarrow \begin{bmatrix} B_1 & B_6 & B_5 \\ B_6 & B_2 & B_4 \\ B_5 & B_4 & B_3 \end{bmatrix}. \tag{19}$$

Correspondingly, the first two suffixes of z_{ijk} are abbreviated into a

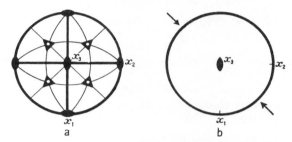

Fig. 13.11 a and b. The same as Figs. 13.10 a and b but for the cubic class $m3$.

single one running from 1 to 6, and we write for the electro-optical effect,
$$\Delta B_i = z_{ij} E_j \quad (i = 1, 2, ..., 6; j = 1, 2, 3). \tag{20}$$
In changing the suffixes in the z_{ijk} *no factors of $\frac{1}{2}$ or 2 appear.*

[Factors of 2 did appear when the suffixes of the piezoelectric moduli d_{ijk} were abbreviated. The reason for the difference is seen by comparing the piezoelectric equation

$$\epsilon_{jk} = d_{ijk} E_i \qquad [\text{Ch. VII, (19)}]$$

with the electro-optical equation

$$\Delta B_{ij} = z_{ijk} E_k.$$

We notice, first of all, that the *last two* suffixes in d_{ijk} are abbreviated, since these correspond to the suffixes of the ϵ_{jk}; while the *first two* suffixes of z_{ijk} are abbreviated, corresponding to the suffixes of ΔB_{ij}. This is not relevant to the question of the 2's, however. 2's come in with the d's because 2's enter the definition of the one-suffix strains. No factors of 2 enter the definition of the one-suffix B_{ij}, and so no factors of 2 come in with the z's.]

The restrictions imposed by crystal symmetry on the z_{ij} are closely similar to those for the piezoelectric moduli d_{ij}. The matrices given for the d_{ij} on pp. 123–4 may be used for the z_{ij} with two changes:

(1) the *first* suffix in z_{ij} runs from 1 to 6, while it is the *second* suffix in d_{ij} that runs from 1 to 6. Therefore, the arrays are correct for z_{ij} if i is read as the *column* number and j as the *row* number, which is the reverse of the usual procedure;

(2) as a consequence of the non-appearance of factors of 2 mentioned above, the double circle symbol is to be interpreted as meaning simply the numerical equal with the sign reversed—the same, in fact, as the single open circle.

(iii) *Reduction in the number of independent photoelastic coefficients.* Since, in equation (14), $\Delta B_{ij} = \Delta B_{ji}$ for all E_k and σ_{kl}, we have

$$\pi_{ijkl} = \pi_{jikl}. \tag{21}$$

Furthermore, since $\sigma_{kl} = \sigma_{lk}$ even with body-torques (p. 87), we may put

$$\pi_{ijkl} = \pi_{ijlk}. \tag{22}$$

This last operation is similar to that used in dealing with the elastic compliances (pp. 132–3).

The relations (21) and (22) reduce the number of independent coefficients from $3^4 = 81$ to 36, and they make possible a similar abbreviated notation to that used for the elastic constants (p. 134). This goes as follows. In the equation for the photoelastic effect, namely,

$$\Delta B_{ij} = \pi_{ijkl}\sigma_{kl} \quad (i,j,k,l = 1,2,3), \tag{23}$$

the suffixes in the ΔB_{ij} are abbreviated by (19), and the suffixes in the σ_{kl} are abbreviated in a precisely similar way. We have then

$$\Delta B_m = \pi_{mn}\sigma_n \quad (m,n = 1,2,...,6). \tag{24}$$

The π_{mn} are related to the π_{ijkl} by the rules:

$$\pi_{mn} = \pi_{ijkl}, \quad \text{when } n = 1, 2, \text{ or } 3;$$
$$\pi_{mn} = 2\pi_{ijkl}, \quad \text{when } n = 4, 5, \text{ or } 6.$$

The factors of 2 appear because of the pairing of the shear stress terms on the right-hand side of (23).

[The above abbreviation may be compared with that for the elastic compliances in the equation

$$\epsilon_{ij} = s_{ijkl}\sigma_{kl}. \qquad \text{[Ch. VIII, (1)]}$$

In transforming the s's, factors of 4 appeared in some terms because of the additional factor of 2 coming from the ϵ's.]

With the elastic constants the existence of a strain energy function uniquely determined by the strains gave the further relations, $c_{ij} = c_{ji}$, and $s_{ij} = s_{ji}$, but in general

$$\pi_{mn} \neq \pi_{nm}. \tag{25}$$

Thus, the number of independent π_{mn} for the triclinic classes remains 36. The number of independent π_{mn} for the other classes reduces on account of symmetry. The reductions are the same as those for the elasticity tensors, except that $\pi_{mn} \neq \pi_{nm}$, and some factors of 2 occur in different places. The forms of the (π_{mn}) matrices are given in Table 15. A key to the notation, which is similar to that already introduced, appears at the head of the table. As usual, we use the standard setting of the reference axes in relation to the symmetry elements (Appendix B), except where otherwise noted. The number of independent components for each class is given in brackets after each matrix.

If we work in terms of strains instead of stresses and use the equation

$$\Delta B_{ij} = p_{ijrs} \epsilon_{rs} \quad (i, j, r, s = 1, 2, 3), \tag{26}$$

the change to the matrix notation is made in a similar way. The suffixes in ΔB_{ij} are abbreviated by (19). In abbreviating the suffixes in ϵ_{rs} we have to remember to introduce factors of $\frac{1}{2}$ in the shear strains [Ch. VII, equation (20)]. (The abbreviation of suffixes in the elastic equation, $\sigma_{ij} = c_{ijkl} \epsilon_{kl}$, is thus an exactly parallel case.) We have then, instead of (26),

$$\Delta B_m = p_{mn} \epsilon_n \quad (m, n = 1, 2, ..., 6),$$

and the p_{mn} are related to the p_{ijrs} by

$$p_{mn} = p_{ijrs}, \quad \text{for all } m \text{ and } n,$$

with no factors of 2 or $\frac{1}{2}$. In general, $p_{mn} \neq p_{nm}$. It is readily shown that

$$p_{mn} = \pi_{mr} c_{rn}, \quad \pi_{mn} = p_{mr} s_{rn}.$$

The forms of the (p_{mn}) matrix in the various classes are similar to those of the (π_{mn}) matrix except that some factors of 2 appear in different places (Table 15).

2.6. The photoelasticity of cubic crystals. It will be seen from Table 15 that the photoelastic behaviour of cubic crystals is not the same as that of isotropic materials. The cubic classes divide into two groups. Four coefficients are needed to define the photoelastic properties in classes where the cube axes are diads; the other group, where the cube axes are tetrads, needs only three coefficients. It is of interest now to examine analytically a few special cases of cubic crystals under stress.

(i) *Uniaxial tension along a cube axis.* Consider a cubic crystal of class *23* or *m3* under a uniaxial tensile stress σ applied parallel to a cube axis. Let Ox_1 be the direction of the stress, and let Ox_2, Ox_3 be the two

TABLE 15

Forms of the π and p photoelastic matrices†

KEY TO NOTATION

In both π and p matrices

-    zero component
-    non-zero component
- equal components
- components numerically equal, but opposite in sign.

In the π matrices

⊙   a component equal to twice the heavy dot component to which it is joined

⊚   a component equal to minus 2 times the heavy dot component to which it is joined

✗ $(\pi_{11} - \pi_{12})$.

In the p matrices

⊙   a component equal to the heavy dot component to which it is joined

⊚   a component equal to minus the heavy dot component to which it is joined

✗ $\tfrac{1}{2}(p_{11} - p_{12})$.

TRICLINIC
Both classes

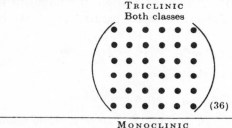

(36)

MONOCLINIC
All classes

Diad ‖ x_2
(standard
orientation)   Diad ‖ x_3

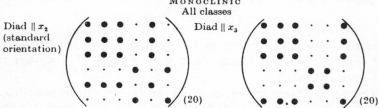

(20)   (20)

ORTHORHOMBIC
All classes

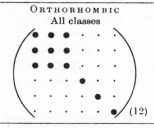

(12)

† When these matrices were originally calculated by Pockels, mistakes were made in 10 out of the 32 classes. These mistakes were copied by all later writers on the subject until Bhagavantam (1942 and 1952) pointed them out.

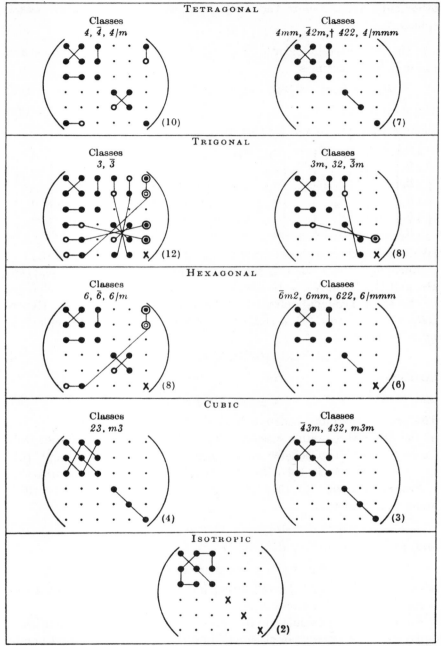

† The same matrix holds for both orientations of class $\bar{4}2m$ ($2 \parallel x_1$ and $m \perp x_1$), for the addition of a centre of symmetry makes the two orientations indistinguishable.

other cube axes. Before the stress was put on, the indicatrix was a sphere,

$$B^0(x_1^2+x_2^2+x_3^2) = 1, \tag{27}$$

and the refractive index was given by $B^0 = 1/(n^0)^2$. Under the stress the indicatrix becomes

$$B_1 x_1^2 + B_2 x_2^2 + B_3 x_3^2 + 2B_4 x_2 x_3 + 2B_5 x_3 x_1 + 2B_6 x_1 x_2 = 1. \tag{28}$$

The matrix equation

$$\Delta B_i = \pi_{ij}\sigma_j, \tag{29}$$

when written out in full for this case, is

$$\begin{pmatrix} \Delta B_1 \\ \Delta B_2 \\ \Delta B_3 \\ \Delta B_4 \\ \Delta B_5 \\ \Delta B_6 \end{pmatrix} = \begin{pmatrix} B_1-B^0 \\ B_2-B^0 \\ B_3-B^0 \\ B_4 \\ B_5 \\ B_6 \end{pmatrix} = \begin{pmatrix} \pi_{11} & \pi_{12} & \pi_{13} & 0 & 0 & 0 \\ \pi_{13} & \pi_{11} & \pi_{12} & 0 & 0 & 0 \\ \pi_{12} & \pi_{13} & \pi_{11} & 0 & 0 & 0 \\ 0 & 0 & 0 & \pi_{44} & 0 & 0 \\ 0 & 0 & 0 & 0 & \pi_{44} & 0 \\ 0 & 0 & 0 & 0 & 0 & \pi_{44} \end{pmatrix} \begin{pmatrix} \sigma \\ 0 \\ 0 \\ 0 \\ 0 \\ 0 \end{pmatrix} = \begin{pmatrix} \pi_{11}\sigma \\ \pi_{13}\sigma \\ \pi_{12}\sigma \\ 0 \\ 0 \\ 0 \end{pmatrix}.$$

(We have used the rules of matrix multiplication given on p. 151.)

Since $B_4 = B_5 = B_6 = 0$ the axes of the indicatrix are simply Ox_1, Ox_2 and Ox_3 (as may also be seen by symmetry). To obtain the changes in the three principal refractive indices we write $B_1 = 1/n_1^2$. Hence, $\Delta B_1 = -(2/n_1^3)\Delta n_1$. To a sufficient approximation we may replace n_1 by n^0 and obtain

$$\Delta n_1 = -\tfrac{1}{2}(n^0)^3\Delta B_1 = -\tfrac{1}{2}(n^0)^3\pi_{11}\sigma. \tag{30}$$

Similarly,

$$\Delta n_2 = -\tfrac{1}{2}(n^0)^3\pi_{13}\sigma, \tag{31}$$

and

$$\Delta n_3 = -\tfrac{1}{2}(n^0)^3\pi_{12}\sigma. \tag{32}$$

The crystal therefore becomes biaxial. In a crystal belonging to one of the other three cubic classes ($\bar{4}3m$, 432, $m3m$) it will be seen from Table 15 that the scheme of coefficients is the same except that $\pi_{12} = \pi_{13}$. Hence, in this case, $\Delta n_2 = \Delta n_3$ and the crystal is uniaxial.

From equations (30), (31), (32) the birefringence for light travelling along x_2 is evidently

$$n_\| - n_\perp = \Delta n_1 - \Delta n_3 = -\tfrac{1}{2}(n^0)^3(\pi_{11}-\pi_{12})\sigma; \tag{33}$$

and, for light travelling along x_3,

$$n_\| - n_\perp = \Delta n_1 - \Delta n_2 = -\tfrac{1}{2}(n^0)^3(\pi_{11}-\pi_{13})\sigma, \tag{34}$$

where $n_\|$ and $n_\perp$ are the refractive indices respectively parallel and perpendicular to the stress direction.

From the above calculation it is easy to see why π_{12} is not identically equal to π_{13} for crystals of classes 23 and $m3$ (as was originally thought by Pockels to be the case). If x_1 is a 4-fold axis, the birefringence in the x_2 and x_3 directions must be the same, by the symmetry principle given on p. 245, and hence $\pi_{12} = \pi_{13}$. In classes *23* and *m3*, however, where x_1 is only a 2-fold axis, x_2 and x_3 are not related by symmetry in the stressed crystal. In fact the symmetry of a crystal of class *23* stressed in this way degenerates to that of class *222*; similarly, *m3* degenerates to *mmm*. Both these are orthorhombic classes, and therefore biaxial.

The reason for some of the equalities in the π's that do exist in classes *23* and *m3* may be readily seen by the direct inspection method (p. 118). A triad axis in the positive octant operates on the axes, and therefore on the suffixes, as follows:

$$1 \to 2, \quad 2 \to 3, \quad 3 \to 1.$$

These are the transformations of suffixes in the *four-suffix*, tensor, notation. In this notation, therefore,

$$11 \to 22, \quad 22 \to 33, \quad 33 \to 11, \quad 23 \to 31, \quad 31 \to 12, \quad 12 \to 23.$$

The same transformations in the *two-suffix* notation are

$$1 \to 2, \quad 2 \to 3, \quad 3 \to 1, \quad 4 \to 5, \quad 5 \to 6, \quad 6 \to 4.$$

Hence,

$$\pi_{11} = \pi_{22} = \pi_{33}, \qquad \pi_{12} = \pi_{23} = \pi_{31}, \qquad \pi_{13} = \pi_{21} = \pi_{32},$$

$$\pi_{44} = \pi_{55} = \pi_{66},$$

as shown in Table 15. The vanishing of the other coefficients is obtained by operating with one of the other triad axes as well.

Formulae (33) and (34) refer to a cubic crystal under uniaxial tension parallel to a cube axis. They give the birefringences along the other two cube axes. In Table 16 we include the corresponding results for some other conditions of stressing and viewing.

TABLE 16

Birefringence of cubic crystals under uniaxial stress for different directions of observation

Direction of uniaxial stress	Direction of observation	Birefringence $(n_{\parallel} - n_{\perp})$	
		Classes 23, m3	Classes $\bar{4}3m$, 432, m3m
[100]	[010]	$-\frac{1}{2}(n^0)^3\sigma(\pi_{11} - \pi_{12})$	$-\frac{1}{2}(n^0)^3\sigma(\pi_{11} - \pi_{12})$
	[001]	$-\frac{1}{2}(n^0)^3\sigma(\pi_{11} - \pi_{13})$	,, ,,
	[011]	$-\frac{1}{2}(n^0)^3\sigma\{\pi_{11} - \frac{1}{2}(\pi_{12} + \pi_{13})\}$	,, ,,
	[01$\bar{1}$]	,, ,,	,, ,,
[111]	All directions $\perp$ to [111]	$-\frac{1}{2}(n^0)^3\sigma\pi_{44}$	$-\frac{1}{2}(n^0)^2\sigma\pi_{44}$

It is evident from formulae (33) and (34) and Table 16 that, by compressing a cubic crystal along a cube axis and measuring the resulting birefringence at right angles to the stress axis, the values of $(\pi_{11}-\pi_{12})$ and $(\pi_{11}-\pi_{13})$ can be found. Such measurements of birefringence are comparatively easy to make. It may be shown, however, that measurements of birefringence alone cannot separate π_{11} from π_{12}, or π_{11} from π_{13}. It is always the differences $(\pi_{11}-\pi_{12})$ or $(\pi_{11}-\pi_{13})$ that enter the formulae. To obtain the absolute values of π_{11}, π_{12}, π_{13} one has to make the considerably more difficult measurement of the absolute change in one of the refractive indices on applying the stress. This requires an interferometer method. Compressing a crystal along [111] gives a convenient way of measuring π_{44}. π_{44}, of course, also determines the changes in refractive index resulting from a shear stress applied about a cube axis.

Some measured values of the π_{ij} and p_{ij} for cubic crystals are given in Table 17. The values are sometimes slightly dependent on the wavelength of the light. Care is needed with the signs.† To illustrate the signs suppose that a crystal of sodium chloride is compressed along a cube axis. From Table 16, $n_{\parallel}-n_{\perp} = -\frac{1}{2}(n^0)^3(\pi_{11}-\pi_{12})\sigma$. From Table 17, $(\pi_{11}-\pi_{12})$ is negative; σ is negative; hence $(n_{\parallel}-n_{\perp}) = (n_e-n_0)$ is negative. Sodium chloride therefore becomes a negative uniaxial crystal when compressed along a cube axis.

EXERCISE 13.1. Prove that, in cubic crystals of classes $\bar{4}3m$, 432 and $m3m$, the change of refractive index produced by a hydrostatic pressure is given by $\rho(dn/d\rho) = \frac{1}{2}n^3 p_0$, where $p_0 = (p_{11}+2p_{12})/3$ and ρ is the density.

3. Second-order effects in general

3.1. Thermodynamics.

We have seen that the electro-optical effect and photoelasticity are second-order effects, in the sense that they arise from a dependence of permittivity on field and stress. A discussion of the thermodynamics of such second-order effects, of which there are many, might be given on the same lines as that developed for first-order effects in Chapter X. However, a full treatment would be too long and detailed to give here, and we shall therefore simply indicate, in outline only, how it could be developed. The suffixes in the various tensors will be omitted. Thus, the notation is not a rigorous one, but it can readily be made rigorous when occasion demands.

† In his definitions of the π_{ij} Pockels (1906) counted compressive stresses as positive and introduced a minus sign on the right-hand side of the defining equation (24). Throughout this book we count tensile stresses as positive. Pockels's π_{ij} are therefore the same as ours. His units for π_{ij} in the *Lehrbuch* (1906) are $\dot{\text{m}}\text{m}^2/\text{kg}$.

TABLE 17

Photoelasticity of cubic crystals

Values of the piezo-optical coefficients for sodium D light
(unit $= 10^{-12}$ m²/newton)   (1 cm²/dyne $= 10$ m²/newton)

Crystal	Class	π_{11}	π_{12}	π_{13}	π_{44}	$\pi_{11}-\pi_{12}$	$\pi_{11}-\pi_{13}$	Ref.
Potassium alum	$m3$	3·7	9·1	8·5	−0·65	−5·43	−4·82	1
Barium nitrate	$m3$	18·11	40·0	35·2	−1·69	−23·84	−17·13	2
Lead nitrate	$m3$	70·21	89·34	82·05	−1·39	−19·13	−11·84	2
Sodium chloride	$m3m$	0·25	1·46		−0·85	−1·21		3
Potassium chloride	$m3m$	. .	. .		−4·31	1·70		3
Fluorite	$m3m$	−0·29	1·16		0·698	−1·45		3
Diamond	$m3m$	−0·43	0·37		−0·27	−0·80		4

Values of the elasto-optical coefficients (dimensionless) for sodium D light

Crystal	Class	p_{11}	p_{12}	p_{13}	p_{44}	$p_{11}-p_{12}$	$p_{11}-p_{13}$	Ref.
Potassium alum	$m3$	0·27	0·35	0·34	−0·0056	−0·0792	−0·0704	1
Barium nitrate	$m3$	2·49	3·40	3·20	−0·0205	−0·992	−0·713	2
Lead nitrate	$m3$	8·50	8·78	8·67	−0·0191	−0·281	−0·174	2
Sodium chloride	$m3m$	0·137	0·178		−0·0108	−0·0408		3
Potassium chloride	$m3m$	. .	. .		−0·0276	0·0595		3
Fluorite	$m3m$	0·0558	0·228		0·0236	−0·1722		3
Diamond	$m3m$	−0·125	0·325		−0·11	−0·45		4

References: (1) Bhagavantam and Suryanarayana (1947), (2) Bhagavantam and Krishna Rao (1953), (3) Pockels (1906), (4) Ramachandran (1947) corrected in Burstein and Smith (1948 b). For other data see Maris (1927), Mueller (1935), West and Makas (1948), Burstein and Smith (1948 a) [corrected in (1948 b)], Bhagavantam and Suryanarayana (1949 b), Bhagavantam and Krishna Rao (1954), Narasimhamurty (1954 a and b).

As in Chapter X, σ, E and T are chosen as independent variables defining the state of the crystal. The photoelastic effect, as we have seen, is essentially due to a change of permittivity with stress. The photoelastic effect is thus proportional to

$$\frac{\partial \kappa}{\partial \sigma} = \frac{\partial}{\partial \sigma}\left(\frac{\partial D}{\partial E}\right). \tag{35}$$

Since D is a function of state, the order of differentiation may be reversed, and we may write

$$\frac{\partial \kappa}{\partial \sigma} = \frac{\partial}{\partial E}\left(\frac{\partial D}{\partial \sigma}\right) = \frac{\partial d}{\partial E}, \tag{36}$$

where d represents the piezoelectric moduli defined by $\partial D/\partial \sigma$. The coefficients giving the change of permittivity with stress are thus numerically equal to the coefficients giving the change of piezoelectric moduli with field. We must note, however, that *the frequency of the field changes would have to be the same in the two cases if equality is to exist.*

The coefficients $\partial d/\partial E$ may be written

$$\frac{\partial d}{\partial E} = \frac{\partial}{\partial E}\left(\frac{\partial D}{\partial \sigma}\right) = \frac{\partial}{\partial E}\left(\frac{\partial \epsilon}{\partial E}\right), \tag{37}$$

where we have used the thermodynamic relation (35) of Chapter X. The change in the permittivity with stress is thus connected with the second derivative of strain with respect to field, $\partial^2\epsilon/\partial E^2$. This derivative represents the phenomenon of *electrostriction* (§ 3.2 below). Electrostriction results from a *quadratic* dependence of ϵ on E; it is thereby distinguished from the converse piezoelectric effect, which is due simply to a *linear* dependence of ϵ on E. We have thus established a link between photoelasticity, electrostriction and the coefficients $\partial d/\partial E$.

Many other connexions may be found, in a similar way, between other second derivatives of ϵ, D and S with respect to σ, E and T. All these second derivatives describe second-order effects. As another example we have the derivatives $\partial^2\epsilon/\partial\sigma^2$, which are the second-order compliances. We also have the derivatives $(\partial^2\epsilon/\partial\sigma\partial T)$, which are coefficients giving the temperature dependence of the compliances $\partial\epsilon/\partial\sigma$; they are equal to the coefficients giving the stress dependence of the thermal expansion coefficients $\partial\epsilon/\partial T$.

In Chapter X first derivatives such as $\partial D/\partial E$, $\partial\epsilon/\partial T$, which describe first-order effects (permittivity, thermal expansion) were expressed as *second* derivatives of the thermodynamic potential Φ. We defined Φ so that

$$d\Phi = -\epsilon\,d\sigma - D\,dE - S\,dT,$$

[Ch. X, equation (32)]

and then, for example, $D = -\dfrac{\partial\Phi}{\partial E}$,

and hence $\kappa = \dfrac{\partial D}{\partial E} = -\dfrac{\partial^2\Phi}{\partial E^2}$.

It follows, therefore, that derivatives like $\partial^2 D/\partial E^2$, which gives the second-order permittivity (and the electro-optical effect), are expressible as the *third* derivatives of Φ:

$$\frac{\partial^2 D}{\partial E^2} = -\frac{\partial^3\Phi}{\partial E^3}. \tag{38}$$

In the same way the coefficients $(\partial^2 D/\partial\sigma\partial E)$, which, as we saw above, give the photoelastic effect and the electrostrictive effect, are expressible as third derivatives of Φ:

$$\frac{\partial^2 D}{\partial \sigma \partial E} = -\frac{\partial^3 \Phi}{\partial \sigma \partial E^2}. \tag{39}$$

3.2. Electrostriction and morphic effects. In the linear (first-order) piezoelectric effect discussed in Chapter VII the polarization is proportional to the stress (direct effect) and the strain is proportional to the field (converse effect). When the stresses and strains are small only the linear effect need be considered. When the stresses and strains are larger, quadratic terms have to be introduced. Thus, for the converse effect we write

$$\epsilon_{jk} = d^0_{ijk} E_i + \gamma_{iljk} E_i E_l. \tag{40}$$

The first term on the right represents the linear piezoelectric effect. d^0_{ijk} is the value of $d_{ijk} = (\partial \epsilon_{jk}/\partial E_i)$ for infinitesimal fields. The second term represents a strain proportional to products of the electric field components; this is the electrostrictive term that gives rise to second derivatives of strain with respect to the field. If the field is reversed in direction, all the components E_i change sign, and so the strain due to the linear terms changes sign; an extension becomes a contraction, and so on. The second term, on the other hand, does not change sign when the field is reversed. γ_{iljk} is a fourth-rank tensor that is evidently symmetrical in i and l and in j and k. Its non-vanishing components in the various classes are the same as those of the photoelastic tensors π_{ijkl} and p_{ijkl}. It follows that, since these components are not all zero even for isotropic materials, electrostriction is a property that is possible in all materials, including glasses and liquids. In this respect the electrostrictive strain differs strikingly from the piezoelectric strain which is linear in the field, for the latter, as we have seen, is destroyed by a centre of symmetry.

Equation (40) may be looked at in another way. It may be written as

$$\epsilon_{jk} = (d^0_{ijk} + \gamma_{iljk} E_l) E_i. \tag{41}$$

In this form the third-rank tensor $\gamma_{iljk} E_l$ appears as a correction to be applied to d^0_{ijk}. In the general case, all the components d^0_{ijk} have to be corrected by small amounts that are proportional to the field.

One further point may be noticed. The application of the field may, or may not, lower the symmetry of the crystal (according to the principle stated on p. 245). In either case the correction terms $\gamma_{iljk} E_l$ form a tensor which must conform to the symmetry of the combination: crystal plus field. If this symmetry is lower than the symmetry of the isolated crystal, the correction terms may give finite values to some of the components of d_{ijk} that were previously zero. In other words, by lowering the symmetry of the crystal the field may *create* new piezoelectric moduli. These new moduli are proportional to the field. They therefore give a strain which is proportional to the square of the field. This particular type of second-order effect is called a *morphic* effect. In general, a morphic effect must be produced whenever an influence, by lowering the symmetry of a crystal, creates new coefficients proportional to the influence.

SUMMARY

Double refraction. The equation of the *indicatrix* is

$$\frac{x_1^2}{n_1^2} + \frac{x_2^2}{n_2^2} + \frac{x_3^2}{n_3^2} = 1, \tag{5}$$

where $n_1 = \sqrt{K_1}$, $n_2 = \sqrt{K_2}$, $n_3 = \sqrt{K_3}$ are the *principal refractive indices*. The indicatrix (5) is identical with the representation quadric for the relative dielectric impermeability:

$$B_1 x_1^2 + B_2 x_2^2 + B_3 x_3^2 = 1, \tag{6}$$

where $B_1 = 1/K_1$, $B_2 = 1/K_2$, $B_3 = 1/K_3$ are the principal relative dielectric impermeabilities.

For a cubic crystal $n_1 = n_2 = n_3$. Hexagonal, tetragonal and trigonal crystals are *uniaxial*: $n_1 = n_2 = n_0$, $n_3 = n_e$; positive if $n_e > n_0$; negative if $n_e < n_0$. Orthorhombic, monoclinic and triclinic crystals are *biaxial*.

The electro-optical effect and photoelasticity. If an electric field and a stress are applied to a crystal, the coefficients of the indicatrix, $B_{ij} x_i x_j = 1$, change by small amounts ΔB_{ij} which are proportional to the field and stress, thus:

$$\Delta B_{ij} = z_{ijk} E_k + \pi_{ijkl} \sigma_{kl}. \tag{14}$$

z_{ijk} are the *electro-optical coefficients* and form a third-rank tensor. π_{ijkl} are the *piezo-optical coefficients* and form a fourth-rank tensor. In terms of the field and the strains:

$$\Delta B_{ij} = z_{ijk} E_k + p_{ijrs} \epsilon_{rs}, \tag{16}$$

where

$$p_{ijrs} = \pi_{ijkl} c_{klrs}, \qquad \pi_{ijkl} = p_{ijrs} s_{rskl}. \tag{17}$$

p_{ijrs} are the *elasto-optical coefficients* and are dimensionless.

The electro-optical effect at constant stress is the sum of (1) the effect at constant strain (the primary effect) and (2) the indirect effect due to the combination of piezoelectricity with photoelasticity (the secondary effect).

Since $B_{ij} = B_{ji}$, $z_{ijk} = z_{jik}$. By abbreviating the suffixes of ΔB_{ij} and the first two suffixes of z_{ijk} into a single one running from 1 to 6 we may write the electro-optical effect as

$$\Delta B_i = z_{ij} E_j \quad (i = 1, 2, ..., 6; j = 1, 2, 3). \tag{20}$$

No factors of 2 or $\frac{1}{2}$ are necessary when abbreviating the z's. There are 18 independent z_{ij} in the general case. This number is reduced by symmetry exactly as for the piezoelectric moduli. The form of the matrix (z_{ij}) is the same as that of the transpose of the piezoelectric matrix (d_{ij}) except for some factors of 2.

Since $B_{ij} = B_{ji}$, $\pi_{ijkl} = \pi_{jikl}$; and since $\sigma_{kl} = \sigma_{lk}$, even with body-torques (p. 87), $\pi_{ijkl} = \pi_{ijlk}$. In the two-suffix notation the photoelastic effect may then be written as

$$\Delta B_m = \pi_{mn} \sigma_n \quad (m, n = 1, 2, ..., 6), \tag{24}$$

where

$$\pi_{mn} = \pi_{ijkl}, \quad \text{when } n = 1, 2, \text{ or } 3;$$

$$\pi_{mn} = 2\pi_{ijkl}, \quad \text{when } n = 4, 5 \text{ or } 6.$$

In general, $\qquad\qquad \pi_{mn} \neq \pi_{nm}$.

In terms of the one-suffix strains the photoelastic effect is

$$\Delta B_m = p_{mn} \epsilon_n \quad (m, n = 1, 2, ..., 6),$$

with $\qquad\qquad p_{mn} = \pi_{mr} c_{rn}, \qquad \pi_{mn} = p_{mr} s_{rn}.$

No factors of 2 or $\frac{1}{2}$ are necessary when abbreviating the p's. In general,

$$p_{mn} \neq p_{nm}.$$

There are 36 independent π_{mn} and p_{mn} in the general case. This number is reduced by crystal symmetry to three in the most symmetrical cubic classes.

The electro-optical and the photoelastic effects arise from derivatives of the form $\partial D^2/\partial E^2$ and $(\partial^2 D/\partial \sigma \partial E)$ respectively. If we write

$$n = n^0 + aE_0 + a'\sigma + bE_0^2 + b'\sigma^2 + b''E_0\sigma + ..., \tag{12}$$

the term aE_0 represents the linear electro-optical effect. No linear electro-optical effect can exist in a crystal with a centre of symmetry. The term $a'\sigma$ represents the linear photoelastic effect, and the term bE_0^2 represents the *Kerr effect*; both these effects can exist in crystals of any symmetry.

By writing down the appropriate derivatives, and using the thermodynamic relations of Chapter X, many connexions may be found between the various second derivatives of ϵ, D and S with respect to σ, E and T. For example, it is found that the coefficients for photoelasticity $(\partial^2 D/\partial E\partial\sigma)$ and electrostriction $(\partial^2\epsilon/\partial E^2)$ are formally related. However, the frequency of the field changes in the two cases would have to be the same for complete equality of the coefficients. Second-order effects such as these are expressible as third derivatives of thermo-dynamic potentials.

In the piezoelectric equation

$$\epsilon_{jk} = d^0_{ijk} E_i + \gamma_{iljk} E_i E_l, \tag{40}$$

the last term represents electrostriction. γ_{iljk} is a fourth-rank tensor similar to π_{ijkl} and p_{ijkl}.

XIV

OPTICAL ACTIVITY

1. Introduction

THE phenomenon of *optical activity* or *optical rotatory power* must be carefully distinguished from the double refraction discussed in Chapter XIII. It occurs in the following way. When plane polarized monochromatic light is transmitted through certain isotropic materials the plane of polarization is found to be rotated, the amount of rotation being proportional to the thickness of the medium traversed. The same phenomenon is found with certain cubic crystals, and with certain uniaxial and biaxial crystals when the light passes along an optic axis, the direction in which ordinary double refraction vanishes. α-quartz is a notable example.

The explanation is as follows. In all these cases, when the incident plane polarized wave enters the medium, it splits up into two circularly polarized waves, one right-handed† and the other left-handed. The situation at the point of entrance, as seen by an observer whose eye the light is entering, is illustrated in Fig. 14.1 a. The vectors OA and OB, representing the **D** vectors of the two circularly polarized vibrations, rotate in opposite senses with equal angular velocity. OA, rotating clockwise, represents the right-handed component; OB, rotating anticlockwise, represents the left-handed component. The two together are formally equivalent to the vector OC, which always remains vertical and represents the incident plane polarized vibration. The medium itself has a certain right-handed or left-handed character, and so the two circularly polarized waves travel through it with slightly different velocities. As a result, if the two circularly polarized components are reckoned to be in phase when they enter the medium, they will be out of phase when they emerge. The situation at the point of exit is shown in Fig. 14.1 b. The **D** vectors, OA' and OB', rotate in opposite senses and with equal angular velocity, but they no longer reach the vertical

---

† In a right circularly polarized wave the electric displacement vector **D** *at a fixed point* rotates anticlockwise as viewed in the direction of propagation. **D** therefore rotates clockwise when viewed by an observer whose eye the light is entering. *At a fixed instant* the ends of the **D** vectors for points on a given wave normal lie on a right-handed helix. The change of **D** as the wave progresses may be visualized by imagining the helix to move rigidly along its axis by pure translation with no rotation.

position at the same time. They combine to give a vector OC' which vibrates in a plane that is rotated through an angle ϕ from the vertical. It will be seen that, when OA' reaches the vertical, OB' is still 2ϕ away from it. 2ϕ is the difference of phase. *The rotation of the plane of polarization is, therefore, half the difference of phase of the two circular components.* In Fig. 14.1 the right-handed vibration is the faster of

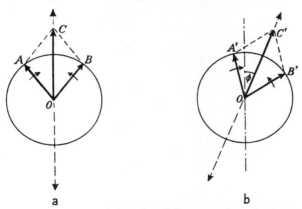

Fig. 14.1. Optical activity when light passes through a crystal in a direction for which the ordinary birefringence vanishes: the situation at (*a*) the entrance point and (*b*) the exit point.

the two, that is, its phase travels faster, and so, as it emerges, it is ahead in phase.† The rotation of the plane of polarization is *clockwise* as seen by an observer whose eye the light is entering; as is customary, we call this a right-handed rotation and define it as *positive* (provided right-handed axes are chosen).‡ One may remember that the sense of the rotation is the same as the sense of the faster of the two circularly polarized components; that is, a right-handed rotation occurs when the right-handed component is the faster.

† Since it is easy to be confused about this the following additional explanation may be helpful. According to the preceding footnote we may picture each wave as a helix moving rigidly, by pure translation, through the medium. The pitches (wavelengths) of the helices are slightly different, and so are the velocities; on the other hand, the number of turns passing a given point in the medium in unit time (the frequency) is the same for each helix. A fixed point on a helix corresponds to a certain phase, or azimuth. At the instant when OA and OB in Fig. 14.1 *a* reach the vertical let us mark the corresponding points on the two helices. The two points are carried along on the helices at different velocities, and the point on the right-handed helix will reach the exit plane first. At the instant it reaches the exit plane, OA' in Fig. 14.1 *b* reaches the vertical. At the later instant when the point on the left-handed helix arrives at the exit plane, OB' reaches the vertical. Therefore, on the exit plane, OA' reaches the vertical before OB', as indicated in Fig. 14.1 *b*.

‡ A right-handed quartz crystal is defined as one that gives a right-handed optical rotation along the optic axis.

It is useful to express the rotation in terms of the refractive indices, n_r and n_l, of the two components. At a given instant, the number of revolutions of the displacement vector of the right-handed wave between entrance and exit (the number of turns of the helix) is d/λ_r, where d is the distance traversed and λ_r is the wavelength in the medium. For the left-handed wave the number of revolutions is d/λ_l. The difference in the number of revolutions is the phase difference. The phase difference, expressed in radians, is therefore

$$2\pi d\left(\frac{1}{\lambda_l} - \frac{1}{\lambda_r}\right) = \frac{2\pi d}{\lambda_0}(n_l - n_r),$$

where λ_0 is the wavelength *in vacuo*. The rotation of the plane of polarization is, accordingly,

$$\phi = \frac{\pi d}{\lambda_0}(n_l - n_r). \tag{1}$$

The rotation per unit path in the medium, known as the *rotatory power*, $\rho = \phi/d$, is given by

$$\rho = \frac{\pi}{\lambda_0}(n_l - n_r). \tag{2}$$

In practice, $(n_l - n_r)$ is very small compared with unity; in most optically active crystals $(n_l - n_r)$ is of the order of 10^{-4} or less. Ordinary double refraction is associated with differences of refractive index of the order of 10^{-3} to 10^{-1}. In spite of the smallness of $(n_l - n_r)$, however, the fact that, in equation (1), λ_0 is very small compared with macroscopic values of d means that appreciable rotations ϕ are still produced. As a numerical example, when sodium light (mean $\lambda_0 = 5893$ A) passes along the optic axis of right-handed α-quartz, $n_l - n_r = 7{\cdot}10 \times 10^{-5}$. Therefore, a plate 1 mm thick cut normal to the optic axis rotates the plane of polarization through an angle

$$\phi = \frac{\pi \times 10^{-1} \times 7{\cdot}10 \times 10^{-5}}{5893 \times 10^{-8}} = 0{\cdot}379 \text{ radians} = 21{\cdot}7°.$$

In agreement with the above explanation of optical rotation it is found that, if a circularly polarized wave is incident on the medium, it emerges still circularly polarized. The essential fact about optical activity in isotropic media, in cubic crystals, and for transmission along the optic axes of uniaxial and biaxial crystals, is this: *the nature of the medium is such that two circularly polarized waves of opposite hand and different velocity may be transmitted through it unchanged in form, that is, in their state of polarization, while a plane polarized wave may not be so transmitted.*

2. Optical activity and birefringence

The treatment in the last section was restricted to isotropic materials, cubic crystals, and crystals in which the light was passing along an optic axis, because in these cases one can study optical activity free from the effects of ordinary birefringence. We must now ask: if this effect can occur when light is transmitted along an optic axis in bire-fringent crystals, what new effects will appear when the light is trans-mitted in other directions? In this case the effects of optical activity

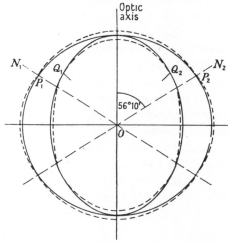

Fig. 14.2. The wave surfaces of α-quartz (not to scale), showing how they are distorted by optical activity. The proportional difference between the radii of the two undistorted surfaces (full curves) at right angles to the optic axis is 6×10^{-3}. The radial distortion of each surface due to optical activity (expressed as a fraction of the mean radius) is $\pm 2\cdot4 \times 10^{-5}$ on the optic axis but much less than this in all directions that are not very close to the optic axis. At right angles to the optic axis the radial distortion is $\pm 2\cdot7 \times 10^{-8}$. ON_1 and ON_2 are wave normal directions for which $k = 0$. P_1, P_2, Q_1, Q_2 are points at which the undistorted and distorted surfaces touch; they give ray directions for which there is no distortion of the wave surfaces. P_1, P_2 on the ordinary wave surface lie on ON_1, ON_2; the tangent planes to the extraordinary wave surface at Q_1, Q_2 are normal to ON_1 and ON_2.

are superimposed on the ordinary birefringence and they may be regarded as a small perturbation of it.

Consider, for example, the wave surfaces of α-quartz, shown, not to scale, in Fig. 14.2. α-quartz is a positive uniaxial crystal and, there-fore, in the absence of optical activity the wave surfaces would be a sphere and an ellipsoid of revolution, as shown by the full lines. The surfaces touch in the direction of the optic axis, for the two wave velocities in this direction would be the same. The two principal refractive indices for quartz are $n_e = 1\cdot553$ and $n_0 = 1\cdot544$. The ratio of the radii of the two wave surfaces at right angles to the optic axis is therefore $1\cdot553/1\cdot544 = 1\cdot006$. The effect of optical activity is slightly to distort these surfaces to the shape shown by the broken lines. Along the optic axis the surfaces no longer touch. Their separation corresponds to the difference of velocity between the two circularly polarized components we have discussed in § 1. It is very small, about 5×10^{-5} times the mean radius, and is exaggerated in the figure.

Before we introduced optical activity, the two wave surfaces of a uniaxial crystal represented the propagation outwards from a point of two separate *plane polarized* waves. Their planes of polarization were at right angles. Now that we have modified the surfaces to allow for optical activity, we find that, along the optic axis, the two waves in question are not plane polarized but circularly polarized. For a general direction in the crystal the new situation is as follows. Associated with any given wave normal there are, just as before, *two* definite waves that travel through the crystal unchanged in form. *These two waves are, in general, elliptically polarized.* The two ellipses that define the state of polarization of the two waves have the same shape, but have opposite senses of rotation. Their major axes, which are at right angles to one another, coincide with the principal vibration directions that would exist for this wave normal if the crystal were not optically active. In the special direction of the optic axis the two waves are circularly polarized, as we have seen. Off the optic axis the ellipticity k (the ratio of the minor to the major axis) varies with direction.

In α-quartz k depends only on the angle between the wave normal and the optic axis, in agreement with the symmetry. Its variation is shown diagrammatically in Fig. 14.3. Along the optic axis $k = 1$, but it very rapidly falls off to small values and becomes zero at an angle of $56°\ 10'$ to the optic axis (Szivessy and Münster 1934) (see also Münster and Szivessy 1935, and Bruhat and Grivet 1935). When the wave normal is perpendicular to the optic axis, k is again finite, although still small, but the sense of rotation of the two ellipses is reversed ($k = -0\cdot00203$).

For optically active *biaxial* crystals the situation is essentially the same. The optical activity can be regarded as producing a small distortion of the wave surface figure. Just as in uniaxial crystals, each wave normal has associated with it two definite elliptically polarized waves, related to one another as described above. The waves are circularly polarized along the two primary optic axes.

To make the description of optical activity more complete we must specify how the amount of distortion of the wave surfaces varies with direction. At the same time we must seek an expression for the ellipticity of the two unchanged waves as a function of direction. We have to distinguish between the direction of propagation, that is, the ray direction, and the wave normal. The separation of the two wave surfaces in Fig. 14.2 in any radial direction gives the difference in ray

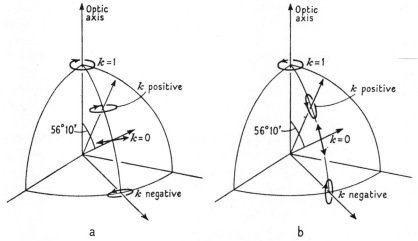

FIG. 14.3. Illustrating the variation of ellipticity k with the direction of the wave normal (shown by the radial arrows) in right-handed quartz (a) for the 'ordinary' wave and (b) for the 'extraordinary' wave. The vibration ellipses are shown as they would appear on the surface of a sphere. The ellipticity is greatly exaggerated for the two directions shown in the diagrams for which k is not equal to zero or unity.

velocity for the two waves having a common ray direction. It is more convenient for the present purpose to specify the two refractive indices for the two waves that have a common wave normal. The possible values of the refractive index n, for a given direction of the wave normal, are the positive roots of the equation

$$(n^2 - n'^2)(n^2 - n''^2) = G^2. \tag{3}$$

n' and n'' are the two refractive indices that the crystal would have in the absence of optical activity; G is a parameter, small compared with unity, that varies with direction and which gives the amount of optical activity or the *gyration*.

Along the optic axes we know that $n' = n'' = \bar{n}$, say. Equation (3) then gives

$$n^2 - \bar{n}^2 = \pm G,$$

and, since G is small, $\qquad\qquad n = \bar{n} \pm \dfrac{G}{2\bar{n}}.$ $\qquad\qquad$ (4)

The difference between the two refractive indices is $G/\bar{n}$ and so, from equation (2), the rotatory power is

$$\rho = \frac{\pi G}{\lambda_0 \bar{n}}. \qquad\qquad (5)$$

We shall not give here any proof of equation (3). The equation is a deduction from theory and has not yet been rigorously tested by experiment. To deduce it one starts from an appropriate atomic model of the crystal and studies its behaviour in the electromagnetic field of a light wave. For the details reference may be made to articles by Szivessy (1928) and Condon (1937) and Born's *Optik* (1933)†, but we warn the reader that the calculation contains some subtleties which have led to errors in some of the published work. Hoek (1941) gives a useful summary of the position.

G, as we have said, varies with the direction of the wave normal. In the theory referred to, G is a quadratic function of the direction cosines l_1, l_2, l_3, of the wave normal with respect to some arbitrarily chosen axes; thus we write

$$G = g_{11} l_1^2 + g_{22} l_2^2 + g_{33} l_3^2 + 2g_{23} l_2 l_3 + 2g_{31} l_3 l_1 + 2g_{12} l_1 l_2, \qquad (6)$$

or $\qquad\qquad\qquad G = g_{ij} l_i l_j \quad (g_{ij} = g_{ji}), \qquad\qquad (7)$

where the g_{ij} are coefficients which describe the optical activity of the crystal. The g_{ij} form a tensor (the symmetric part of Born's gyration tensor) about whose nature we shall have more to say later.

3. The principle of superposition

Let us now consider what happens when a plane wave strikes a parallel-sided plate of optically active crystal at normal incidence. It splits up into the two elliptically polarized components which have the normal to the plate as their wave normal. If n_1 and n_2 are the refractive indices for these two waves, the phase difference between them after passing through unit thickness of the crystal is

$$\Delta = (2\pi/\lambda_0)(n_1 - n_2).$$

n_1 and n_2 are to be found from equation (3). (3) is a quadratic equation for n^2 and the roots are n_1^2 and n_2^2. Now

$$(n_1 - n_2)^2 = n_1^2 + n_2^2 - 2\sqrt{(n_1^2 n_2^2)},$$

† But see pp. 316–317 for later developments. Szivessy (1928) takes ρ *positive* for a *left-rotating* crystal (with right-handed axes) contrary to the usual definition, which we are adopting. We are taking G to have the same sign as ρ.

and so we may use the expressions for the sum and the product of the roots of (3) to obtain

$$(n_1-n_2)^2 = n'^2 + n''^2 - 2\sqrt{(n'^2 n''^2 - G^2)}.$$

Hence, since $G \ll n'n''$, to a sufficient approximation we have

$$\Delta^2 = \frac{4\pi^2}{\lambda_0^2}\left\{(n'-n'')^2 + \frac{G^2}{n'n''}\right\}. \tag{7\,a}$$

If $\bar{n}$ now stands for a mean refractive index $\bar{n} = \sqrt{(n'n'')}$, this equation may be written as
$$\Delta^2 = \delta^2 + (2\rho)^2, \tag{8}$$

where
$$\delta = \frac{2\pi(n'-n'')}{\lambda_0}, \qquad \rho = \frac{\pi G}{\lambda_0 \bar{n}}. \tag{9}$$

Equation (8) expresses Δ, the phase difference between the two elliptically polarized components for the particular wave normal we are considering, in terms of δ, which is the phase difference in the absence of optical activity, and 2ρ, which is the phase difference that would be given by a rotatory power ρ in the absence of ordinary birefringence. We see at once that optical activity always increases and never decreases the birefringence (as an example see Fig. 14.2

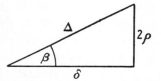

Fig. 14.4. Illustrating the principle of superposition for optical activity and ordinary birefringence.

in which the wave surfaces are pushed apart by the optical activity). The relation (8) is illustrated by the right-angled triangle in Fig. 14.4. The angle β in this triangle is connected with the ellipticity k (p. 264) by the relation
$$k = \tan(\tfrac{1}{2}\beta); \tag{10}$$

equation (10) is a result of the full theory which we shall not prove here. From the geometry of the triangle

$$\tan\beta = 2\rho/\delta, \tag{11}$$

and hence, from (9),
$$\tan\beta = \frac{G}{\bar{n}(n'-n'')}. \tag{12}$$

The ratio k, for any wave normal, is thus completely determined, through equations (10) and (12), by the values of n', n'', and G for this wave normal.

We have already seen that the distortion of the wave surfaces caused by the optical activity is determined by n', n'' and G through equation (3). Our description of the phenomenon is thus formally complete.

Equation (8) suggests an important *principle of superposition* for optical activity and ordinary birefringence. We shall merely quote it and leave the proof to the reader. If any wave is incident normally on the crystal plate that we are considering, it splits up into the two elliptically polarized components. The ratio of the minor to the major axis for each component is k. After suffering the phase difference Δ per unit path, the components combine on leaving the crystal to produce a wave whose state of polarization is, of course, different from that of the original wave. It may be proved that the state of polarization of the emerging wave is exactly the same as if, instead of passing through the optically active birefringent crystal, it had alternately suffered pure double refraction and pure optical rotation in the proportions given by the triangle of Fig. 14.4. A more precise statement is as follows. The wave behaves as if it had passed through a succession of thin plates. The plates are alternately purely doubly refracting and purely optically active. For unit path, the total amount of pure double refraction is equivalent to a phase difference of δ, and the total amount of pure rotation is equivalent to a rotation of ρ. δ and ρ are related to Δ and k by the equations:

$$2\rho = \Delta \sin\beta, \quad \delta = \Delta \cos\beta, \quad \text{where} \quad \tan(\tfrac{1}{2}\beta) = k.$$

4. The size of the effect

For transmission along an optic axis, the difference of refractive indices for the two circularly polarized components may be about 10^{-4}, as we have seen. Thus, by equation (4), $G/\bar{n}$ will be equal to about 10^{-4}. Therefore, by equation (12), and assuming that $G/\bar{n}$ remains of this order of magnitude in other directions, $\tan\beta$ will be very small unless the ordinary birefringence $(n'-n'')$ is also very small, that is, unless we are very close to an optic axis. For example, in a favourable case the birefringence in a direction far from an optic axis might be 10^{-2}. Then $\tan\beta = 10^{-4}/10^{-2} = 10^{-2}$. Hence, $k = \tan\tfrac{1}{2}\beta = 1/200$. The ellipticity k of the two unchanged waves will therefore fall off very rapidly from the value of 1 which it takes on an optic axis. This is the reason why optical activity is so difficult to measure in directions which are not very close to an optic axis: the two unchanged waves are very nearly plane polarized.

Moreover, it can be seen from equation (7 a) that for directions not close to an optic axis the extra birefringence due to optical activity is of order $G^2/\{\bar{n}^2(n'-n'')\}$. With the above figures this is $10^{-8}/10^{-2} = 10^{-6}$. Thus, so far as birefringence is concerned, the effect of optical activity as one leaves an optic axis is not merely overwhelmed by the ordinary birefringence, but at the same time becomes considerably less in absolute magnitude—with the above figures it decreases from 10^{-4} to 10^{-6}. The numerical data for α-quartz given in the legend to Fig. 14.2 provide a further illustration of this point.

5. The tensor character of $[g_{ij}]$

The next topic we wish to take up is how crystal symmetry affects the gyration tensor $[g_{ij}]$, introduced in equation (6). For this purpose we have first to find out the tensor character of $[g_{ij}]$, that is, how it transforms on change of axes. This, in turn, is connected with the character of G and of ρ.

We begin by considering an optically active cubic crystal, for which the rotatory power is the same in all directions. Now rotatory power may be right-handed or left-handed, and a sign convention is needed which is valid for both right-handed and left-handed axes of reference. We have already set up a convention in § 1, by defining a right-handed rotation as positive, provided that right-handed axes are chosen. The general definition for right-handed or left-handed axes is as follows. First set up a set of axes and refer the rotatory power to them. If the sense of the rotation is the same as the hand of the axes (that is, both right-handed or both left-handed), then ρ is defined to be positive; if different then negative. The sign thus depends on the hand of the axes. It follows that the rotatory power ρ of the crystal will change sign if we change the hand of the reference axes. We call a physical quantity that behaves in this way a *pseudo-scalar*. Its transformation law may be written

$$\rho' = \pm\rho,$$

where the $+$ sign refers to transformations that do not change the hand of the axes, and $-$ sign to those that do.

Proceeding now to the anisotropic case, we see that G introduced in equation (3) is a pseudo-scalar, for its value for light passing along an optic axis is proportional to the rotatory power ρ for that direction.

Now consider the transformation of equation (7),

$$G = g_{ij} l_i l_j. \tag{7}$$

On changing from the axes x_i to new axes x_i' we know that l_i and l_j transform as (polar) vectors, thus:

$$l_i = a_{ki} l_k', \qquad l_j = a_{mj} l_m';$$

and that G transforms as a pseudo-scalar, thus:

$$G = \pm G',$$

the sign depending on whether the hand of the axes is changed or not.
Hence (7) becomes

$$\pm G' = g_{ij} a_{ki} l'_k a_{mj} l'_m,$$

or

$$G' = g'_{km} l'_k l'_m,$$

where

$$g'_{km} = \pm a_{ki} a_{mj} g_{ij}. \tag{13}$$

Thus, if equation (7) is to retain its form for all possible choices of
axes, the g_{ij} must transform according to the law (13). It will be seen
that (13) is identical with the transformation law of a second-rank
tensor, except for the $\pm$ signs. Physical quantities that transform
according to (13) are called *axial second-rank tensors*. The ordinary
second-rank tensors that we have met hitherto may be called *polar
second-rank tensors* when it is necessary to make an explicit distinction.
The distinction between polar and axial second-rank tensors is analogous
to the distinction between polar and axial vectors (p. 39). Table 18
collects the transformation laws for polar and axial tensors of ranks
zero, one and two.

TABLE 18

Transformation laws for polar and axial tensors

Quantity	Transformation law	Example
Scalar	$\phi' = \phi$	temperature T
Pseudo-scalar . . .	$\phi' = \pm\phi$	rotatory power ρ
Polar vector . . .	$p'_i = a_{ij} p_j$	temperature gradient $\partial T/\partial x_i$
Axial vector . . .	$r'_i = \pm a_{ij} r_j$	vector product of any two polar vectors
Polar second-rank tensor .	$T'_{ij} = a_{ik} a_{jl} T_{kl}$	permittivity tensor κ_{ij}
Axial second-rank tensor .	$R'_{ij} = \pm a_{ik} a_{jl} R_{kl}$	gyration tensor g_{ij}

6. The effect of crystal symmetry on the g_{ij}

We may now use the results of § 5 to study the effect of crystal symmetry on the components of $[g_{ij}]$. The principle of the method is as
follows. Transform $[g_{ij}]$ by one of the symmetry elements of the
crystal, remembering to introduce a minus sign if the transformation
changes the hand of the axes. Then the new components must be
identical with the old ones. The direct inspection method (p. 118) is
usually the most convenient way of making the transformation. Consider, for example, a crystal with a centre of symmetry. The transformation of axes is

$$1 \to -1, \quad 2 \to -2, \quad 3 \to -3,$$

and the hand is changed. So we have

$$g'_{ij} = -g_{ij}.$$

But $\qquad g'_{ij} = g_{ij}$; hence $g_{ij} = 0.$

Therefore, *a crystal with a centre of symmetry cannot be optically active.*

As a further example, consider class m. With m perpendicular to x_2 the transformation of axes is

$$1 \to 1, \quad 2 \to -2, \quad 3 \to 3,$$

and the hand is changed. Therefore, the components transform as follows:

$$g_{11} \to -g_{11}, \quad g_{12} \to g_{12}, \quad g_{31} \to -g_{31},$$
$$g_{22} \to -g_{22}, \quad g_{23} \to g_{23},$$
$$g_{33} \to -g_{33};$$

and the tensor reduces to

$$\begin{bmatrix} 0 & g_{12} & 0 \\ g_{12} & 0 & g_{23} \\ 0 & g_{23} & 0 \end{bmatrix}. \tag{14}$$

Proceeding through the non-centrosymmetrical classes in this way we find the arrays of non-vanishing components of $[g_{ij}]$ that are shown in Table 19. The axes are chosen according to the conventions listed in Appendix B. Since all the arrays are symmetrical across the leading diagonal, the lower left-hand part is omitted. The number of independent components is shown in brackets after each tensor.

The arrays in Table 19 contain several points of interest. Crystals of classes m, $mm2$, $\bar{4}$ and $\bar{4}2m$ can theoretically show optical activity even though these classes are not enantiomorphous.† Classes $\bar{4}$ and $\bar{4}2m$ are uniaxial, and it can be seen by inspection of symmetry that the rotation along the optic axis must be zero. The same conclusion may be reached by noticing that, for both these classes, $g_{33} = 0$. Optical activity in classes $\bar{4}$ and $\bar{4}2m$ can occur, therefore, only along directions where there is also birefringence, and so there would be experimental difficulties in detecting it. Class m, on the other hand, is biaxial. There are two possibilities: (1) the plane of symmetry contains the two optic axes, or (2) it bisects the angle between them. In case (1) the symmetry would forbid any rotation along the optic axes, and so manifestations of optical activity must be sought in other

† A class is said to be *enantiomorphous* when crystals belonging to it can exist in two different forms which are mirror images of one another.

TABLE 19

Forms of the gyration tensor $[g_{ij}]$

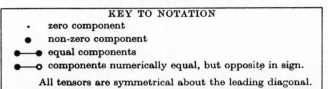

KEY TO NOTATION

. zero component
● non-zero component
●—● equal components
●—○ components numerically equal, but opposite in sign.

All tensors are symmetrical about the leading diagonal.

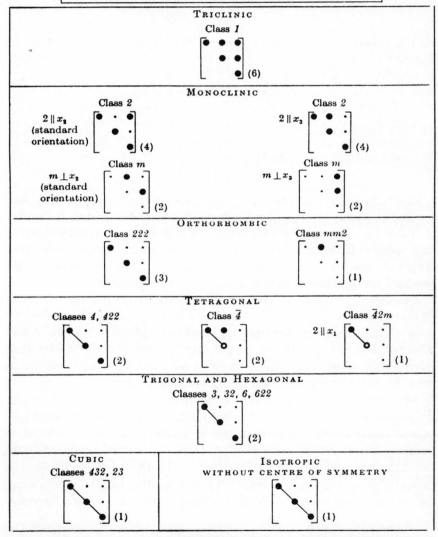

In the following non-centrosymmetric classes all the components of $[g_{ij}]$ vanish:

4mm, 43m, 3m, 6mm, 6, 6m2.

directions; in case (2) the symmetry allows rotation along the optic axes but it must be equal in magnitude and opposite in sign for the two axes. One reported example of case (1) exists, but none of case (2) (Szivessy 1928, p. 830).

Class *2* likewise gives two possibilities: the optic axes may or may not be related by the diad axis. If they are so related, the rotations along each must be the same; if they are not so related, the rotations along each may be different. Examples of both cases exist (Wooster 1938, p. 155).

In class *222* the two optic axes are always related by the diad axes, and so the rotations about each of them must be the same. Several examples exist (Wooster 1938, p. 155).

For α-quartz (class *32*), which is the only optically active and bire-fringent crystal that has been fully investigated, the theory we have given is confirmed; the measured values of the g_{ij} for $\lambda_0 = 5100$ A are (Szivessy and Münster 1934)†

$$g_{11} = g_{22} = \mp 5 \cdot 82 \times 10^{-5}, \qquad g_{33} = \pm 12 \cdot 96 \times 10^{-5}.$$

With a right-handed choice of axes, the upper sign refers to right-handed and the lower sign to left-handed quartz.

EXERCISE 14.1. What restrictions on the optical rotation along the optic axes exist in class *mm2* ?

EXERCISE 14.2. How do the principal axes of $[g_{ij}]$ lie in relation to the sym-metry plane in class *m* ? What is the form of the tensor referred to its principal axes ?

EXERCISE 14.3. Verify the numerical statements made in the legend for **Fig. 14.2** by using the other numerical data for α-quartz given in this chapter.

SUMMARY

Optical activity in isotropic media, cubic crystals, and uniaxial and biaxial crystals when the light passes along an optic axis. In these cases there are two plane waves that are transmitted unchanged in their state of polarization. The two waves are circularly polarized and of opposite hand. They travel with different velocities, corresponding to refractive indices n_r and n_l. A plane-polarized incident wave splits into the two circularly polarized com-ponents, and emerges as a plane-polarized wave with the plane of polarization

† We have changed the signs because these authors count a left-rotating crystal as positive, contrary to our convention (cf. footnote, p. 266).

rotated through an angle ϕ. The corresponding difference of phase of the two circularly polarized components is 2ϕ. The rotatory power is

$$\rho = \phi/d = (\pi/\lambda_0)(n_l - n_r), \tag{2}$$

where d is the path in the medium and λ_0 is the wavelength *in vacuo*. If the rotation is *clockwise* as seen by an observer whose eye the light is entering, it is a *right-handed* rotation. The sign of ρ depends on the hand of the reference axes. If the sense of the rotation is the same as the hand of the axes, ρ is positive; if different, then negative.

Optical activity in uniaxial and biaxial crystals in directions inclined to the optic axes. In a crystal that is both optically active and ordinarily birefringent there are, for any wave normal, two waves that are transmitted unchanged in their state of polarization. The waves are, in general, elliptically polarized. The ellipses defining the states of polarization have the same shape but opposite senses of rotation. Their major axes are mutually perpendicular, and coincide with the principal vibration directions that would exist for this wave normal if there were no optical activity. The refractive indices for the two waves are the positive roots of the equation

$$(n^2 - n'^2)(n^2 - n''^2) = G^2, \tag{3}$$

where n', n'' are the two refractive indices in the absence of optical activity. G is the *gyration* and is a measure of the optical activity for the direction in question. G as a function of direction is given by

$$G = g_{ij} l_i l_j \quad (g_{ij} = g_{ji}), \tag{7}$$

where the g_{ij} are the components of the *gyration tensor*.

For transmission through a parallel-sided plate we have, for unit thickness, the equations

$$\Delta^2 = \delta^2 + (2\rho)^2, \tag{8}$$

$$\delta = (2\pi/\lambda_0)(n' - n''), \qquad \rho = (\pi G)/(\lambda_0 \bar{n}), \qquad \bar{n} = \surd(n'n'').$$

These equations express Δ (the phase difference between the two elliptically polarized components) in terms of δ (the phase difference in the absence of optical activity) and 2ρ (the phase difference that would be given by a rotatory power ρ without ordinary birefringence). k, the ratio of minor to major axis of the ellipses defining the privileged vibrations, is given through the parameter β by

$$k = \tan\tfrac{1}{2}\beta, \qquad \tan\beta = \frac{G}{\bar{n}(n' - n'')}. \tag{10}, (12)$$

Equation (8) suggests the *principle of superposition* for optical activity and ordinary birefringence stated on p. 268.

ρ and G are *pseudo-scalars* with the transformation law:

$$\rho' = \pm\rho, \qquad\qquad G' = \pm G,$$

the sign depending on whether the hand of the axes changes or not. $[g_{ij}]$ is an *axial second-rank tensor* with the transformation law:

$$g'_{ij} = \pm a_{ik} a_{jl} g_{kl} \quad \text{(from Table 18)}.$$

Optical activity cannot exist in a centrosymmetric crystal. It exists in 15 of the 21 non-centrosymmetric classes.

APPENDIX A

SUMMARY OF VECTOR NOTATION AND FORMULAE

In this book vectors are printed in bold-face type, thus, $\mathbf{p}$. The components of $\mathbf{p}$ referred to axes Ox_1, Ox_2, Ox_3 are p_1, p_2, p_3. We write

$$\mathbf{p} = [p_1, p_2, p_3],$$

and often denote $\mathbf{p}$ by p_i or $[p_i]$.

The *magnitude*, or *length*, of $\mathbf{p}$ is denoted by p:

$$p^2 = p_1^2 + p_2^2 + p_3^2 = p_i p_i.$$

A *unit vector* is one of unit length.

The *scalar product* of $\mathbf{p}$ and $\mathbf{q}$ is denoted by $\mathbf{p} . \mathbf{q}$:

$$\mathbf{p} . \mathbf{q} = p_i q_i = pq \cos \theta,$$

where θ is the angle between $\mathbf{p}$ and $\mathbf{q}$.

The *vector product* of $\mathbf{p}$ and $\mathbf{q}$ is denoted by $\mathbf{p} \wedge \mathbf{q}$:

$$\mathbf{p} \wedge \mathbf{q} = (pq \sin \theta)\mathbf{l},$$

where $\mathbf{l}$ is a unit vector perpendicular to $\mathbf{p}$ and $\mathbf{q}$ such that $\mathbf{p}$, $\mathbf{q}$, $\mathbf{l}$ form a right-handed set. The components of $\mathbf{p} \wedge \mathbf{q}$ referred to right-handed axes are

$$[p_2 q_3 - p_3 q_2, \; p_3 q_1 - p_1 q_3, \; p_1 q_2 - p_2 q_1].$$

The *gradient* of a scalar ϕ which varies with position is a vector denoted by $\operatorname{grad} \phi$:

$$\operatorname{grad} \phi = \left[\frac{\partial \phi}{\partial x_1}, \frac{\partial \phi}{\partial x_2}, \frac{\partial \phi}{\partial x_3} \right].$$

The *divergence* of a vector $\mathbf{p}$ which varies with position is a scalar denoted by $\operatorname{div} \mathbf{p}$:

$$\operatorname{div} \mathbf{p} = \frac{\partial p_1}{\partial x_1} + \frac{\partial p_2}{\partial x_2} + \frac{\partial p_3}{\partial x_3} = \frac{\partial p_i}{\partial x_i}.$$

The *curl* of a vector $\mathbf{p}$ which varies with position is a vector denoted by $\operatorname{curl} \mathbf{p}$, whose components referred to right-handed axes are

$$\left[\frac{\partial p_3}{\partial x_2} - \frac{\partial p_2}{\partial x_3}, \; \frac{\partial p_1}{\partial x_3} - \frac{\partial p_3}{\partial x_1}, \; \frac{\partial p_2}{\partial x_1} - \frac{\partial p_1}{\partial x_2} \right].$$

APPENDIX B

THE SYMMETRY OF CRYSTALS AND CONVENTIONS FOR THE CHOICE OF AXES

THIS appendix summarizes the meanings of various terms and symbols used in connexion with crystal symmetry; it also describes the conventions used in this book for the choice of reference axes. In notation and terminology we follow closely the *International tables for X-ray crystallography*, volume I (1952).

A macroscopic view of crystal symmetry is adopted in the main body of this book, but in the present account it is more convenient to begin by taking a microscopic, atomic, point of view.

Lattice and unit cell. A *lattice* is an infinite array of evenly spaced points which are all similarly situated; points are regarded as similarly situated if the rest of the lattice appears the same, and in the same orientation, when viewed from them. An *ideal crystal* is defined to be a body in which the atoms are arranged in a lattice; by this is meant (*a*) that the atomic arrangement appears the same, and in the same orientation, when viewed from all the lattice points, and (*b*) the atomic arrangement viewed from any point that is not a lattice point is different from the atomic arrangement viewed from a lattice point. If, conversely, one is given an ideal crystal, the lattice may be constructed by first marking an arbitrary point as origin and then marking all points that are similarly situated as regards their atomic environment. The form and orientation (but not, of course, the position) of the lattice so found is independent of which point in the crystal is chosen as origin. An ideal crystal, thus defined, is infinite in extent; real crystals are not only bounded, but depart from the ideal crystal by possessing occasional imperfections.

The lattice may be considered as being the points of intersection of three sets of evenly spaced parallel planes. These planes divide up the crystal into identical elementary parallelepipeds called *primitive unit cells*. Thus, a primitive unit cell is a parallelepiped having lattice points at its corners only. In a given lattice the primitive unit cell is not unique, for the sets of planes may be chosen in an infinite number of ways, and to each choice corresponds a primitive unit cell. It is often more convenient to choose a larger unit cell which has lattice points at the centres of its faces, or at its body centre, or occasionally at other positions, in addition to the points at its corners. Such cells are called *multiply primitive unit cells*. A unit cell, whether primitive or multiply primitive, is defined by the lengths and directions of three non-parallel edges; these are denoted by the vectors **a, b, c**.

Lattice planes, crystal faces and directions. Choose a lattice point as origin and draw axes Ox, Oy, Oz parallel to **a, b, c** respectively (Fig. B.1). It may be verified that the plane whose intercepts on the axes are a/h, b/k, c/l, where h, k, l are integers without a common factor, passes through lattice points and is one of a set of evenly spaced parallel planes which, collectively, pass through all the points of the lattice. This set of *lattice planes* is denoted by the symbol (hkl). h, k, l are the (*Miller*) *indices* of the set of planes.

When crystals are bounded by plane faces the faces are parallel to lattice planes. The symbol (hkl) is given a second meaning as denoting a *face* parallel to the planes (hkl). There can be two such faces, one on each side of the crystal. The face on the same side of the origin as the plane making intercepts a/h, b/k, c/l is denoted (hkl); the opposite face is denoted $(\bar{h}\,\bar{k}\,\bar{l})$, the bars above the letters signifying minus signs. The *law of rational indices* states that all planes which can occur as faces of a crystal have intercepts on the axes which, when expressed as multiples of certain unit lengths along the axes (proportional to a, b and c), have ratios that are rational numbers (a rational number is one which can be written as p/q, where p and q are integers). The law was originally deduced from

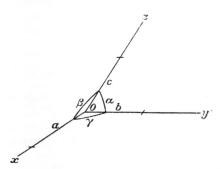

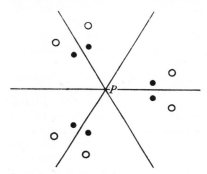

FIG. B.1. Showing the axes Ox, Oy, Oz, the lengths of the sides of the unit cell a, b, c, and the interaxial angles α, β, γ.

FIG. B.2

observations on crystal faces, but in the present atomic approach it is a direct result of the lattice structure of crystals.

The vector drawn from the origin to any lattice point at a corner of a unit cell may be written as $U\mathbf{a}+V\mathbf{b}+W\mathbf{c}$, where U, V, W are integers. The direction so defined is denoted by the symbol $[UVW]$. Since only the ratio $U:V:W$ is significant for the direction, it is usual to choose the three integers so that they have no common factor. In crystals of the cubic system (defined below) the direction $[UVW]$ is normal to the lattice planes (UVW), but this result is not generally true for other crystals. In all cases, a direction $[UVW]$, being parallel to rows of lattice points, is parallel to a family or *zone* of faces; $[UVW]$ is then the direction of the *zone axis*. The *Weiss zone law*, which is readily proved, states that the face (hkl) is a member of the zone of faces whose axis is $[UVW]$ if

$$hU+kV+lW = 0.$$

Symmetry elements. Most crystals possess symmetry in addition to the repetitions expressed by the crystal lattice. This symmetry is best described by analysing it into *symmetry elements*. As an illustration of what is meant, Fig. B.2 shows a pattern made up of two different kinds of atoms represented by full and open circles. If the array of atoms is rotated through 120°, or one-third of a revolution, about an axis through P and perpendicular to the plane of the paper, the pattern is clearly unchanged. Likewise, further rotations through 120° will produce no change. This property of the pattern is described by saying that it possesses a *three-fold rotation axis*, or a *rotation triad axis*, through the point P perpendicular to the plane of the paper. The three full lines in Fig. B.2

denote *mirror planes* or *planes of symmetry* perpendicular to the plane of the paper. The patterns on the two sides of a plane of symmetry are related as the object and its image in a plane mirror. A list of symmetry elements into which the symmetry of any array can be analysed, together with the operation associated with each element, is as follows:

(i) *centre of symmetry*: taking an origin of coordinates at the centre of symmetry the operation is to move each point (x, y, z) to the position $(-x, -y, -z)$. This operation is known as *inversion*.

(ii) *mirror plane*: an operation in which each point is moved to the position of its mirror image in a certain plane.

(iii) *glide plane*: an operation in which each point is moved to the position of its mirror image in a certain plane, and then translated parallel to this plane.

(iv) *n-fold rotation axis*: a rotation of $2\pi/n$ about a given axis, where n is a positive integer.

(v) *n-fold screw axis*: a rotation of $2\pi/n$ about a given axis, followed by a translation along this axis.

(vi) *n-fold inversion axis*: a rotation of $2\pi/n$ about a given axis, followed by inversion through a given point on the axis.

These symmetry operations are not all independent. For example, a 1-fold inversion axis is the same as a centre of symmetry; a 2-fold inversion axis is equivalent to a plane of symmetry normal to the axis; a 3-fold inversion axis is equivalent to a 3-fold rotation axis together with a centre of symmetry.

The presence of a lattice structure prevents the occurrence in crystals of certain symmetry elements in the above list. It may be proved that the only rotation and screw axes possible in crystals are those with $n = 1, 2, 3, 4$ or 6 (in the historical development of the subject the non-existence of rotation axes with $n = 5$ or $n > 6$ was deduced from the law of rational indices). Likewise, the lattice structure imposes certain restrictions on the translations associated with the screw axes and the glide planes.

Space-groups. A self-consistent arrangement of symmetry elements in a lattice is known as a *space group*. The operation of any element of the group must leave the pattern of symmetry elements unaltered. It may be shown by detailed enumeration of the possibilities that there are only 230 essentially different space groups.

Point groups and crystal classes. In the study of the physical properties of crystals we are not concerned with the relative *positions* of the symmetry elements, but only with their *orientations*. This poses the question: what combinations of symmetry elements are possible (a) without regard to the relative positions of the symmetry elements, as distinct from their relative orientations, and (b) without regard to the translations associated with glide planes and screw axes? The answer to this question determines the kinds of symmetry which crystals can show in their macroscopic properties, and in the orientation of their faces.

The possible macroscopic symmetry elements in crystals reduce to the following:

(i) centre of symmetry,
(ii) mirror plane,
(iii) 1-, 2-, 3-, 4- or 6-fold rotation axes,
(iv) 1-, 2-, 3-, 4- or 6-fold inversion axes.

In a combination of these elements we may regard all the members as passing through a single point; the possible combinations of macroscopic symmetry elements are accordingly called *point groups*.†

It may be shown by inspection of the 230 space groups, or from first principles, that there are just 32 different point groups. Crystals are divided into 32 *crystal classes* according to the point-group symmetry they possess.

Enumeration of the 32 point groups and crystal classes. The symmetry elements of the 32 point groups are shown in Table 21 (pp. 284–8) by means of stereographic projections. For the purposes of Table 21 the stereographic projection, which is widely used in crystallography, may be defined as follows: (i) the origin of the point group is supposed to be at the centre of a reference sphere; (ii) the orientations of the symmetry elements are determined by their intersections with the surface of this reference sphere; (iii) each diagram in Table 21 represents the projection on to a diametral plane of one hemisphere of the reference sphere from the opposite point on the far side of the sphere (as, for example, in the projection of the northern hemisphere of the Earth from the South Pole on to the equatorial plane). The symbolism used for showing the symmetry elements on the stereograms (stereographic projections) is given in column 2 of Table 20.

In Table 21 each class is given a symbol, placed above the stereogram, which summarizes its essential symmetry elements according to the notation of column 3 of Table 20. The symbol does not specifically include all the symmetry elements of a given class, but the presence of those included in the symbol necessarily implies the presence of all the others (the symbol occasionally includes more information than is strictly necessary for the definition of the point group). Thus, for example, $4/mmm$ means that there is a 4-fold rotation axis perpendicular to a mirror plane ($4/m$), together with a mirror plane containing the axis ($4/mm$), and a further mirror plane containing the axis and at 45° to the latter plane ($4/mmm$). Reference to Table 21 shows that these are not the only symmetry elements present in this point group; however, it is not difficult to show that their presence necessarily implies the existence of the other mirror planes and the diads that are indicated in the stereogram. The full rules for constructing the symbol for a given point group will not be discussed here, since they are not needed for understanding the rest of the book. The stereograms and symbols in Table 21 contain all the essential information. For the purposes of the table only, the symbols for classes with a centre of symmetry (centrosymmetrical classes) are enclosed in boxes.

The notation just described for the crystal classes is based upon one devised by Mauguin and Hermann. The older notation of Schoenflies is still found in some books and papers, but we shall not use it in this book; it is shown for reference in column 2 of Table 25 on p. 294.

The symbols $\{hkl\}$ and $\langle UVW \rangle$. The symbol $\{hkl\}$ means all crystal faces or sets of lattice planes which can be obtained from the face or planes (hkl) by repeated operation of the symmetry elements of a given point group. The set of faces so found constitutes a *form*: one speaks of 'faces of the form $\{hkl\}$'.

---

† The possible symmetry elements of bounded geometrical forms are the same as the macroscopic symmetry elements of crystals, except that, since there is no lattice, there are no restrictions on the values of n for the rotation and inversion axes. The symmetry elements of bounded geometrical forms all *necessarily* pass through a single point.

In a similar way, the symbol $\langle UVW \rangle$ means all directions which can be reached from $[UVW]$ by repeated operation of the symmetry elements of a given point group.

Crystal systems. The 32 crystal classes are conventionally grouped into seven crystal systems, as shown in Table 21. The requirement for membership of a given crystal system is that the symmetry of the class should possess a

TABLE 20

Symbolism for the symmetry elements of the 32 point groups

Symmetry element	Symbol on stereogram	International symbol
centre of symmetry	no symbol	$\bar{1}$
mirror plane	full line (great circle)	m
Rotation axes		
1-fold (monad).	no symbol	1
2-fold (diad)	◆	2
3-fold (triad)	▲	3
4-fold (tetrad)	◆	4
6-fold (hexad)	⬢	6
Inversion axes		
1-fold (inverse monad) ≡ centre of symmetry	no symbol	$\bar{1}$
2-fold (inverse diad) ≡ mirror plane normal to the axis	as for mirror plane	$\bar{2}\ (\equiv m)$
3-fold (inverse triad) ≡ 3-fold rotation axis plus a centre of symmetry	⬭	$\bar{3}$
4-fold (inverse tetrad) (includes a rotation diad axis)	◈	$\bar{4}$
6-fold (inverse hexad) ≡ a rotation triad axis plus a plane normal to it	⬣	$\bar{6}\ (\equiv 3/m)$

certain characteristic. The names of the systems, together with the characteristics required for membership, are as follows:

(i) *triclinic*: no symmetry other than a 1-fold axis (rotation or inverse);

(ii) *monoclinic*: a single 2-fold axis (rotation or inverse);

(iii) *orthorhombic*: three mutually perpendicular 2-fold axes (rotation or inverse), but no axes of higher order;

(iv) *tetragonal*: a single 4-fold axis (rotation or inverse);

(v) *cubic*: four 3-fold axes arranged like the body diagonals of a cube;

(vi) *trigonal*: a single 3-fold axis (rotation or inverse);

(vii) *hexagonal*: a single 6-fold axis (rotation or inverse).

Choice of axes Ox, Oy, Oz. The reason for the above grouping into systems is that it is possible to refer all the classes within a single system to a similar set of axes Ox, Oy, Oz. Moreover, when the axes have been thus chosen, according to certain conventions set out below, a unit cell may be outlined in which the ratio $a:b:c$ has features which are characteristic for the system. This often necessitates the choice of a multiply primitive unit cell, and, indeed, it was precisely to allow for this contingency that we reserved the right to use unit cells that were not primitive. The following list gives the conventions for the choice of Ox, Oy, Oz in relation to the characteristic symmetry elements of each system; it also gives any relations that exist between a, b and c, and between the interaxial angles α, β and γ, as defined in Fig. B.1.

(i) *triclinic*: $a \neq b \neq c$, $\alpha \neq \beta \neq \gamma$.

(ii) *monoclinic*: Oy parallel to the 2-fold axis;
$$a \neq b \neq c, \quad \alpha = \gamma = 90° \neq \beta.$$

(iii) *orthorhombic*: Ox, Oy, Oz parallel to the 2-fold axes;
$$a \neq b \neq c, \quad \alpha = \beta = \gamma = 90°.$$

(iv) *tetragonal*: Oz parallel to the 4-fold axis;
$$a = b \neq c, \quad \alpha = \beta = \gamma = 90°.$$

(v) *cubic*: Ox, Oy, Oz parallel to the edges of the cube whose body diagonals are the 3-fold axes;
$$a = b = c, \quad \alpha = \beta = \gamma = 90°.$$

(vi) *trigonal*: Oz parallel to the 3-fold axis;
$$a = b \neq c, \quad \alpha = \beta = 90°, \quad \gamma = 120°.$$

(vii) *hexagonal*: Oz parallel to the 6-fold axis;
$$a = b \neq c, \quad \alpha = \beta = 90°, \quad \gamma = 120°.$$

The trigonal and hexagonal systems, having similar sets of axes, are sometimes regarded as forming a single system. In some systems (for example, the tetragonal) the above rules do not specify all the axial directions, and in the more symmetrical classes of these systems further conventions are introduced. The directions of Ox, Oy, Oz are inserted on the stereograms in Table 21 whenever the conventions fix them completely in relation to the symmetry elements.

Miller-Bravais and rhombohedral axes.† The axes we have given in (vi) and (vii) of the above list for the trigonal and hexagonal systems do not display the 3-fold, and still less the 6-fold, symmetry characteristic of these systems. For this reason a fourth, redundant, axis denoted Ou is often added in the Ox, Oy plane at 120° to Ox and Oy, with an axial length equal to a. The axes are then called *Miller–Bravais axes*. The symbol for a face or set of lattice planes is obtained just as before, except that it now contains four indices, thus: $(hkil)$, where the i refers to the Ou axis. h, k, i are not independent, and it may be shown that $h+k+i = 0$. The advantage of using the fourth axis is that crystal faces of the same form, say $\{hkil\}$, have sets of indices which are, apart from signs, permutations of one another; this then becomes a common property for all the systems.

The problem of assigning symbols to directions in the trigonal and hexagonal systems does not arise in the text of this book; but since it is usually neglected

† The material of this section is not needed for an understanding of the main text but is included here for completeness.

entirely in the standard textbooks the following remarks may be helpful.† The best symbolism to adopt depends on the use to which the symbols are to be put. For many purposes it is best to use only the three axes Ox, Oy, Oz and to write either $[UV*W]$ or $[UVW]$ for the direction of the vector $U\mathbf{a} + V\mathbf{b} + W\mathbf{c}$. The three numbers U, V, W may then be treated precisely like the three numbers of the direction symbol $[UVW]$ in any other crystal system. In particular, the Weiss zone law holds for them when they are used with the face indices h, k, l. An objection to this procedure is that directions related by the point-group symmetry do not always have sets of indices which are, apart from signs, permutations of one another.

Another notation is in use which avoids this last objection by the use of four indices. It must be very carefully distinguished from the three-index notations described above (that is, $[UVW]$ or $[UV*W]$). A direction, like a face, is indicated in this alternative notation by a set of four numbers $[U'V'T'W']$ of which the sum of the first three is zero: $U' + V' + T' = 0$. The direction $[U'V'T'W']$ means the direction of the vector

$$U'\mathbf{a} + V'\mathbf{b} + T'\mathbf{t} + W'\mathbf{c},$$

where $\mathbf{t}$ is a vector of length a in the direction of the Ou axis. (It may be verified that the four indices are then the normal projections of the vector on the four axes, expressed as multiples of $\frac{3}{2}a$, $\frac{3}{2}a$, $\frac{3}{2}a$ and c respectively.) In the four-index notation the Ox axis has the symbol $[2\,\overline{1}\,\overline{1}\,0]$; while in a three-index notation it is $[10*0]$ or $[100]$; thus the symbols of the same direction in the two notations may look quite different. In the four-index notation the statement of the Weiss zone law is as follows: the face $(hkil)$ is a member of the zone of faces whose axis is $[U'V'T'W']$ if

$$hU' + kV' + iT' + lW' = 0.$$

In some crystals of the trigonal system it is possible to choose a primitive rhombohedral unit cell; the corresponding axes Ox, Oy, Oz, called *rhombohedral* or *Miller axes*, are all equally inclined to the 3-fold axis with

$$a = b = c, \qquad \alpha = \beta = \gamma.$$

A rhombohedral unit cell may also be chosen, but usually less advantageously, in other crystals of the trigonal system and in crystals of the hexagonal system, but in these cases such a unit cell would not be primitive.

Choice of axes Ox_1, Ox_2, Ox_3. For writing out tensors and matrices which represent physical properties of crystals we use another set of axes Ox_1, Ox_2, Ox_3, which are always mutually perpendicular. The choice of Ox_1, Ox_2, Ox_3 in relation to Ox, Oy, Oz in the various crystal systems is again a matter of convention, and in this book we follow *Standards on piezoelectric crystals* (1949). For our purpose the conventions may be summarized as:

monoclinic: $Ox_2 \parallel Oy$;

tetragonal, trigonal and hexagonal: $Ox_3 \parallel Oz$, $Ox_1 \parallel Ox$;

orthorhombic and cubic: $Ox_1 \parallel Ox$, $Ox_2 \parallel Oy$, $Ox_3 \parallel Oz$.

Ox_1, Ox_2, Ox_3 always form a right-handed set; this is so for the left-handed, as well as for the right-handed, crystals in an enantiomorphous class.‡ Ox_1, Ox_2, Ox_3 are inserted in the stereograms whenever the above rules fix them completely in

---

† A paper by Donnay (1947) gives a discussion and further references.

‡ A class is said to be *enantiomorphous* when crystals belonging to it can exist in two different forms which are mirror images of one another.

relation to the symmetry elements. These are the only conventions necessary in order to specify the general forms of the crystal-property matrices in the 32 classes. However, before numerical values of the physical constants (for example, the piezoelectric coefficients) for a given crystal can be quoted with unambiguous meanings, further conventions, both for the orientation and for the sense of the axes Ox_1, Ox_2, Ox_3, are needed. A wide range of choice is possible, but the system proposed in *Standards on piezoelectric crystals* (1949) seems likely to gain general acceptance.

Departures from convention. In the matrix tables for crystal properties which appear throughout the book we have sometimes explicitly admitted alternative settings of axes Ox_1, Ox_2, Ox_3. The most important of these is:

$$\text{monoclinic:} \quad Ox_3 \| Oy,$$

a choice which is followed by a number of authors.

Optically isotropic, uniaxial and biaxial crystals. The crystal systems are sometimes grouped into three categories on an optical basis. As is shown in Chapter XIII, §§ 1.1 and 1.2, the optical properties of transparent crystals may be described by reference to an ellipsoid called the indicatrix. For triclinic, monoclinic and orthorhombic crystals the three axes of the indicatrix are unequal in length. The surface therefore possesses two circular central sections (Fig. 13.3, p. 238), and the normals to these sections are the *primary optic axes*. These three systems are accordingly said to be *biaxial*. For tetragonal, trigonal and hexagonal crystals the indicatrix is an ellipsoid of revolution and has only one circular central section. The normal to this section, which is the axis of revolution, is the *optic axis*. These three systems are therefore said to be *uniaxial*. For the cubic system the indicatrix is a sphere and the system is said to be *anaxial* or *optically isotropic*. Although it is founded on the optical properties, this grouping of the systems has a wider relevance, because many other crystal properties may be described by an ellipsoid, or, more generally, by a quadric surface, namely all those properties represented by symmetrical second-rank tensors (Ch. I, §§ 4 and 5).

Biaxial	*Uniaxial*	*Anaxial*
triclinic	tetragonal	cubic
monoclinic	trigonal	
orthorhombic	hexagonal	

GENERAL REFERENCES

A good introductory account of crystal symmetry is:

PHILLIPS, F. C. (1946), *An introduction to crystallography*, London: Longmans, Green.

The standard reference work on the results of crystallographic symmetry theory is:

International tables for X-ray crystallography, volume I, 1952, published for the International Union of Crystallography by the Kynoch Press, Birmingham.

A complete and concise, but mathematically advanced, derivation of all the point and space groups is:

F. SEITZ, 'A matrix-algebraic development of the crystallographic groups,' *Z. f. Krist.* (i) **88**, 433–59 (1934) (point groups); (ii) **90**, 289–313 (1935) (lattices); (iii) **91**, 336–66 (1935) (space groups); (iv) **94**, 100–30 (1936) (space groups).

TABLE 21

Symmetry elements and conventions for the choice of axes in the 32 crystal classes

Symbols of centrosymmetrical classes are enclosed in boxes

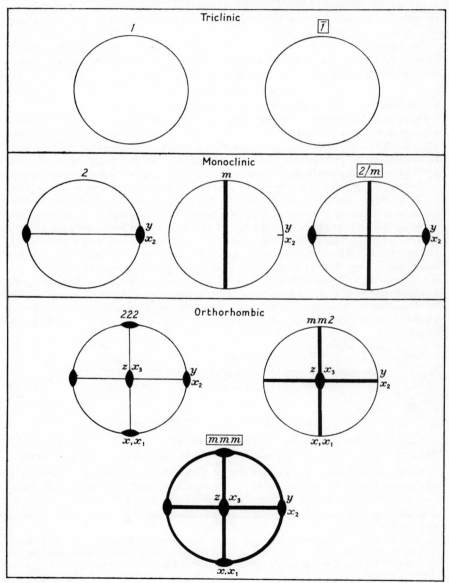

TABLE 21.1

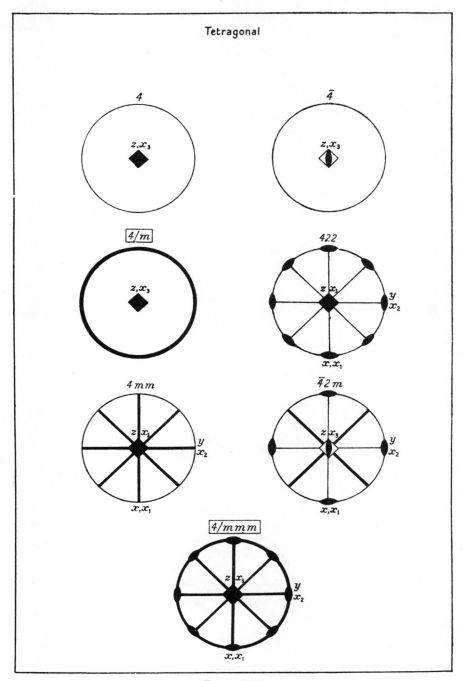

Tetragonal

TABLE 21.2

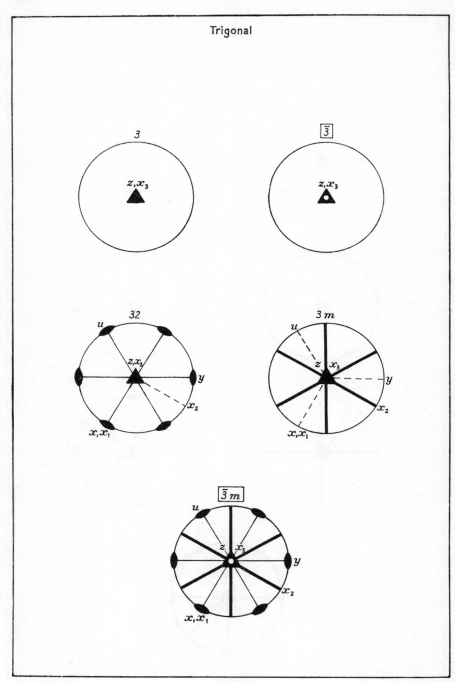

Trigonal

TABLE 21.3

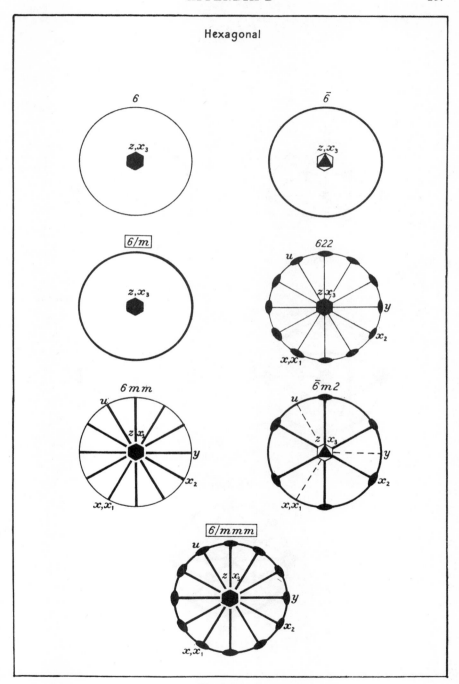

Hexagonal

TABLE 21.4

Cubic

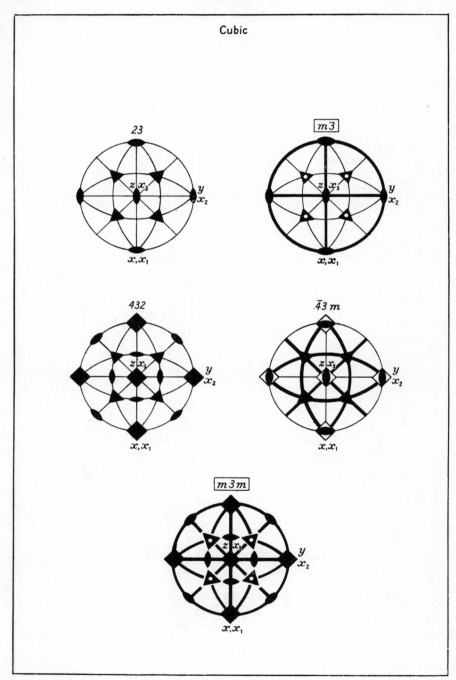

TABLE 21.5

APPENDIX C

SUMMARY OF CRYSTAL PROPERTIES

TABLE 22 a

Tensors representing crystal properties mentioned in this book

All vectors and tensors in groups (ii) to (vi) are polar, except for the vectors **H**, **B** and **I**, which are taken as axial, [note (2), p. 291].

Chapter, equation and page number	Property or coefficients		Defining equation [note (1)]	General relations between components
	(i) Scalar relating two scalars			
X, (1), 170	heat capacity C		$\Delta S = (C/T)\,\Delta T$	..
	(ii) First-rank tensor relating a scalar and a vector			
IV, (21), 78	pyroelectricity	1	$\Delta P_i = p_i\,\Delta T$ ⎫	
	[note (6)]			..
X, (37), 180	electrocaloric effect	1	$\Delta S = p_i\,E_i$ ⎭	
X, (46), 183	heat of polarization	1	$\Delta S = t_i\,\Delta P_i$ ⎫	
X, (46), 183	field due to temperature change	1	$E_i = -t_i\,\Delta T$ ⎭	..
IV, — , 78	polarization by hydrostatic pressure	1	$P_i = -d_{ijj}\,p$	..
	(iii) Second-rank tensor relating two vectors			
IV, (6), 69	permittivity	2	$D_i = \kappa_{ij}\,E_j$	symmetrical
IX, (20 a), 156	[dielectric impermeability]	2	$E_i = \beta_{ij}\,D_j$	,,
IV, (5), 69	dielectric susceptibility	2	$P_i = \kappa_0\,\chi_{ij}\,E_j$	,,
III, (7), 55	permeability	2	$B_i = \mu_{ij}\,H_j$	,,
III, (5), 54	magnetic susceptibility	2	$I_i = \mu_0\,\psi_{ij}\,H_j$	,,
XI, (17), 204	electrical conductivity	2	$j_i = \sigma_{ik}\,E_k$	,,
XI, (18), 204	[electrical resistivity]	2	$E_i = \rho_{ik}j_k$	,,
XI, (3), 195	thermal conductivity	2	$h_i = -k_{ij}(\partial T/\partial x_j)$	,,
XI, (8), 196	[thermal resistivity]	2	$\partial T/\partial x_i = -r_{ij}\,h_j$	,,
XII, (36), 225	thermoelectricity	3	$\partial\bar{\mu}/\partial x_i = -\Sigma_{ik}(\partial T/\partial x_k)$	
XII, (43), 226			$(j^e = 0)$ *or* 'flow of entropy' $= j_i^e\,\Sigma_{ik}$	not symmetrical

Chapter, equation and page number	Property or coefficients		Defining equation [note (1)]	General relations between components
(iv) Second-rank tensor relating a scalar and a second-rank tensor				
VI, (13), 106	thermal expansion	2	$\epsilon_{ij} = \alpha_{ij}\Delta T$	symmetrical
X, (19), 176	piezocaloric effect	2	$\Delta S = \alpha_{ij}\sigma_{ij}$	
X, (24), 177	thermal pressure	2	$\sigma_{ij} = -f_{ij}\Delta T$	,,
X, (24), 177	heat of deformation	2	$\Delta S = f_{ij}\epsilon_{ij}$	
VIII, (24), 145	strain by hydrostatic pressure	2	$\epsilon_{ij} = -s_{ijkk}\,p$	symmetrical $(s_{ijkk} = s_{jikk})$
XII, (46), 227	Peltier coefficients	3	$\Pi_{ik} = (T/e)\Sigma_{ik}$	not symmetrical
(v) Third-rank tensor relating a vector and a second-rank tensor				
VII, (3), 111	direct piezoelectric effect	4	$P_i = d_{ijk}\,\sigma_{jk}$	$d_{ijk} = d_{ikj}$
VII, (19), 115	converse piezoelectric effect	4	$\epsilon_{jk} = d_{ijk}\,E_i$	
X, (47), 183	a piezoelectric effect	4	$P_i = e_{ijk}\,\epsilon_{jk}$	$e_{ijk} = e_{ikj}$
X, (47), 183	a piezoelectric effect	4	$\sigma_{jk} = -e_{ijk}\,E_i$	
XIII, (14), 244	electro-optical effect	5	$\Delta B_{ij} = z_{ijk}\,E_k$	$z_{ijk} = z_{jik}$
(vi) Fourth-rank tensor relating two second-rank tensors				
VIII, (1), 132	elastic compliances	6	$\epsilon_{ij} = s_{ijkl}\,\sigma_{kl}$	$s_{ijkl} = s_{jikl} = s_{ijlk} = s_{klij}$
VIII, (2), 132	[elastic stiffnesses]	6	$\sigma_{ij} = c_{ijkl}\,\epsilon_{kl}$	$c_{ijkl} = c_{jikl} = c_{ijlk} = c_{klij}$
XIII, (16), 244	elasto-optical coefficients	7	$\Delta B_{ij} = p_{ijkl}\,\epsilon_{kl}$	$p_{ijkl} = p_{jikl} = p_{ijlk}$
XIII, (14), 244	piezo-optical coefficients	7	$\Delta B_{ij} = \pi_{ijkl}\,\sigma_{kl}$	$\pi_{ijkl} = \pi_{jikl} = \pi_{ijlk}$
XIII, (40), 257	electrostriction	7	$\epsilon_{jk} = \gamma_{iljk}\,E_i E_l$	$\gamma_{iljk} = \gamma_{lijk} = \gamma_{ilkj}$
(vii) Axial second-rank tensor giving the variation of a pseudo-scalar with direction				
XIV, (7), 266	optical activity (gyration tensor)	8	$G = g_{ij}\,l_i\,l_j$	symmetrical

TABLE 22 *b*

Tensors mentioned in this book and not representing crystal properties

Second-rank polar tensor relating two polar vectors

Chapter, equation and page number	Tensor	Defining equation [note (1)]	General relations between components
V, (7), 88	stress	$p_i = \sigma_{ij}\,l_j$	symmetrical [note (5)]
VI, (9), 99	strain+rotation	$u_i - (u_0)_i = e_{ij}\,l_j$	$[e_{ij}]$ not symmetrical
VI, (10), 99	strain, rotation	$u_i - (u_0)_i = \epsilon_{ij}x_j + \varpi_{ij}x_j$	$[\epsilon_{ij}]$ symmetrical $[\varpi_{ij}]$ antisymmetrical

Notes on Tables 22 a and 22 b

(1) The simple proportionality between the variables shown in the defining equations holds only for sufficiently small values of the variables. For larger values a differential definition is adopted, for example, for permittivity,

$$\kappa_{ij} = \partial D_i / \partial E_j.$$

For this reason the equation given in the tables differs sometimes from that given in the main text.

The division of the tensor properties into the groups (i) to (vii) is to some extent arbitrary. For example, the electrical resistivity in group (iii) relates a second-rank tensor, namely the products of the current density components, with a scalar, namely the rate of production of Joule heat, by the equation

$$\dot{q} = \rho_{ik} j_i j_k;$$

it could be so defined, and thus placed in group (iv). The arbitrariness lies in choosing the defining equation. In our scheme, when the defining equation has been chosen, the group is determined. The rank of the tensor, on the other hand, is not arbitrary.

(2) Whether **E** and **H** and related vectors are polar or axial is partly a matter of convention (footnote, p. 54).

(3) A property enclosed in square brackets [. . .] is the reciprocal of the property appearing directly above it. All properties, of course, have reciprocals, but only those reciprocals mentioned in the main text are inserted in Table 22 a.

(4) Properties linked by a brace } are related thermodynamically.

(5) The stress tensor is assumed to be symmetrical throughout both tables (see p. 87).

(6) Numbers in bold type refer to Appendix D.

TABLE 23

Some tensors in crystal physics not in Tables 22 a and 22 b

Coefficients of self-diffusion (Boas and Mackenzie 1950)
Pyromagnetism and piezomagnetism (Voigt 1910) (existence of piezomagnetism not firmly established)
Magnetostriction
Second-order elastic coefficients (p. 256) (Birch 1947; Fumi 1951, 1952 b; Hearmon 1953)
    ,,   ,, photoelastic coefficients
    ,,   ,, electro-optical coefficients (Kerr effect) (p. 243)
Effect of pressure on electrical conductivity (Cookson 1935)
Temperature coefficients of thermal expansion coefficients
Tensors giving other second-order and higher-order effects (pp. 254-7).

TABLE 24

Some anisotropic crystal properties not directly represented by tensors

The appearance of a property in this table does not mean that it may not ultimately be expressible in terms of tensors (for example, refractive index is connected with the tensor dielectric constant) but merely that the property itself does not appear to be a tensor.

Cleavage strength
Dielectric strength
Yield stress for plastic deformation; strain-hardening, and fracture in plastic crystals
Refractive index (pp. 235–7)
Surface properties including:

    Friction
    Surface hardness
    Surface energy
    Reflection of polarized light
    Rate of crystal growth and solution
    Physico-chemical properties: tarnishing rate, etch pits etc.

For a general review of anisotropy in metals see the article by Boas and Mackenzie (1950).

APPENDIX D

THE NUMBER OF INDEPENDENT COEFFICIENTS IN THE 32 CRYSTAL CLASSES

In Table 25, p. 294, are listed, for each of the 32 crystal classes, the number of independent coefficients needed to specify each crystal property completely. The crystal properties are divided into groups numbered 1, 2,.... The particular properties in any one group may be found by referring back to column 2 of Table 22 a, Appendix C, where each property is labelled with the number of the group to which it belongs. The complete matrix for each crystal property and crystal class will be found on the page whose number is given in parenthesis at the head of each column.

For the number and listing of the independent coefficients for other possible tensor properties, including some as yet undiscovered, see p. 122 and the following papers: Bhagavantam and Suryanarayana (1949 a), Jahn (1949), Fumi (1951, 1952 a, b, c, d) and Fieschi and Fumi (1953). A particular sixth-rank tensor is discussed by Mason (1951) [see also *errata* in Mason (1952)].

TABLE 25

Class symbol		Number of independent coefficients						
Inter-national	Schoen-flies	1 (p. 79)	2 (p. 23)	3 (p. 227)	4 and 5 (pp. 123–4, 247–8)	6 (pp. 140–1)	7 (pp. 250–1)	8 (pp. 271–2)
1	C_1	3	6	9	18	21	36	6
$\bar{1}$	$C_i(S_2)$	0	6	9	0	21	36	0
2	C_2	1	4	5	8	13	20	4
m	$C_s(C_{1h})$	2	4	5	10	13	20	2
$2/m$	C_{2h}	0	4	5	0	13	20	0
222	$D_2(V)$	0	3	3	3	9	12	3
$mm2$	C_{2v}	1	3	3	5	9	12	1
mmm	$D_{2h}(V_h)$	0	3	3	0	9	12	0
$\frac{4}{4}$	C_4	1	2	3	4	7	10	2
$\bar{4}$	S_4	0	2	3	4	7	10	2
$4/m$	C_{4h}	0	2	3	0	7	10	0
422	D_4	0	2	2	1	6	7	2
$4mm$	C_{4v}	1	2	2	3	6	7	0
$\bar{4}2m$	$D_{2d}(V_d)$	0	2	2	2	6	7	1
$4/mmm$	D_{4h}	0	2	2	0	6	7	0
3	C_3	1	2	3	6	7	12	2
$\bar{3}$	$C_{3i}(S_6)$	0	2	3	0	7	12	0
32	D_3	0	2	2	2	6	8	2
$3m$	C_{3v}	1	2	2	4	6	8	0
$\bar{3}m$	D_{3d}	0	2	2	0	6	8	0
6	C_6	1	2	3	4	5	8	2
$\bar{6}$	C_{3h}	0	2	3	2	5	8	0
$6/m$	C_{6h}	0	2	3	0	5	8	0
622	D_6	0	2	2	1	5	6	2
$6mm$	C_{6v}	1	2	2	3	5	6	0
$\bar{6}m2$	D_{3h}	0	2	2	1	5	6	0
$6/mmm$	D_{6h}	0	2	2	0	5	6	0
23	T	0	1	1	1	3	4	1
$m3$	T_h	0	1	1	0	3	4	0
432	O	0	1	1	0	3	3 •	1
$\bar{4}3m$	T_d	0	1	1	1	3	3	0
$m3m$	O_h	0	1	1	0	3	3	0
Isotropic without centre of symmetry		0	1	1	0	2	2	1
Isotropic with centre of symmetry		0	1	1	0	2	2	0
Number of crystal classes possessing each property		10	32	32	20	32	32	15

APPENDIX E

MATRICES FOR EQUILIBRIUM PROPERTIES IN THE 32 CRYSTAL CLASSES

TABLE 26 shows how the matrices that occur in Chapter X, equations (44) and (45), are affected by crystal symmetry in each of the 32 classes. For convenience of reference we repeat the equations here.

$$
\begin{aligned}
\epsilon_1 &= s_{11}^{E,T}\sigma_1 + s_{12}\sigma_2 + s_{13}\sigma_3 + s_{14}\sigma_4 + s_{15}\sigma_5 + s_{16}\sigma_6 &&+ d_{11}^T E_1 + d_{21}E_2 + d_{31}E_3 &&+ \alpha_1^E \Delta T \\
\epsilon_2 &= s_{12}\sigma_1 + s_{22}\sigma_2 + s_{23}\sigma_3 + s_{24}\sigma_4 + s_{25}\sigma_5 + s_{26}\sigma_6 &&+ d_{12}E_1 + d_{22}E_2 + d_{32}E_3 &&+ \alpha_2 \Delta T \\
\epsilon_3 &= s_{13}\sigma_1 + s_{23}\sigma_2 + s_{33}\sigma_3 + s_{34}\sigma_4 + s_{35}\sigma_5 + s_{36}\sigma_6 &&+ d_{13}E_1 + d_{23}E_2 + d_{33}E_3 &&+ \alpha_3 \Delta T \\
\epsilon_4 &= s_{14}\sigma_1 + s_{24}\sigma_2 + s_{34}\sigma_3 + s_{44}\sigma_4 + s_{45}\sigma_5 + s_{46}\sigma_6 &&+ d_{14}E_1 + d_{24}E_2 + d_{34}E_3 &&+ \alpha_4 \Delta T \\
\epsilon_5 &= s_{15}\sigma_1 + s_{25}\sigma_2 + s_{35}\sigma_3 + s_{45}\sigma_4 + s_{55}\sigma_5 + s_{56}\sigma_6 &&+ d_{15}E_1 + d_{25}E_2 + d_{35}E_3 &&+ \alpha_5 \Delta T \\
\epsilon_6 &= s_{16}\sigma_1 + s_{26}\sigma_2 + s_{36}\sigma_3 + s_{46}\sigma_4 + s_{56}\sigma_5 + s_{66}\sigma_6 &&+ d_{16}E_1 + d_{26}E_2 + d_{36}E_3 &&+ \alpha_6 \Delta T \\
D_1 &= d_{11}^T\sigma_1 + d_{12}\sigma_2 + d_{13}\sigma_3 + d_{14}\sigma_4 + d_{15}\sigma_5 + d_{16}\sigma_6 &&+ \kappa_{11}^{\sigma,T}E_1 + \kappa_{12}E_2 + \kappa_{13}E_3 &&+ p_1^\sigma \Delta T \\
D_2 &= d_{21}\sigma_1 + d_{22}\sigma_2 + d_{23}\sigma_3 + d_{24}\sigma_4 + d_{25}\sigma_5 + d_{26}\sigma_6 &&+ \kappa_{12}E_1 + \kappa_{22}E_2 + \kappa_{23}E_3 &&+ p_2 \Delta T \\
D_3 &= d_{31}\sigma_1 + d_{32}\sigma_2 + d_{33}\sigma_3 + d_{34}\sigma_4 + d_{35}\sigma_5 + d_{36}\sigma_6 &&+ \kappa_{13}E_1 + \kappa_{23}E_2 + \kappa_{33}E_3 &&+ p_3 \Delta T \\
\Delta S &= \alpha_1^E\sigma_1 + \alpha_2\sigma_2 + \alpha_3\sigma_3 + \alpha_4\sigma_4 + \alpha_5\sigma_5 + \alpha_6\sigma_6 &&+ p_1^\sigma E_1 + p_2 E_2 + p_3 E_3 &&+ \\
& && && (C^{\sigma,E}/T)\Delta T
\end{aligned}
$$

Ch. X, (44)

$$
\begin{aligned}
\boldsymbol{\epsilon} &= \mathbf{s}^{E,T}\boldsymbol{\sigma} + \mathbf{d}_t^T\mathbf{E} + \boldsymbol{\alpha}^E\Delta T \\
\mathbf{D} &= \mathbf{d}^T\boldsymbol{\sigma} + \boldsymbol{\kappa}^{\sigma,T}\mathbf{E} + \mathbf{p}^\sigma\Delta T \\
\Delta S &= \boldsymbol{\alpha}_t^E\boldsymbol{\sigma} + \mathbf{p}_t^\sigma\mathbf{E} + (C^{\sigma,E}/T)\Delta T
\end{aligned}
$$

Ch. X, (45)

Table 26 collects together the matrices given in the text on pp. 23, 79, 123–4, 140–1. All the coefficients on the right-hand side of the above equations are presented as one matrix $[(6+3+1)\times(6+3+1)] = [10\times10]$ thus:

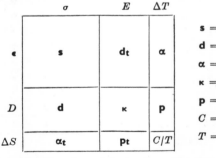

s = elastic compliances

d = piezoelectric moduli

α = thermal expansion coefficients

κ = permittivities

p = pyroelectric coefficients

C = heat capacity

T = absolute temperature

At the side of each matrix is given the number of independent coefficients for each property in the order, **s**, **d**, **α**, **κ**, **p**, C/T, with the total number at the bottom. The setting of the reference axes x_1, x_2, x_3, relative to the symmetry elements follows the conventions given in Appendix B. All the 10×10 matrices are symmetrical about the leading diagonal. All other equalities and relations between the elements demanded by the point-group symmetry are shown by the symbolism explained at the head of the table; this is the same as the symbolism used in the main text.

TABLE 26

Matrices for equilibrium properties in the 32 crystal classes

KEY TO NOTATION

- · zero component ● non-zero component ●—● equal components
●—○ components numerically equal, but opposite in sign
  ⊙ a component equal to twice the heavy dot component to which it is joined
  ⊚ a component equal to minus 2 times the heavy dot component to which it is joined
✗ $2(s_{11} - s_{12})$

Each complete 10×10 matrix is symmetrical about the leading diagonal.

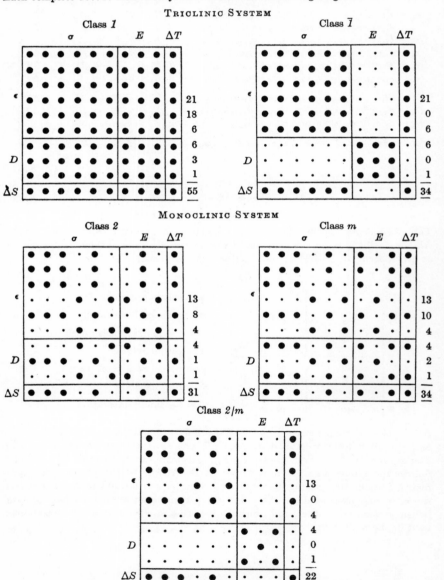

ORTHORHOMBIC SYSTEM

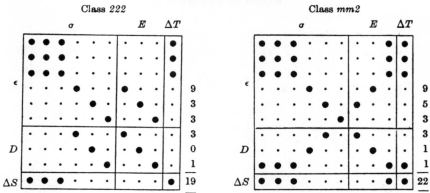

Class *222*

Class *mm2*

Class *mmm*

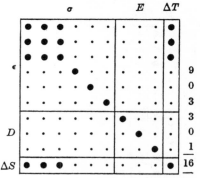

TETRAGONAL SYSTEM

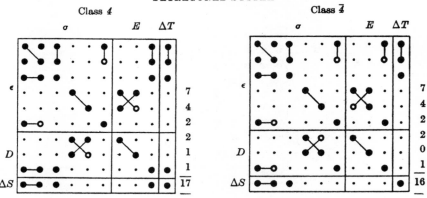

Class *4*

Class *4̄*

TETRAGONAL SYSTEM (*continued*)

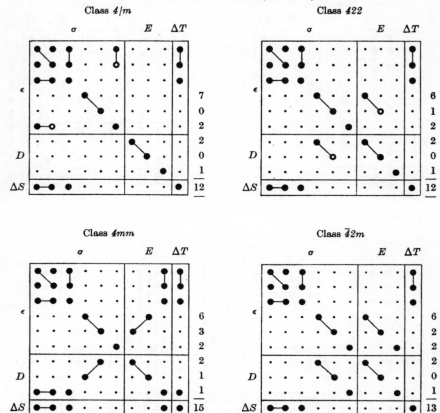

Class *4/m*

Class *422*

Class *4mm*

Class *42m*

Class *4/mmm*

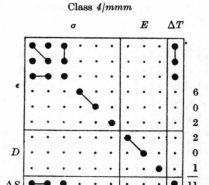

TRIGONAL SYSTEM

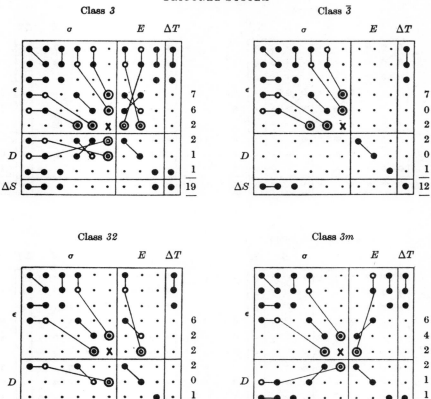

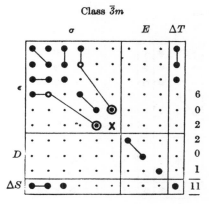

APPENDIX E

HEXAGONAL SYSTEM

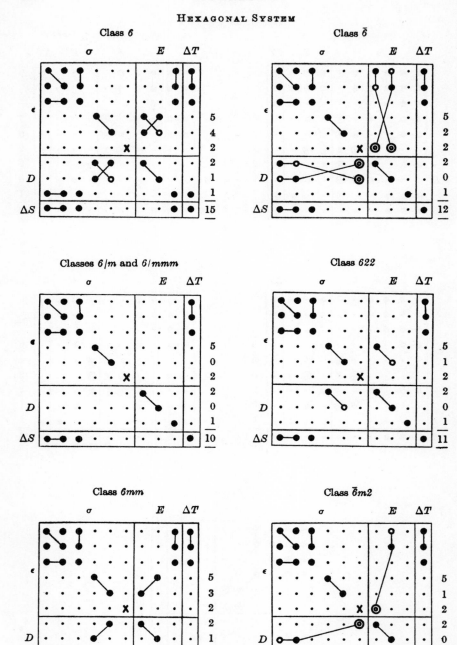

Class 6

Class 6̄

Classes 6/m and 6/mmm

Class 622

Class 6mm

Class 6̄m2

CUBIC SYSTEM

Classes *23* and *4̄3m*   Classes *m3*, *432* and *m3m*

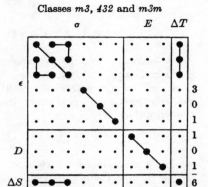

ISOTROPIC

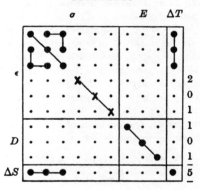

APPENDIX F

MAGNETIC AND ELECTRICAL ENERGY

THE following remarks on magnetic and electrical energy supplement those made in Chapter III, § 2, Chapter IV, § 4, and Chapter X, § 3.

It may be shown from Maxwell's equations that when a change in a static electromagnetic field is brought about by shifting charges and altering currents the work done is

$$dW = \int_{\substack{\text{all} \\ \text{space}}} (\mathbf{E}\,d\mathbf{D} + \mathbf{H}\,d\mathbf{B})\,d\tau, \tag{1}$$

where $d\tau$ is an element of volume.

Now in an isothermal reversible process the work done equals the increase $d\Psi''$ in the free energy Ψ'':

$$dW = d\Psi'';$$

and so, if the thermodynamic system under consideration is taken to be the polarized or magnetized body plus the charges and currents which are used to control the field, (1) can be equated to $d\Psi''$ for a reversible isothermal change in this composite system. Now Ψ'' depends not only on the condition of the body but also on the state of the external charges and currents; it is therefore convenient to split Ψ'' into two parts, one of which, Ψ, depends only on the state of the body, and to call this part the 'free energy of the body'. In the related case of the elastic energy of a body (Ch. VIII, § 3) there is no difficulty in defining the free energy of the body, for all the energy is localized (or, at any rate, assumed to be localized) in the body. But here, because of the long-range nature of electrical and magnetic forces, the division of the energy into a part possessed by the body and a remainder is somewhat arbitrary. One way is to define

$$\Psi = \Psi'' - \tfrac{1}{2} \int_{\substack{\text{space} \\ \text{outside} \\ \text{body}}} (\kappa_0 \mathbf{E}^2 + \mu_0 \mathbf{H}^2)\,d\tau.$$

Then, since outside the body $\mathbf{D} = \kappa_0 \mathbf{E}$ and $\mathbf{B} = \mu_0 \mathbf{H}$, for any reversible field change,

$$d\Psi = \int_{\text{body}} (\mathbf{E}\,d\mathbf{D} + \mathbf{H}\,d\mathbf{B})\,d\tau, \tag{2}$$

where the integral is taken over the volume occupied by the body.

An alternative is to define

$$\Psi = \Psi'' - \tfrac{1}{2} \int_{\substack{\text{all} \\ \text{space}}} (\kappa_0 \mathbf{E}^2 + \mu_0 \mathbf{H}^2)\,d\tau.$$

Then, for any reversible field change,

$$d\Psi = \int_{\text{body}} (\mathbf{E}\,d\mathbf{P} + \mathbf{H}\,d\mathbf{I})\,d\tau. \tag{3}$$

Either $E_i\,dP_i$ or $E_i\,dD_i$ may therefore be used with equal propriety in an expression for $d\Psi$ or dU (p. 179). The same thermodynamic conclusions will be reached whichever definition is adopted, for the two expressions for $d\Psi$, (2) and (3), differ by a perfect differential,

$$\tfrac{1}{2}d\left\{\int\limits_{\text{body}} (\kappa_0\,\mathbf{E}^2+\mu_0\,\mathbf{H}^2)\,d\tau\right\}.$$

It should be pointed out that if the size and shape of the crystal change as the field is changed, as they will do if the crystal is piezoelectric, certain points of space which were originally inside the crystal will be outside it after the change, and vice versa. At such points $\mathbf{P}$ and $\mathbf{I}$, and therefore $\mathbf{D}$ and $\mathbf{B}$, will change discontinuously; to say that $\mathbf{E}\,d\mathbf{D}$ is the increase in free energy per unit volume of the crystal is only strictly correct, therefore, when the size and shape of the crystal are held fixed. However, it is not difficult to show that, if the expression $\mathbf{E}\,d\mathbf{D}$ is assumed to be correct even when there are strains present, the errors in the thermodynamical relations so derived are small correction terms.

For fuller discussion of the points involved here reference may be made to the book by Becker and Döring (1939) and papers by Guggenheim (1936) and Koenig (1937).

APPENDIX G

THE DIFFERENCE BETWEEN THE CLAMPED AND FREE ISOTHERMAL PERMITTIVITIES

To illustrate the manipulations necessary to derive equations (49) to (55) on p. 186 we derive equation (51) here. This equation expresses the difference between the clamped and free permittivities, and so gives the piezoelectric contribution to the permittivity:

$$\kappa_{ij}^{\epsilon} - \kappa_{ij}^{\sigma} = -d_{ikl}d_{jmn}c_{klmn}^{E} \quad (T \text{ constant}). \qquad \text{Ch. X, (51)}$$

Let T be constant. Then, from Chapter X, (27) and (28),

$$d\epsilon_{mn} = \left(\frac{\partial \epsilon_{mn}}{\partial \sigma_{pq}}\right)_E d\sigma_{pq} + \left(\frac{\partial \epsilon_{mn}}{\partial E_j}\right)_\sigma dE_j \left.\right\} \qquad (1)$$

$$dD_i = \left(\frac{\partial D_i}{\partial \sigma_{kl}}\right)_E d\sigma_{kl} + \left(\frac{\partial D_i}{\partial E_j}\right)_\sigma dE_j \qquad (2)$$

Put $d\epsilon_{mn} = 0$ in (1) and multiply by $(\partial \sigma_{kl}/\partial \epsilon_{mn})_E$:

$$0 = \left(\frac{\partial \sigma_{kl}}{\partial \epsilon_{mn}}\right)_E \left(\frac{\partial \epsilon_{mn}}{\partial \sigma_{pq}}\right)_E d\sigma_{pq} + \left(\frac{\partial \sigma_{kl}}{\partial \epsilon_{mn}}\right)_E \left(\frac{\partial \epsilon_{mn}}{\partial E_j}\right)_\sigma dE_j. \qquad (3)$$

From Chapter IX, equation (21) (or otherwise), the first term on the right-hand side is equal to

$$\delta_{kp}\delta_{lq} d\sigma_{pq} = d\sigma_{kl}.$$

Therefore, substituting the value of $d\sigma_{kl}$ so obtained from (3) into (2), and dividing by dE_j, we have

$$\left(\frac{\partial D_i}{\partial E_j}\right)_\epsilon - \left(\frac{\partial D_i}{\partial E_j}\right)_\sigma = -\left(\frac{\partial D_i}{\partial \sigma_{kl}}\right)_E \left(\frac{\partial \sigma_{kl}}{\partial \epsilon_{mn}}\right)_E \left(\frac{\partial \epsilon_{mn}}{\partial E_j}\right)_\sigma.$$

This is equivalent to the required equation (51), Chapter X.

APPENDIX H

PROOF OF THE INDICATRIX PROPERTIES FROM MAXWELL'S EQUATIONS (see pp. 235–7)

IN m.k.s. units Maxwell's equations are

$$\left.\begin{array}{l} \operatorname{curl} \mathbf{H} = \mathbf{j} + \dot{\mathbf{D}} \\ \operatorname{curl} \mathbf{E} = -\dot{\mathbf{B}} \end{array}\right\}; \tag{1}$$

a dot denotes differentiation with respect to time. Transparent crystals have low conductivities and are not ferromagnetic. Their permeabilities therefore differ only slightly from that of a vacuum, and we put

$$\mathbf{j} = 0, \qquad \mathbf{B} = \mu_0 \mathbf{H}. \tag{2}$$

Equations (1) then become

$$\left.\begin{array}{l} \operatorname{curl} \mathbf{H} = \dot{\mathbf{D}} \\ \operatorname{curl} \mathbf{E} = -\mu_0 \dot{\mathbf{H}} \end{array}\right\}. \tag{3}$$
$$\tag{4}$$

We wish to examine the properties of plane-polarized electromagnetic waves travelling through a crystal. We therefore try as a solution to these equations

$$\mathbf{E} = \mathbf{E}_0 \exp\left\{ i\omega\left(t - \frac{\mathbf{r} \cdot \mathbf{l}}{v}\right) \right\}, \dagger \tag{5}$$

where $\mathbf{E}_0$ is a constant vector, ω and v are constants, $\mathbf{l}$ is a unit vector and $\mathbf{r}$ is the position vector $[x_1, x_2, x_3]$. This represents a plane wave; for, at a given time t, $\mathbf{E}$ is constant over the surface $\mathbf{r} \cdot \mathbf{l} = \text{constant}$, which is a plane normal to $\mathbf{l}$. v is evidently the velocity of the wave front along the wave normal $\mathbf{l}$. We recognize $\mathbf{E}_0$ as the amplitude and $\omega/2\pi$ as the frequency of the wave.

Substituting this expression for $\mathbf{E}$ into (4) gives

$$-\mu_0 \dot{\mathbf{H}} = \operatorname{curl}\left[\mathbf{E}_0 \exp\left\{ i\omega\left(t - \frac{\mathbf{r} \cdot \mathbf{l}}{v}\right) \right\} \right]. \tag{6}$$

Now the curl of the product of a scalar (ϕ, say) with a vector ($\mathbf{A}$, say) decomposes as follows:

$$\operatorname{curl} \phi \mathbf{A} = \phi \operatorname{curl} \mathbf{A} + (\operatorname{grad} \phi) \wedge \mathbf{A}.$$

Hence, applying this to (6), and remembering that $\mathbf{E}_0$ does not depend on position, so that $\operatorname{curl} \mathbf{E}_0 = 0$, we have

$$\mu_0 \dot{\mathbf{H}} = \mathbf{E}_0 \wedge \operatorname{grad} \exp\left\{ i\omega\left(t - \frac{\mathbf{r} \cdot \mathbf{l}}{v}\right) \right\}. \tag{7}$$

Now $\operatorname{grad}(\mathbf{r} \cdot \mathbf{l}) = \mathbf{l}$, as may be seen by writing out the components; hence (7) becomes

$$\mu_0 \dot{\mathbf{H}} = \mathbf{E}_0 \wedge \left[-\frac{i\omega}{v} \exp\left\{ i\omega\left(t - \frac{\mathbf{r} \cdot \mathbf{l}}{v}\right) \right\} \operatorname{grad}(\mathbf{r} \cdot \mathbf{l}) \right],$$

or

$$\mu_0 \mathbf{H} = \frac{i\omega}{v}(\mathbf{l} \wedge \mathbf{E}_0) \exp\left\{ i\omega\left(t - \frac{\mathbf{r} \cdot \mathbf{l}}{v}\right) \right\}.$$

---

† This equation involves complex quantities although the physical quantities involved are real. *Either* the real *or* the imaginary part of the equation may be taken as the physical statement. See, for example, Joos (1934), pp. 43 et seq.

Integrating with respect to t we have

$$\mu_0\,\mathbf{H} = \frac{1}{v}(\mathbf{l}\wedge\mathbf{E}_0)\exp\left\{i\omega\left(t - \frac{\mathbf{r}.\mathbf{l}}{v}\right)\right\} = \frac{1}{v}(\mathbf{l}\wedge\mathbf{E}). \tag{8}$$

We may write

$$\mathbf{H} = \mathbf{H}_0\exp\left\{i\omega\left(t - \frac{\mathbf{r}.\mathbf{l}}{v}\right)\right\}, \tag{9}$$

where

$$\mu_0\,\mathbf{H}_0 = \frac{1}{v}(\mathbf{l}\wedge\mathbf{E}_0). \tag{10}$$

(9) shows that $\mathbf{H}$ is represented by a plane wave, in phase with $\mathbf{E}$. (8) contains the important result that $\mathbf{H}$ is perpendicular to the plane of $\mathbf{l}$ and $\mathbf{E}$. This is illustrated in Fig. H.1.

We do not yet know the angle between $\mathbf{E}$ and $\mathbf{l}$, and we have not yet used

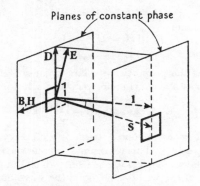

FIG. H.1. The relation between $\mathbf{D}$, $\mathbf{E}$, $\mathbf{B}$, $\mathbf{H}$, $\mathbf{l}$ and $\mathbf{S}$ in a plane-polarized light wave passing through a birefringent crystal. $\mathbf{l}$ is the wave-normal and $\mathbf{S}$ gives the ray direction.

equation (3). Substituting (9) into (3), just as (5) was substituted into (4), we find

$$\dot{\mathbf{D}} = \text{curl}\left[\mathbf{H}_0\exp\left\{i\omega\left(t - \frac{\mathbf{r}.\mathbf{l}}{v}\right)\right\}\right]$$

$$= -\mathbf{H}_0\wedge\text{grad}\exp\left\{i\omega\left(t - \frac{\mathbf{r}.\mathbf{l}}{v}\right)\right\}$$

$$= -\frac{i\omega}{v}(\mathbf{l}\wedge\mathbf{H}_0)\exp\left\{i\omega\left(t - \frac{\mathbf{r}.\mathbf{l}}{v}\right)\right\}.$$

Integrating with respect to t we have

$$\mathbf{D} = -\frac{1}{v}(\mathbf{l}\wedge\mathbf{H}_0)\exp\left\{i\omega\left(t - \frac{\mathbf{r}.\mathbf{l}}{v}\right)\right\}, \tag{11}$$

or, from (9),

$$\mathbf{D} = -\frac{1}{v}(\mathbf{l}\wedge\mathbf{H}). \tag{12}$$

We may therefore insert $\mathbf{D}$ in Fig. H.1 perpendicular to $\mathbf{l}$ and $\mathbf{H}$. *With respect to $\mathbf{D}$ and $\mathbf{H}$ we thus have pure transverse waves.* $\mathbf{D}$, $\mathbf{E}$ and $\mathbf{l}$ all lie in one plane, for they are all perpendicular to $\mathbf{H}$.

Substituting for **H** in (12) from (8) we find

$$\mathbf{D} = -\frac{1}{\mu_0 v^2}\{\mathbf{1}\wedge(\mathbf{1}\wedge\mathbf{E})\}. \tag{13}$$

Now, by the rule for the expansion of a triple vector product,

$$\mathbf{1}\wedge(\mathbf{1}\wedge\mathbf{E}) = \mathbf{1}(\mathbf{1}.\mathbf{E})-\mathbf{E}(\mathbf{1}.\mathbf{1}) = \mathbf{1}(\mathbf{1}.\mathbf{E})-\mathbf{E}.$$

Therefore (13) is equivalent to

$$\boxed{\mu_0 v^2\mathbf{D}-\mathbf{E}+\mathbf{1}(\mathbf{1}.\mathbf{E}) = 0.} \tag{14}$$

Equation (14) is the conclusion we reach by starting from Maxwell's equations (1), together with (2), and taking the case of plane waves. It is independent of any connexion between **D** and **E** demanded by the properties of the medium. We now assume that **D** and **E** are connected by the permittivity tensor. When referred to the principal axes of the permittivity tensor, so that $E_1 = D_1/\kappa_1$ etc., (14) becomes, in component form,

$$\mu_0 v^2 D_1-\frac{D_1}{\kappa_1}+l_1(\mathbf{1}.\mathbf{E}) = 0,$$

and two similar equations. Hence

$$D_1 = \frac{l_1(\mathbf{1}.\mathbf{E})}{\left(\dfrac{1}{\kappa_1}-\mu_0 v^2\right)}, \tag{15}$$

and two similar equations.

Since **D** is normal to **1**, the scalar product **D**.**1** vanishes. We now express this product in terms of the components of **D** given by (15) and the two similar equations. Dividing by $(\mathbf{1}.\mathbf{E})$, which does not, in general, vanish, we obtain

$$\frac{l_1^2}{\left(\dfrac{1}{\kappa_1}-\mu_0 v^2\right)}+\frac{l_2^2}{\left(\dfrac{1}{\kappa_2}-\mu_0 v^2\right)}+\frac{l_3^2}{\left(\dfrac{1}{\kappa_3}-\mu_0 v^2\right)} = 0. \tag{16}$$

This is an equation giving the wave velocity v in terms of the direction cosines l_1, l_2, l_3 of the wave normal. We notice that it is a quadratic equation for v^2. Hence, there are, in general, two values of v^2 for every set of l_1, l_2, l_3. If we fix attention only on positive values of v, this gives two distinct waves of different velocity for every direction of **1**. By putting $[l_1, l_2, l_3] = [1, 0, 0]$ we find the two velocities associated with the x_1 axis. They are

$$v = 1/\sqrt(\mu_0\kappa_2) \text{ and } 1/\sqrt(\mu_0\kappa_3).$$

We therefore define the *three principal velocities* by

$$v_1 = 1/\sqrt(\mu_0\kappa_1), \qquad v_2 = 1/\sqrt(\mu_0\kappa_2), \qquad v_3 = 1/\sqrt(\mu_0\kappa_3),$$

and write (16) in the simpler form

$$\boxed{\frac{l_1^2}{(v_1^2-v^2)}+\frac{l_2^2}{(v_2^2-v^2)}+\frac{l_3^2}{(v_3^2-v^2)} = 0.} \tag{17}$$

Since $c = 1/\sqrt(\mu_0\kappa_0)$, the three principal velocities may be written

$$v_1 = c/\sqrt{K_1}, \qquad v_2 = c/\sqrt{K_2}, \qquad v_3 = c/\sqrt{K_3};$$

or

$$v_1 = c/n_1, \qquad v_2 = c/n_2, \qquad v_3 = c/n_3,$$

where n_1, n_2, n_3 are the three principal refractive indices (pp. 236–7).

Proof of the indicatrix construction. We now prove that the two wave velocities associated with a given wave normal are given by the indicatrix construction described on p. 236.

Let $v^2 = v'^2$ and $v^2 = v''^2$ be the two roots of equation (17). Then, if (17) is written as $f(v^2) = 0$, we have

$$f(v'^2) = 0 \quad \text{and} \quad f(v''^2) = 0. \tag{18}$$

From (15), the direction cosines of **D** are proportional to

$$\frac{l_1}{v_1^2 - v^2}, \qquad \frac{l_2}{v_2^2 - v^2}, \qquad \frac{l_3}{v_3^2 - v^2}. \tag{19}$$

When v'^2 and v''^2 are substituted for v^2 in (19) we obtain the direction cosines of the two directions of **D** associated with the wave normal $[l_1, l_2, l_3]$. Call these sets of direction cosines (a', b', c') and (a'', b'', c'') respectively. To prove that the two directions in question are mutually perpendicular we note that

$$a'a'' + b'b'' + c'c'' = \frac{l_1^2}{(v_1^2 - v'^2)(v_1^2 - v''^2)} + \frac{l_2^2}{(v_2^2 - v'^2)(v_2^2 - v''^2)} + \frac{l_3^2}{(v_3^2 - v'^2)(v_3^2 - v''^2)}$$

$$= \frac{f(v'^2) - f(v''^2)}{v'^2 - v''^2} = 0, \quad \text{by (18)}.$$

Now transform to new axes, as follows,

	x_1	x_2	x_3
x_1'	a'	b'	c'
x_2'	a''	b''	c''
x_3'	l_1	l_2	l_3

x_3' is then the direction of the wave normal, and x_1' and x_2' are the two directions of **D**. We have now to prove that x_1' and x_2' are the principal axes of the central section of the indicatrix normal to x_3', and that the lengths of the semi-axes of this ellipse are the two refractive indices c/v' and c/v''.

The equation of the indicatrix (p. 237) is

$$B_1 x_1^2 + B_2 x_2^2 + B_3 x_3^2 = 1,$$

and we know that $B_1 = (v_1/c)^2$, etc. Transformed to the new axes the equation is

$$B_{ij}' x_i' x_j' = 1, \tag{20}$$

where

$$B_{ij}' = a_{ik} a_{jl} B_{kl}. \tag{21}$$

The elliptical section normal to x_3' is found by putting $x_3' = 0$:

$$B_{11}' x'^2 + 2B_{12}' x_1' x_2' + B_{22}' x_2'^2 = 1. \tag{22}$$

We have, from (21),

$$B_{12}' = a_{1k} a_{2l} B_{kl} = a_{11} a_{21} B_{11} + a_{12} a_{22} B_{22} + a_{13} a_{23} B_{33}$$

$$= a'a'' B_1 + b'b'' B_2 + c'c'' B_3. \tag{23}$$

By using the expressions for the direction cosines given by (19) and by putting B_1 proportional to v_1^2, and so on, we see that the right-hand side of (23) is proportional to

$$\frac{l_1^2 v_1^2}{(v_1^2 - v'^2)(v_1^2 - v''^2)} + \frac{l_2^2 v_2^2}{(v_2^2 - v'^2)(v_2^2 - v''^2)} + \frac{l_3^2 v_3^2}{(v_3^2 - v'^2)(v_3^2 - v''^2)}.$$

Inspection shows that this expression is identically equal to

$$\frac{v'^2 f(v'^2) - v''^2 f(v''^2)}{v'^2 - v''^2},$$

which is zero, by (18). Hence $B'_{12} = 0$. It follows that x'_1 and x'_2 are the principal axes of the elliptical section.

We must now show that the lengths of the semi-axes are the two refractive indices. The length of the semi-axis parallel to x'_1 is $1/\sqrt{B'_{11}}$, and we want to show that this is equal to c/v'. In other words, we have to prove that $c^2 B'_{11} = v'^2$.

We have
$$B'_{11} = a_{1k} a_{1l} B_{kl} = a_{11}^2 B_{11} + a_{12}^2 B_{22} + a_{13}^2 B_{33}$$
$$= a'^2 B_1 + b'^2 B_2 + c'^2 B_3,$$

or
$$c^2 B'_{11} = \frac{\dfrac{l_1^2 v_1^2}{(v_1^2 - v'^2)^2} + \dfrac{l_2^2 v_2^2}{(v_2^2 - v'^2)^2} + \dfrac{l_3^2 v_3^2}{(v_3^2 - v'^2)^2}}{\dfrac{l_1^2}{(v_1^2 - v'^2)^2} + \dfrac{l_2^2}{(v_2^2 - v'^2)^2} + \dfrac{l_3^2}{(v_3^2 - v'^2)^2}}.$$

If we try putting $c^2 B'_{11} = v'^2$ in this equation, we obtain, after multiplying up and collecting the terms,

$$\frac{l_1^2}{(v_1^2 - v'^2)^2}(v_1^2 - v'^2) + \frac{l_2^2}{(v_2^2 - v'^2)^2}(v_2^2 - v'^2) + \frac{l_3^2}{(v_3^2 - v'^2)^2}(v_3^2 - v'^2) = 0.$$

Equation (17) shows that this relation is satisfied. Hence
$$c^2 B'_{11} = v'^2.$$

Similarly,
$$c^2 B'_{22} = v''^2;$$

and so our result is proved.

The *ray direction* in a wave is the direction in which a bounded part of the wave front would travel (see p. 239). It is the direction of the flow of energy, and is therefore given by the *Poynting vector*†

$$\mathbf{S} = \mathbf{E} \wedge \mathbf{H}. \tag{24}$$

$\mathbf{S}$ is therefore perpendicular to $\mathbf{E}$ and $\mathbf{H}$ and may be inserted in Fig. H.1. Notice that the angle between $\mathbf{D}$ and $\mathbf{E}$ is the same as that between $\mathbf{S}$ and $\mathbf{l}$.

† See, for example, Abraham and Becker (1937).

BIBLIOGRAPHY

Books and papers referred to in the text

ABRAHAM, M., and BECKER, R. (1937). *The classical theory of electricity and magnetism*, English translation, Glasgow: Blackie.

AITKEN, A. C. (1948). *Determinants and matrices*, Edinburgh: Oliver and Boyd.

BATES, L. F. (1937). *Rep. Prog. Phys.* **3**, 185.

—— (1951). *Modern magnetism*, 3rd edition, Cambridge.

BECKER, R., and DÖRING, W. (1939). *Ferromagnetismus*, Art. 6, Berlin: Springer.

BELL, R. J. T. (1937). *An elementary treatise on coordinate geometry of three dimensions*, 2nd edition, London: Macmillan.

BHAGAVANTAM, S. (1942). *Proc. Ind. Acad. Sci.* A, **16**, 359–65.

—— (1952). *Acta Cryst.* **5**, 591–3.

—— (1955). *Proc. Ind. Acad. Sci.* A, **41**, 72–90.

—— and KRISHNA RAO, K. V. (1953). *Acta Cryst.* **6**, 799–801.

—— —— (1954). *Current Science*, **23**, 257–8.

—— and SURYANARAYANA, D. (1947). *Proc. Ind. Acad. Sci.* A, **26**, 97–109.

—— —— (1949 a). *Acta Cryst.* **2**, 21–26.

—— —— (1949 b). *Acta Cryst.* **2**, 26–30.

—— and VENKATARAYUDU, T. (1951). *Theory of groups and its application to physical problems*, 2nd edition, Waltair: Andhra University.

BIRCH, F. (1947). *Phys. Rev.* **71**, 809–24.

BOAS, W., and MACKENZIE, J. K. (1950). *Prog. Met. Phys.* **2**, 90–120.

BOND, W. L. (1943). *Bell System Tech. Journ.* **22**, 1–72.

BORN, M. (1933). *Optik*, Springer, 413–20.

BÖTTCHER, C. J. F. (1952). *Theory of electric polarisation*, ch. v, Elsevier: Amsterdam.

BRUHAT, G., and GRIVET, P. (1935). *J. Phys. Radium*, series 7, **6**, 12–26.

BURSTEIN, E., and SMITH, P. L. (1948 a). *Phys. Rev.* **74**, 229–30.

—— —— (1948 b). *Phys. Rev.* **74**, 1880–1.

CADY, W. G. (1946). *Piezoelectricity*, New York: McGraw-Hill.

CALLEN, H. B. (1948). *Phys. Rev.* **73**, 1349–58.

CARPENTER, R. O'B. (1950). *J. Opt. Soc. Amer.* **40**, 225–9.

CASIMIR, H. B. G. (1945). *Rev. Mod. Phys.* **17**, 343–50.

CONDON, E. U. (1937). *Rev. Mod. Phys.* **9**, 432–57.

COOKSON, T. W. (1935). *Phys. Rev.* **47**, 194–5.

DE GROOT, S. R. (1951). *Thermodynamics of irreversible processes*, Amsterdam: North-Holland Publishing Company.

DEVONSHIRE, A. F. (1954. *Advances in Physics*, **3**, 85–130.

DOMENICALI, C. A. (1953). *Phys. Rev.* **92**, 877–81.

DONNAY, J. D. H. (1947). *Amer. Mineralogist*, **32**, 52–58 (and 477–8).

EDDINGTON, A. S. (1923). *The mathematical theory of relativity*, Cambridge, §§ 20, 21.

EHRENFEST, P., and RUTGERS, A. J. (1929). *Proc. Acad. Amsterdam*, **32**, 698–706; 883–93.

EISENHART, L. P. (1939). *Coordinate geometry*, Boston: Ginn.

EPSTEIN, P. S. (1397). *Textbook of thermodynamics*, New York: Wiley.

FERRAR, W. L. (1941). *Algebra, a text-book of determinants, matrices and algebraic forms*, Oxford: Clarendon Press.

FIESCHI, R., and FUMI, F. G. (1953). *Il Nuovo Cimento*, **10**, 865–82 (in English).

FUMI, F. G. (1951). *Phys. Rev.* **83**, 1274–5.
—— (1952 *a*). *Acta Cryst.* **5**, 44–48.
—— (1952 *b*). *Phys. Rev.* **86**, 561.
—— (1952 *c*). *Acta Cryst.* **5**, 691–4.
—— (1952 *d*). *Il Nuovo Cimento*, **9**, 739–56 (in English).
GOENS, E. (1933). *Ann. d. Phys.* **16**, 793–809.
GUGGENHEIM, E. A. (1936). *Proc. Roy. Soc.* A, **155**, 49, 70.
HARTREE, D. R. (1952). *Numerical analysis*, Oxford: Clarendon Press.
HARTSHORNE, N. H., and STUART, A. (1950). *Crystals and the polarising microscope*, 2nd. edition, London: Arnold.
HEARMON, R. F. S. (1946). *Rev. Mod. Phys.* **18**, 409–40.
—— (1953). *Acta Cryst.* **6**, 331–40.
——(1956). *Advances in Physics*, **5**, 323–82.
HECKMANN, G. (1925). *Ergebnisse der exakten Naturwissenschaften*, **4**, 140.
HIGMAN, B. (1955). *Applied group-theoretic and matrix methods*, Oxford: Clarendon Press.
HOEK, H. (1941). *Physica*, **8**, 209–25.
HUNTINGTON, H. B. (1958). *The elastic constants of crystals. Solid State Physics*, **7**, 213–351, New York: Academic Press.
International Critical Tables (1929). New York: McGraw-Hill.
International tables for X-ray crystallography (1952). Vol. I, published for the International Union of Crystallography by the Kynoch Press, Birmingham.
JAHN, H. A. (1937). *Z. Kristallogr.* **98**, 191–200.
—— (1949). *Acta Cryst.* **2**, 30–33.
JEFFREYS, H. (1931). *Cartesian tensors*, Cambridge.
JOOS, G. (1934). *Theoretical physics*, English translation, Glasgow: Blackie, pp. 355 et seq.
KOENIG, F. O. (1937). *J. Phys. Chem.* **41**, 597.
Landolt-Börnstein Physikalisch-Chemische Tabellen, 1923, 1936, Berlin: Springer.
LONSDALE, K. (1937). *Rep. Prog. Phys.* **4**, 368–89.
MARIS, H. B. (1927). *J. Opt. Soc. Amer.* **15**, 194.
MASON, W. P. (1946). *Phys. Rev.* **69**, 173–94.
—— (1951). *Phys. Rev.* **82**, 715–23.
—— (1952). *Phys. Rev.* **85**, 1065.
McGREEVY, T. (1953). *The M.K.S. system of units*, London: Pitman.
MUELLER, H. (1935). *Phys. Rev.* **47**, 947–57.
MÜNSTER, C., and SZIVESSY, G. (1935). *Phys. Z.* **36**, 101–6.
NARASIMHAMURTY, T. S. (1954 *a*). *Current Science*, **23**, 149–50.
—— (1954 *b*). *Proc. Ind. Acad. Sci.* A, **40**, 167–75.
ONSAGER, L. (1931 *a*). *Phys. Rev.* **37**, 405–26.
—— (1931 *b*). *Phys. Rev.* **38**, 2265–79.
PHILLIPS, F. C. (1946). *An introduction to crystallography*, London: Longmans, Green.
POCKELS, F. (1906). *Lehrbuch der Kristalloptik*, Leipzig: Teubner.
RAMACHANDRAN, G. N. (1947). *Proc. Ind. Acad. Sci.* A, **25**, 208–19.
RAMAN, Sir C. V. and KRISHNAMURTI, D. (1955). *Proc. Ind. Acad. Sci.* A, **42**, 111–29.
SAS, R. K., and PIDDUCK, F. B. (1947). *The metre-kilogram-second system of electrical units*, London: Methuen.
SCARBOROUGH, J. B. (1950). *Numerical mathematical analysis*, 2nd edition, p. 399, Baltimore: the Johns Hopkins Press.
SCHMID, E., and BOAS, W. (1950). *Plasticity of crystals*, English translation, London: Hughes, p. 193.

SEARS, F. W. (1946). *Principles of physics II, Electricity and magnetism*, Cambridge, Mass.: Addison–Wesley.

SLATER, J. C. and FRANK, N. H. (1947). *Electromagnetism*, New York: McGraw-Hill.

SORET, C. (1893). *Arch. de Genève*, **29**, 322.

—— (1894). *Arch. de Genève*, **32**, 631.

Standards on piezoelectric crystals (1949). Proc. Institute of Radio Engineers, **37**, 1378–95.

STRATTON, J. A. (1941). *Electromagnetic theory*, New York: McGraw-Hill.

SZIVESSY, G. (1928). *Handbuch der Physik*, **20**, 804–37.

—— and MÜNSTER, C. (1934). *Ann. Phys.* **20**, 703–36.

TIFFEN, R. and STEVENSON, A. C. (1956). *Quart. J. Mech. Appl. Math.* **9**, 306–312.

TUTTON, A. E. H. (1922). *Crystallography and practical crystal measurement*, London: Macmillan, vol. 2, pp. 1308 et seq.

VAN DYKE, K. S., and GORDON, G. D. (1950). *A manual of piezoelectric data*, supplementary volume, 2nd edition, privately circulated.

VOIGT, W. (1903). *Gött. Nachr.* 87.

—— (1910). *Lehrbuch der Kristallphysik*, 1st edition (reprinted in 1928 with an additional appendix), Leipzig: Teubner.

WEST, C. D., and MAKAS, A. S. (1948). *J. Chem. Phys.* **16**, 427.

WOODS, F. S. (1934). *Advanced calculus*, new edition, Boston, Mass.: Ginn.

WOOSTER, W. A. (1938). *A text-book on crystal physics* (reprinted 1949), Cambridge.

SUPPLEMENTARY REFERENCES
AND NOTES (1985)

(a) *Comprehensive reference works and tables of properties*:

Landolt-Börnstein, New series ed. K.-H. Hellwege, Group III: *Crystal and solid state physics*, Vol. 11 (1979). *Elastic, piezoelectric, pyroelectric, piezooptic, electrooptic constants, and nonlinear dielectric susceptibilities of crystals.* Berlin: Springer-Verlag.

Contains definitive articles on linear and non-linear elastic, pyroelectric, dielectric and optical effects in crystals and on interactions between these effects. The tables of numerical values of properties and the bibliographies (some 4700 references) aim to be complete up to 1977.

Landolt-Börnstein, as above, Vol. 16a (1981) and 16b (1982). *Ferroelectrics and related substances.*

These volumes contain a literature survey and numerical data up to 1979 and 1980 respectively.

(b) *General works on crystal physics and the relation of properties to crystal symmetry:*

Atomic structure and properties of solids (1972). Proceedings of the International School of Physics 'Enrico Fermi', Course 52, Varenna 1971 (edited by E. Burstein), New York and London: Academic Press. Contains reviews on many aspects of crystal physics.

BHAGAVANTAM, S. (1966). *Crystal symmetry and physical properties*, London and New York: Academic Press.

BILLINGS, A. R. (1969). *Tensor properties of materials – generalized compliance and conductivity*, London: Wiley-Interscience.

BIRSS, R. R. (1963). Macroscopic symmetry in space-time. *Rep. Prog. Phys.* **26**, 307–360.

——(1964). *Symmetry and magnetism*, Amsterdam: North Holland.

BOAS, W., and MACKENZIE, J. K. (1950). Anisotropy in metals, *Prog. Met. Phys.* **2**, 90–120.

BRADLEY, C. J. and CRACKNELL, A. P. (1972). *The mathematical theory of symmetry in solids*, Oxford: Clarendon Press. A comprehensive treatment.

CADY, W. G. (1946). *Piezoelectricity*, New York: McGraw-Hill. Mainly concerned with piezoelectricity, but includes much information on related topics in crystal physics.

CALLEN, H. (1968). Crystal symmetry and macroscopic laws. *Amer. J. Phys.* **36**, 735–748. Didactic treatment assuming some knowledge of group theory.

CALLEN, H., CALLEN, E. and KALVA, Z. (1970). Crystal symmetry and macroscopic laws. II. *Amer. J. Phys.* **38** 1278–1284. A sequel to (Callen 1968).

CRACKNELL, A. P. (1969). Group theory and magnetic phenomena in solids. *Rep. Prog. Phys.* **32**, 633–707.

DE FIGUEIREDO, I. M. B. and RAAB, R. E. (1980). A pictorial approach to macroscopic space-time symmetry, with particular reference to light scattering. *Proc. R. Soc. Lond.* A**369**, 501–516. The existence of any particular electromagnetic effect that may be postulated is tested by applying space and time symmetry operations to an entire macroscopic experiment: an extension of earlier work cited.

FUMI, F. G. and RIPAMONTI, C. (1980). Tensor properties and rotational symmetry of crystals. I. *Acta. Cryst.* A36, 535–551; II. *Acta. Cryst.* A36, 551–558 (Erratum in A37, 137); III. (1983) *Acta. Cryst.* A39, 245–251 (Erratum in A39, 594–595). These papers extend the results of Fumi (1952d) and Fieschi and Fumi (1953), referred to on p. 122, and introduce a new method for dealing with the 'difficult' 3-fold axis parallel to Oz (see below), results being given for tensors up to rank 8. Paper I usefully reviews existing methods of studying the effect of symmetry on tensor properties.

On pp. 121–2 and 139 we point out the problem of applying the direct inspection method to the trigonal and hexagonal systems. Fumi (1952c) shows that one can in fact derive by this method the results for all the trigonal and hexagonal crystal classes (including class 6), with the axes chosen according to the usual conventions, provided one knows in advance the results for class 3. Thus class 3 is the only class outside the reach of the method.

JERPHAGNON, J., CHEMLA, D. and BONNEVILLE, R. (1978). The description of the physical properties of condensed matter using irreducible tensors. *Advances in Physics*, 27, 609–650. This is an approach to crystal properties that makes use of the decomposition of a cartesian tensor into parts that are independent of the coordinate system (rather as a localised distribution of electric charge may be expressed as a sum of a pole, dipole, quadrupole etc.). See also Yih-O Tu (1968) below.

JURETSCHKE, H. J. (1974). *Crystal physics. Macroscopic physics of anisotropic solids*, Reading, Mass.: Benjamin (also 1975 London: Addison Wesley).

MASON, W. P. (1966). *Crystal physics of interaction processes*, New York and London: Academic Press.

NARASIMHAMURTY, T. S. (1981). *Photoelastic and electro-optic properties of crystals*, New York and London: Plenum Press. Also useful for elasticity. Contains over 1600 references.

NELSON, D. F. (1979). *Electric, optic, and acoustic interactions in dielectrics*, Wiley-Interscience. In particular, Chaps. 13 and 14 for the photoelastic and electrooptical effects, and the tables in the appendix.

SHUBNIKOV, A. V. and KOPTSIK, V. A. (1974). *Symmetry in science and art* (English translation by G. D. Archard, edited by D. Harker), New York and London: Plenum Press. See especially Chap. 12.

SMITH, C. S. (1958). Macroscopic symmetry and properties of crystals. *Solid State Physics*, 6, 175–249, New York: Academic Press. Includes, for a number of specific properties, an atomic model to show how crystalline anisotropy might arise.

VOIGT, W. (1898). *Die fundamentalen physikalischen Eigenschaften der Krystalle*, Leipzig: Teubner. This and the following one are the books that laid the foundations of the subject.

——(1910). *Lehrbuch der Kristallphysik*, first edition (reprinted in 1928 with an additional appendix), Leipzig: Teubner.

WALPOLE, L. J. (1984). Fourth-rank tensors of the thirty-two crystal classes: multiplication tables. *Proc. Roy. Soc. Lond.* A, 391, 149–179.

WOOSTER, W. A. (1938). *A text-book on crystal physics* (reprinted 1949), Cambridge.

——(1973). *Tensors and group theory for the physical properties of crystals*. Oxford: Clarendon Press.

——and BRETON, A. (1970). Experimental crystal physics, 2nd edition, Oxford: Clarendon Press. A course for teaching.

YIH-O TU (1968). The decomposition of an anisotropic elastic tensor. *Acta Cryst.* A24, 273–282.

(c) *Transport properties and magnetism:*

AKGÖZ, Y. C. and SAUNDERS, G. A. (1975). Space-time symmetry restrictions on the form of transport tensors: I. Galvanomagnetic effects. *J. Phys. C.:* *Solid State Phys.* 8, 1387–1396. II. Thermomagnetic effects. The same, 2962–2970.

CRACKNELL, A. P. (1973). Symmetry properties of the transport coefficients of magnetic crystals. *Phys. Rev.* B, 7, 2145–2154.

HURD, C. M. (1974). Galvanomagnetic effects in anisotropic metals. *Advances in Physics*, 23, 315–433.

KLEINER, W. H. (1966). Space-time symmetry of transport coefficients. *Phys. Rev.* 142, 318–326.

——(1967). Space-time symmetry restrictions on transport coefficients. II. Two theories compared. *Phys. Rev.* 153, 726–727. III. Thermogalvanomagnetic coefficients. *Phys. Rev.* 182, 705–709 (1969).

OPECHOWSKI, W. and GIUCCONE, R. (1965). Magnetic symmetry. In *Magnetism*, Vol. 2A (edited by G. T. Rado and H. Suhl), Chap. 3, New York: Academic Press.

POURGHAZI, A., SAUNDERS, G. A. and AKGÖZ, Y. C. (1976). Symmetry and anti-symmetry restrictions on the form of transport tensors for magnetic crystals. *Phil. Mag.* 33, 781–784.

SHTRIKMAN, S. and THOMAS, H. (1965). Remarks on linear magneto-resistance and magneto-heat-conductivity. *Solid State Comm.* 3, 147–150.

SMITH, A. C., JANAK, J. F. and ADLER, R. B. (1967). *Electronic conduction in solids*, New York: McGraw-Hill. An intermediate level textbook that includes a careful discussion of the Onsager relations and of galvanomagnetic, thermoelectric and thermomagnetic effects in relation to crystal symmetry.

(d) *Couple stresses, strain-gradient elasticity and polarization-gradient theories:*

The classical treatment of stress given in Ch. V assumes that a small surface element within a stressed body transmits only a force per unit area, σ_{ij}. A more general continuum theory of elasticity recognises that a small surface element also transmits a couple per unit area, γ_{ij}, called the couple stress. Just as equation (2a), p. 85, relates the spatial derivatives of σ_{ij} to the body forces, so the spatial derivatives of γ_{ij} are related to the anti-symmetrical part of the stress tensor and to the body-torques G_i (if any) thus:

$$\frac{\partial \gamma_{ij}}{\partial x_j} - t_i + G_i = 0,$$

where t_i is the axial vector with components $[\sigma_{23} - \sigma_{32}, \sigma_{31} - \sigma_{13}, \sigma_{12} - \sigma_{21}]$. This extension does not undermine the basis of the classical linear theory, as has sometimes been asserted, by changing the number of independent elastic constants in Hooke's Law, but rather confirms its status as a good first approximation. When couple stresses are admitted the expression for the elastic energy density (p. 137) contains, in addition to the strain ϵ_{ij}, the gradient of the rotation $\bar{\omega}_{ij}$, defined on pp. 97, 99 (e.g. Rajagopal 1960, Krishnan and Rajagopal 1961). This is called the rotation-gradient theory. The rotation-gradient (nine components with one relation between them) comprises only eight of the eighteen components of the strain-gradient, or rather of the eighteen independent linear combinations of these components. A theory that includes all the components

(Toupin 1962) introduces, to accompany the ten extra strain-gradient components, ten components of self-equilibrating double forces per unit area, called hyper-stresses. These contribute neither resultant force nor couple, per unit area, across a surface in the material, but do nevertheless contribute to the potential energy.

This theory that takes account of the first gradient of the strain suggests a further generalisation that includes, in the energy density, strain and all its spatial derivatives (gradients) up to infinite order. Because the energy density contains these quantities quadratically this is still a linear theory, in the sense that the stress and hyperstresses depend linearly on the strain and strain-gradients; accordingly, solutions to specific problems can still be built up by superposition. (It is thus quite different from non-linear conventional elasticity, which involves higher-order elastic coefficients (p. 256) and which is solely concerned with strain.) When one is considering elastic waves in crystals the higher-order strain-gradients become progressively more important as the wave-length becomes shorter. Such problems can also be approached from a quite different starting point, by using a lattice theory of interatomic forces, and it is interesting to note that a complete correspondence has been demonstrated between the two points of view for the case when the lattice is of the simple type having one atom per unit cell; the continuum field of strain-gradients up to infinite order contains the same information as the deformed lattice (Krum-hansl 1965).

It is characteristic of strain-gradient continuum theory, sometimes called non-local elasticity theory, that material constants with the dimensions of length appear and it thus yields length scales typical of the structure of the material (in fact they are of the order of the interatomic distances). The theory shows that a free crystal will exhibit a spontaneous strain, confined to a region near its surface and extending to a depth given by these length scales. Thus surface tension or surface energy per unit area, with a surface relaxation effect, is in-trinsic to such continuum strain-gradient theories. There is nothing equivalent to this in classical elasticity theory. It is noteworthy that, to obtain a surface energy of deformation for centrosymmetric crystals, it is necessary to include strain and both its first and second gradients in the expression for the energy density (Mindlin 1965). A surface energy of deformation does appear when strain and only its first gradient are included, but the associated material con-stant is a third-rank tensor and so provides an effect only in non-centrosymmetric crystals.

The same line of thought leads to a generalisation of piezoelectric theory in which not only polarization but polarization-gradient is included in the ex-pression for the stored electrical energy. To understand this one should reflect that the polarization at a point in a medium is determined, not just by the field at that point, but by the field in a small region around it (spatial dispersion). Conversely, the field at a point is determined by the polarization in a small region, and therefore by the polarization at the point together with its spatial derivatives. With this generalisation, characteristic effects again appear that are associated with free surfaces, in this case not only a spontaneous deformation but also a polarization localised near the surface.

Gradient theories are also capable of explaining both optical activity (Ch. XIV) (Landau and Lifshitz 1960, Agranovich and Ginzburg 1966, 1971) and acoustical activity. Acoustical activity is the rotation of the direction of mech-anical displacement along the path of a transverse elastic wave. According to these theories optical activity is due to an interaction between the polarization and the polarization gradient. Acoustical activity, on the other hand, depends on interactions of the strain with both the polarization and the polarization gradient, and is absent if either interaction is missing (Mindlin 1972).

If the crystal lattice has a basis i.e. a group of two or more atoms per unit cell, it can be resolved into two or more interpenetrating simple lattices, and these can vibrate at high frequency relative to one another. Although these vibrations, the 'optical' modes, can be of very long, or infinite, wavelength, they cannot be accommodated within classical elasticity; they are conventionally treated by a lattice theory. However, by introducing the concepts of a 'compound continuum' and 'inner elasticity' (Cousins 1978a, b), it is still possible to treat them by continuum theory. In a similar way, if the unit cell contains one or more molecules, the rotation and strain of the molecules themselves will not, in general, be the same as the rotation and strain of the lattice itself. This micro-rotation is taken into account in what is called the 'Cosserat continuum', and the theory of 'micropolar elasticity'. Minagawa et al (1980) give a recent application and references.

These insights into the value of gradient theories were obtained mostly in the period 1960–65; before that time there was much confusion, especially on the role of couple-stresses. The subject is excellently reviewed by Mindlin (1972).

We should also mention a connection between gradient theories of elasticity and the continuum theory of dislocations in crystals; this is the theory in which discrete dislocations are regarded as spread out through the lattice to form a continuum (Kroner 1981).

AGRANOVICH, V. M. and GINZBURG, V. L. (1966). *Spatial dispersion in crystal optics and the theory of excitons*, London: Wiley-Interscience.

——(1971). Crystal optics with spatial dispersion. *Progress in Optics*, **9**, 235–280.

COUSINS, C. S. G. (1978a). Inner elasticity. *J. Phys. C*. **11**, 4867–4879.

——(1978b). The symmetry of inner elastic constants. *J. Phys. C.*, **11**, 4881–4900.

KRISHNAN, R. S. and RAJAGOPAL, E. S. (1961). The atomistic and the continuum theories of crystal elasticity. *Ann. Phys.* **8**, 121–136.

KRONER, E. (1981). Continuum theory of defects. In *Physics of Defects* (ed. R. Balian, M. Kléman, J.-P. Poirier), Les Houches, Amsterdam: North-Holland, pp. 215–315.

KRUMHANSL, J. A. (1965). Generalised continuum field representations for lattice vibrations. In *Lattice dynamics* (edited by R. F. Wallis), Oxford: Pergamon Press, pp. 627–634.

LANDAU, L. D. and LIFSHITZ, E. M. (1960). *Electrodynamics of continuous media* (English translation by J. B. Sykes and J. S. Bell), Oxford: Pergamon. §§82, 83 of this excellent book provide for optical activity a satisfying macroscopic gradient theory, rather than the atomic theory invoked on p. 266.

MINAGAWA, S., ARAKAWA, K. and YAMADA, M. (1980). Diamond crystals as Cosserat continua with constrained rotation. *Phys. Stat. Sol. (a)* **57**, 713–718.

MINDLIN, R. D. (1965). Second gradient of strain and surface-tension in linear elasticity. *Int. J. Solids Struct.*, **1**, 417–438.

——(1972). Elasticity, piezoelectricity and crystal lattice dynamics. *Journal of Elasticity*, **2**, 217–282.

RAJAGOPAL, E. S. (1960). The existence of interfacial couples in infinitesimal elasticity. *Ann. Phys.* **6**, 192–201.

TOUPIN, R. A. (1962). Elastic materials with couple-stresses. *Arch. Rational Mech. Anal.* **11**, 385–414.

(e) Product properties of composite materials:

In a composite material two phases may interact to produce a property not possessed by either phase on its own. For example, if one phase is magnetostrictive and the other is piezoelectric, a magnetic field produces a distortion of the magnetostrictive phase, which in turn distorts the piezoelectric phase, which generates an electric field. Viewed macroscopically the material is homogeneous and shows a magneto-electric effect; it arises as a product (as distinct from a sum or average) of the two single-phase properties.

VAN SUCHTELEN, J. (1972). Product properties: a new application of composite materials. *Philips Res. Repts.* **27**, 28–37.

NEWNHAM, R. E., SKINNER, D. P. and CROSS, L. E. (1978). Connectivity and piezoelectric-pyroelectric composites. *Mat. Res. Bull.* **13**, 525–536.

(f) Displacement-gradient photoelasticity (rotooptic effect):

When a crystal vibrates in an acoustic mode that possesses a shear component its elements are not only strained but also rotated. The observed change in birefringence is only partly due to strain, by the photoelastic effect (Ch. XIII), for, if the crystal is naturally birefringent, an additional contribution will arise from the rotation. This extra effect of the rotation is, of course, calculable from the known natural birefringence and does not represent a new crystal property. Nevertheless, it is convenient to describe the combined effect (the rotooptic effect) by a new tensor which relates birefringence to displacement gradient rather than to strain. This effect of rotation has nothing to do with the rotation-gradient theory of elasticity mentioned above. Relevant papers are:

NELSON, D. F. and LAX, M. (1970). New symmetry for acousto-optic scattering. *Phys. Rev. Lett.* **24**, 378–380.

——(1971). Theory of the photoelastic interaction. *Phys. Rev. B.*, **3**, 2778–2794.

NELSON, D. F., LAZAY, P.D. and LAX, M. (1972). Brillouin scattering in anisotropic media: calcite. *Phys. Rev. B.*, **6**, 3109–20.

VACHER, R. and BOYER, L. (1972). Brillouin scattering: a tool for the measurement of elastic and photoelastic constants. *Phys. Rev. B.*, **6**, 639–673.

VACHER, R., BOISSIER, M. and BOYER, L. (1973). Brillouin scattering measurements of the algebraic values of the photoelastic constants in a birefringent crystal. *Phys. Status Solidi (a)* **18**, 523–529.

See also the book by Nelson (1979) in section (b).

(g) Other aspects of crystal optics:

AGRANOVICH, V. M. and GINZBURG, V. L. (1971). See section (d) above.

BLOEMBERGEN, N. (1977). *Nonlinear optics*, Reading, Mass: Benjamin. First published in 1965, the book deals with the higher-order interactions between light and matter produced by intense laser beams. This 1977 reprint adds a guide to more recent literature.

HAYES, W. and LOUDON, R. (1978). *Scattering of light by crystals*, Wiley-Interscience. (See e.g. Tables 1.2, 4.1, 4.2, 8.2 and 8.3 for effects of symmetry.)

LANDAU, L. D. and LIFSHITZ, E. M. (1960). See section (d) above.

NARASIMHAMURTY, T. S. (1981). See section (b) above.

NELSON, D. F. (1979). See section (b) above.

O'LOANE, J. K. (1980). Optical activity in small molecules, nonenantiomorphous crystals and nematic liquid crystals. *Chem. Rev.* **80**, 41–61. A valuable historical review with many references.

PINNOW, D. A. (1972). Elasto-optic materials. In *Laser Handbook* (edited by F. T. Arecchi and E. O. Schulz-DuBois), Vol. 1, pp. 995–1008, Amsterdam: North-Holland.

POCKELS, F. (1906). *Lehrbuch der Kristalloptik*, Leipzig: Teubner. A classic work.

SAPRIEL, J. (1976). *Acousto-optics* (English translation by S. Francis and B. Kelly), Wiley-Interscience.

TABOR, W. J. (1972). Magneto-optic materials. In *Laser Handbook* (edited by F. T. Arecchi and E. O. Schulz-DuBois), Vol. 1, pp.1009–1027, Amsterdam: North-Holland.

WEMPLE, S. H. (1972). Electro-optic materials. In *Laser Handbook* (edited by F. T. Arecchi and E. O. Schulz-DuBois), Vol. 1, pp. 975–994, Amsterdam: North-Holland.

(h) Pyroelectricity, ferroelectricity, thermal expansion, elasticity and elastic waves:

FEDOROV, F. I. (1968). *Theory of elastic waves in crystals* (English translation by J. E. S. Bradley), New York: Plenum Press.

GRINDLAY, J. (1970). *An introduction to the phenomenological theory of ferroelectricity*, Oxford: Pergamon.

HEARMON, R. F. S. (1961). *An introduction to applied anisotropic elasticity*, Oxford University Press.

HUNTINGTON, H. B. (1958). The elastic constants of crystals. *Solid State Physics*, 7, 213–351. New York: Academic Press.

KRISHNAN, R. S., SRINIVASAN, R. and DEVANARAYANAN, S. (1979). *Thermal expansion of crystals*, Faridabad, Haryana: Thomson Press (India). Contains more than 2100 references.

LANDAU, L. D., and LIFSHITZ, E. M. (1959). *Theory of elasticity* (English translation by J. B. Sykes and W. H. Reid), London: Pergamon.

LANG, S. B. (1974). *Source book of pyroelectricity*, London: Gordon and Breach.

LEKNITSKY, S. G. (1963). *Theory of elasticity of an anisotropic elastic body* (English translation by P. Fern), San Francisco: Holden-Day.

LINES, M. E. (1979). Elastic properties of magnetic materials. *Physics Reports*, 55, 133–181.

MUSGRAVE, M. J. P. (1970). *Crystal acoustics. Introduction to the study of elastic waves and vibrations in crystals*, San Francisco: Holden-Day

SIMMONS, G. and WANG, H. (1971). *Single crystal elastic constants and calculated aggregate properties: a handbook*, 2nd edition, Cambridge, Mass.: M.I.T. Press. Numerical data.

WONG, H. C. and GRINDLAY, J. (1974). The elastic dielectric. *Advances in Physics*, 23, 261–313. Treats non-linear response.

ZHELUDEV, I. S. (1971). *Physics of crystalline dielectrics*, 2 vols. (English translation by A. Tybulewicz), New York and London: Plenum Press.

SOLUTIONS TO THE EXERCISES WITH NOTES

EXERCISE 1.1, p. 15. No, because they do not transform like vector components.

1.2, p. 22. Neumann's Principle applies to the *symmetry* of the physical property. The symmetry of X-ray patterns does obey Neumann's Principle, and depends on the point group. The space group is not deduced from the symmetry of the X-ray patterns but from characteristic absences among the spots.

1.3, p. 31.

[2]:
$$\begin{bmatrix} 25 \times 10^7 & 0 & 0 \\ 0 & 4 \times 10^7 & 0 \\ 0 & 0 & 16 \times 10^7 \end{bmatrix} \text{ ohm}^{-1}\,\text{m}^{-1};$$

x_i' must be the principal axes. [5]: Using equation (32), with the direction cosines referred to the x_i' axes, gives $\sigma = 7 \times 10^7$ ohm^{-1} m^{-1}. (Alternatively, notice the symmetry between Ox_2 and OP, and hence obtain $\sigma = \sigma_{22}$.) [7]: Magnitude $\sqrt{76} \times 10^7$ amps/m²; direction cosines $[0, -1/\sqrt{76}, 5\sqrt{(3/76)}]$, i.e. in $x_2 x_3$ plane at an angle of 6° 36′ with Ox_3 and 83° 24′ with $-Ox_2$. [9]: They should be parallel.

2.1, p. 35. We have, using the vector transformation law, equation (6), and the substitution property of δ_{jk},

$$p_i p_i = a_{ji} p_j' a_{ki} p_k' = \delta_{jk} p_j' p_k' = p_k' p_k'.$$

2.2, p. 46.
$$\begin{vmatrix} S_{11} & S_{12} \\ S_{12} & S_{22} \end{vmatrix} = S_{11} S_{22} - S_{12}^2 = OC^2 - r^2$$

in Fig. 2.4 b, p. 44, where r is the radius of the circle. This is the square of the length of the tangent from O to the circle and is therefore invariant.

2.4, p. 47. (a) $\begin{bmatrix} 10 & 0 & 0 \\ 0 & 20 & 0 \\ 0 & 0 & 43 \end{bmatrix}$ (by rotation through 19° about x_3, from x_2 towards x_1).

(b) $\begin{bmatrix} -9 & 0 & 0 \\ 0 & 3 & 0 \\ 0 & 0 & 10 \end{bmatrix}$ (by rotation through 30° about x_3, from x_1 towards x_2).

(c) $\begin{bmatrix} 0 & 0 & 0 \\ 0 & 4 & 0 \\ 0 & 0 & 9 \end{bmatrix}$ (by rotation through 45° about x_3, from x_2 towards x_1).

(d) $\begin{bmatrix} 10 & 0 & 0 \\ 0 & 12 & 0 \\ 0 & 0 & 0 \end{bmatrix}$ (by rotation through 26° 34′ about x_3, from x_3 towards x_1).

2.5, p. 47. Rotate through 53° 38′ or −82° 16′ about x_3, from x_1 to x_2.

3.1, p. 60. The work done is

(1) $\frac{1}{2}\mu_{11} H_1^2 + \frac{1}{2}\mu_{22} H_2^2 + \mu_{12} H_1 H_2$

and (2) $\frac{1}{2}\mu_{22} H_2^2 + \frac{1}{2}\mu_{11} H_1^2 + \mu_{21} H_2 H_1$.

3.2, p. 64. $\psi = 1\cdot3 \times 10^{-5}$ (rationalized units).

3.3, p. 64. ψ in direction of field is $-25\cdot6 \times 10^{-6}$; $\theta = 53\cdot7°$.

3.4, p. 65. Force $= 1\cdot47 \times 10^{-5}$ newtons ($= 1\cdot47$ dynes).

Direction of force $= (0\cdot29, 0\cdot81, -0\cdot52)$. Remember that for a magnetostatic field in free space, div $\mathbf{H} = 0$ and curl $\mathbf{H} = 0$.

Couple $= 3\cdot1 \times 10^{-8}$ m.-newton, about the direction $(-0\cdot573, 0\cdot817, 0\cdot082)$, right-handed if the axes are right-handed (this is approximately the couple given by the weight of a 3-mg mass acting on a lever arm of 1 mm).

5.1, p. 92. $\sigma_{ij} = \sigma\delta_{ij} + \sigma_{ij}^*$, where $\sigma_{ij}^* = \sigma_{ij} - \tfrac{1}{3}\delta_{ij}\sigma_{kk}$, and $\sigma = \tfrac{1}{3}\sigma_{kk}$. $\sigma\delta_{ij}$ is a hydrostatic stress. σ_{ij}^* has the property that the sum of the components on the leading diagonal (σ_{ii}^*) vanishes. We transform σ_{ij}^* so that the three leading diagonal terms vanish, as follows. Let σ_{11}^* and σ_{22}^* be of opposite sign. Rotate about Ox_3 until $\sigma_{11}^* = 0$. (Mohr circle construction). Now rotate about the new Ox_1 axis

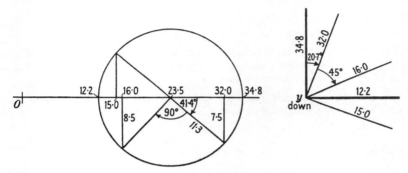

Fig. S.1. Illustrating the solution to Exercise 6.5. Values shown are those of $10^6 \times \alpha$ (α in $^\circ\,\mathrm{C}^{-1}$). One diagram shows the Mohr circle construction; the other shows the thermal expansions for the various directions, the principal expansion directions being given by the heavy lines. The rules used in drawing the two diagrams are those suggested on p. 46.

until $\sigma_{22}^* = 0$. Then σ_{11}^* remains zero and σ_{33}^* must also be zero. The hydrostatic stress $\sigma\delta_{ij}$ is unaffected by the transformation. The result is therefore proved.

6.1, p. 106. Principal strain directions are obtained by rotating the axes $22\cdot5^\circ$ about Ox_2 from $+Ox_3$ towards $+Ox_1$. Their magnitudes are

$$(9\cdot24, 6, 0\cdot76) \times 10^{-6}.$$

ϖ_{ij} represents a rotation of $\sqrt5 \times 10^{-6}$ radians about direction $(0, 2/\sqrt5, 1/\sqrt5)$, right-handed if axes are right-handed.

6.2, p. 108. Angle $= 2\Delta T \sqrt{(\alpha_{31}^2 + \alpha_{23}^2)}$.

6.3, p. 108. *Hint.* Resolve the radius vector and the displacement of its extremity into components along the principal axes.

6.4, p. 109. The principal axes are parallel and perpendicular to the z-axis $(\pm 2^\circ)$; $\alpha_1, \alpha_3 = 29\cdot0\ (\perp z),\ 7\cdot5\ (\|\,z) \times 10^{-6}$ per $^\circ$ C.

6.5, p. 109. $41\cdot0, 34\cdot8, 12\cdot2 \times 10^{-6}$ per $^\circ$ C; $90^\circ, 20\cdot7^\circ, 69\cdot3^\circ$ (Fig. S.1, this page).

6.6, p. 109. $\alpha_1 = 9\cdot9 \times 10^{-6}$, $\alpha_2 = 17\cdot3 \times 10^{-6}$, $\alpha_3 = 34\cdot7 \times 10^{-6}$ per $^\circ$ C.

8.1, p. 138. See Table 9, p. 141.

8.2, p. 147. Consider a unit cube. From the form of the compliance matrix, shear stresses on the {100} faces cause no change in volume. The normal stresses on the {100} faces are $l_1^2 T$, $l_2^2 T$, $l_3^2 T$, where (l_1, l_2, l_3) are the direction cosines of the direction of the tension. The volume change given by these components is

$$\epsilon_1 + \epsilon_2 + \epsilon_3 = (s_{11} + 2s_{12})(l_1^2 T + l_2^2 T + l_3^2 T) = (s_{11} + 2s_{12})T.$$

9.1, p. 152. $\mathbf{AB} = \begin{pmatrix} a_1 b_1 + c_1 b_2 \\ a_2 b_1 + c_2 b_2 \\ a_3 b_1 + c_3 b_2 \end{pmatrix}$.

9.2, p. 152. $\begin{pmatrix} 5 & 19 \\ -9 & 25 \\ -5 & 37 \end{pmatrix}$; $\begin{pmatrix} 4 & 2 & 4 & 3 \\ 5 & 0 & -10 & 0 \end{pmatrix}$.

9.3, p. 152.   (11) and $\begin{pmatrix} 8 & 12 \\ 2 & 3 \end{pmatrix}$.

11.1, p. 200. If α is the angle between the axis of the rod and the 3-fold axis (x_3), $\tan \alpha = \sqrt{(r_3/r_1)}$; $\alpha = 37 \cdot 2°$; $\sin \theta_{max} = (r_1 - r_3)/(r_1 + r_3)$; $\theta_{max} = 15 \cdot 7°$. (The use of the Mohr circle construction avoids the need for differentiation.)

13.1, p. 254. From the principle on p. 245, the strain ellipsoid in a cubic crystal under hydrostatic pressure must be a sphere. We have

$$\Delta B_1 = \Delta B_2 = \Delta B_3 = (p_{11} + 2p_{12})\epsilon, \qquad \Delta B_4 = \Delta B_5 = \Delta B_6 = 0.$$

Therefore $$\Delta n = -\frac{n^3}{2}(p_{11} + 2p_{12})\frac{\Delta v}{3v} = \frac{n^3}{2} \cdot \frac{p_{11} + 2p_{12}}{3} \cdot \frac{\Delta\rho}{\rho}.$$

14.1, p. 273. Zero rotation if optic axes lie in plane perpendicular to x or y; equal and opposite rotations if optic axes lie in plane perpendicular to z.

14.2, p. 273. One axis lies in the symmetry plane; the other two are at 45° to the symmetry plane. Referred to its principal axes the tensor is

$$\begin{bmatrix} +\sqrt{(g_{12}^2 + g_{23}^2)} & 0 & 0 \\ 0 & -\sqrt{(g_{12}^2 + g_{23}^2)} & 0 \\ 0 & 0 & 0 \end{bmatrix},$$

where the third principal axis is taken to be the one lying in the symmetry plane.

INDEX OF NAMES

INDEX OF SUBJECTS

The summaries at the ends of the chapters contain further references to the subjects listed.